PENGUIN BOO.

UNRULY WATI

'When confronted with tragedy, the modernist project has always retreated behind the sober voice of science. There is no more vivid description of this encounter than Sunil Amrith's wonderful new book, nor a better example of combining sympathy for the main protagonists – the planners, the engineers, the meteorologists – with a sustained sense of how, with the best of intentions, things can go horribly wrong'
Abhijit V. Banerjee, co-author of *Poor Economics: A Radical Rethinking of the Way to Fight Global Poverty*

'In this groundbreaking work, Sunil Amrith deftly and imaginatively steers us towards an understanding of both water's worldly historical importance and its sublime capacity to exceed the human scale. Between its haunting opening pages and chilling epilogue, Amrith's sensitive, deeply engaging and densely woven narrative reminds us that the present water crisis is the legacy of a colonial past – not of the peculiarities of Asian people and climate. This is a politically urgent book that shows the need to tell more expansive histories to help us address climate risks that transcend national borders'
Priya Satia, author of *Empire of Guns: The Violent Making of the Industrial Revolution*

'Across Asia, water is power. Sunil Amrith's *Unruly Waters* is a gripping work, both timely and necessary, that captures the forces at work in the struggle to control Asia's water. From cultural influences of colonial empire engineering to atmospheric chemistry in a time of climate change, Amrith reveals all that is at stake for half the planet's population' Meera Subramanian, author of *A River Runs Again*

'A compelling history of India over the last 200 years mostly describing how its people and rulers have dealt with the weather' *Kirkus Reviews*

ABOUT THE AUTHOR

Sunil Amrith is the Mehra Family Professor of South Asian Studies at Harvard University. He is also the author of *Crossing the Bay of Bengal: The Furies of Nature and the Fortunes of Migrants*. He has been a Research Fellow at Trinity College, Cambridge and in 2017 was awarded a MacArthur 'Genius' Fellowship.

SUNIL AMRITH

Unruly Waters

How Mountain Rivers and Monsoons
Have Shaped South Asia's History

PENGUIN BOOKS

PENGUIN BOOKS

UK | USA | Canada | Ireland | Australia
India | New Zealand | South Africa

Penguin Books is part of the Penguin Random House group of companies
whose addresses can be found at global.penguinrandomhouse.com.

First published in the United States of America by Basic Books, an imprint
of Perseus Books, LLC, a subsidiary of Hachette Book Group, Inc 2018
First published in Great Britain by Allen Lane 2018
Published in Penguin Books 2020

003

Printed and bound in Great Britain by Clays Ltd, Elcograf S.p.A.

A CIP catalogue record for this book is available from the British Library

ISBN: 978-0-141-98263-2

www.greenpenguin.co.uk

MIX
Paper | Supporting
responsible forestry
FSC   FSC® C018179
www.fsc.org

Penguin Random House is committed to a
sustainable future for our business, our readers
and our planet. This book is made from Forest
Stewardship Council® certified paper.

For Theodore and Lydia

CONTENTS

LIST OF MAPS

x

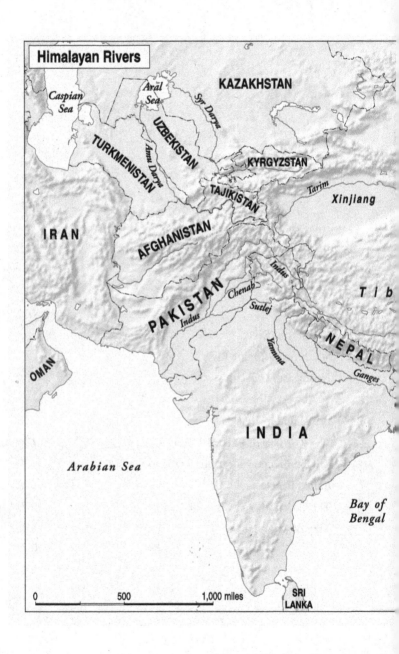

Himalayan Rivers

Caspian
Sea

Aral
Sea

Syr Darya

KAZAKHSTAN

TURKMENISTAN

Amu Darya

UZBEKISTAN

KYRGYZSTAN

TAJIKISTAN

Tarim

Xinjiang

IRAN

AFGHANISTAN

Indus

PAKISTAN

Chenab

Sutlej

Indus

Yamuna

T i b

N E P A L

Ganges

OMAN

INDIA

Arabian Sea

Bay of
Bengal

0   500   1,000 miles

SRI
LANKA

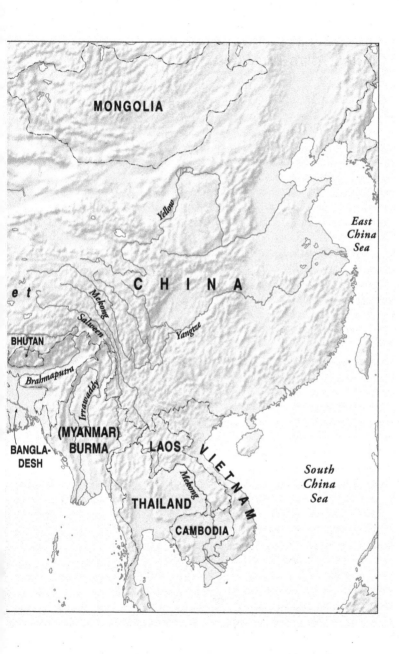

MONGOLIA

Yellow

East
China
Sea

C H I N A

e t

Mekong

Salween

Yangtze

BHUTAN

Brahmaputra

Irrawaddy

(MYANMAR)
BURMA

LAOS

V I E T N A M

BANGLA-
DESH

South
China
Sea

Mekong

THAILAND

CAMBODIA

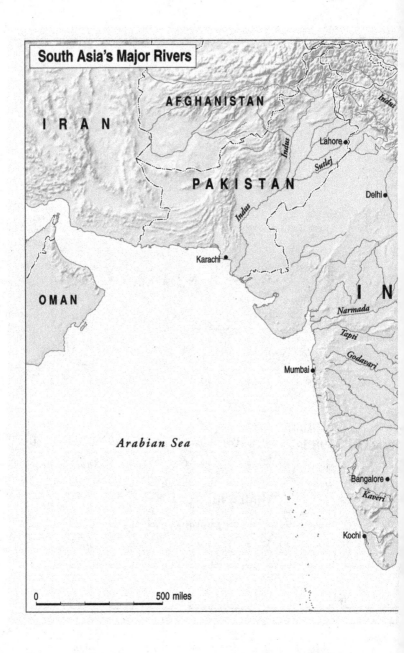

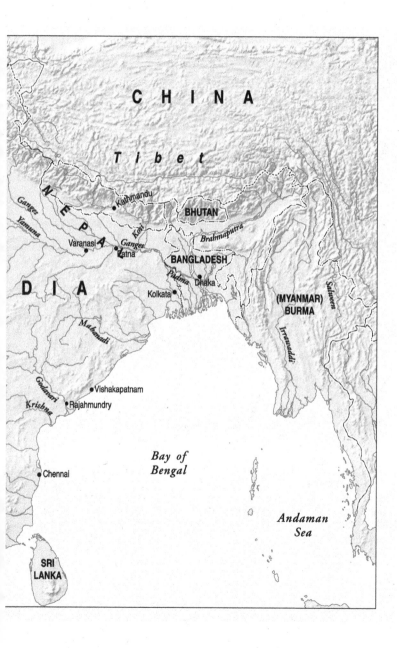

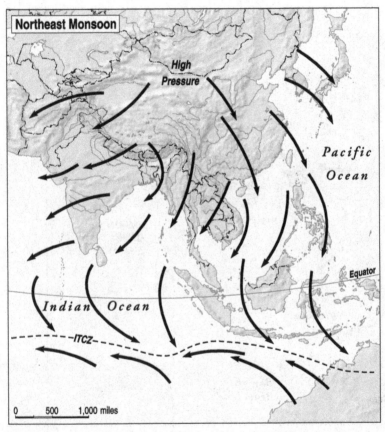

Map showing the winds during the northeast monsoon, which blows from December to March.

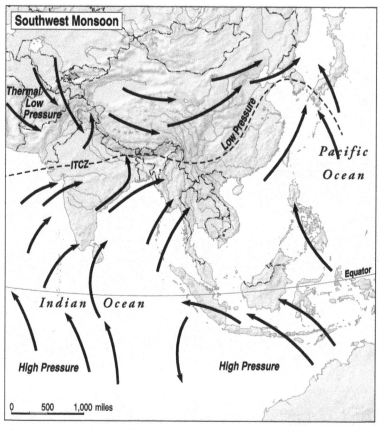

Map showing the winds during the southwest monsoon, from June to September.

A NOTE ON NAMES AND TERMINOLOGY

MANY OF THE PLACES I WRITE ABOUT IN THIS BOOK HAVE BEEN known by different names at different points in time. As a rule, I have used the names that correspond to the period I am writing about—to cite a few examples, I use Bombay, Madras, Calcutta, Poona, and Rangoon when I am discussing the colonial period and the early decades after independence; I switch to Mumbai, Chennai, Kolkata, Pune, and Yangon, respectively, when I am talking about more recent history, as those names were formally changed in the 1990s. I adopt a similar strategy when it comes to country names: for example, I use Ceylon and Malaya when discussing the colonial period, and Sri Lanka and Malaysia when writing about the post-independence era.

For clarity I have transliterated words from South Asian languages in a way that reflects common practice in the region rather than employing the formal diacritical marks favored by scholars of South Asian languages.

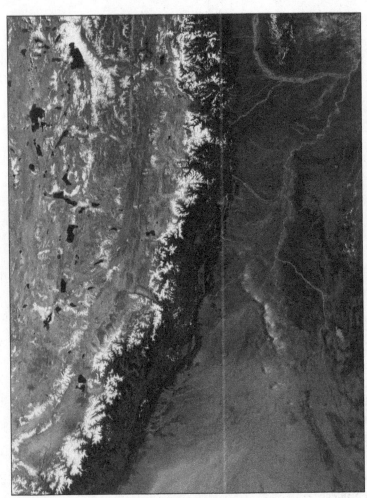

A NASA satellite image from October 27, 2002, showing a Himalayan mountain range and the rivers that descend from the Tibetan Plateau into North India. CREDIT: Jacques Descloitres, MODIS Rapid Response Team, NASA/GSFC

ONE

THE SHAPE OF MODERN ASIA

OOKING DOWN FROM ORBIT, THE LENS OF A NASA SATELLITE LANDS upon this patch of Earth. In the upper half of the picture lies the curve of a Himalayan mountain range, fringed by the iridescent lakes of the Tibetan plateau.

The satellite picture is a snapshot of a single moment on October 27, 2002. But there are layers of history embedded within it. It shows us the outcome of a process that unfolded in deep time. Approximately 50 million years ago, the Himalayas were created by the collision of what would become the Indian peninsula, which had detached from Madagascar, with the Eurasian landmass. The island buckled under the edge of Eurasia, pushed up the Tibetan Plateau, and eradicated a body of water later named the Tethys Sea. "Geology, looking further than religion," E. M. Forster wrote in *A Passage to India*, "knows of a time when neither the river [Ganges] nor the Himalayas that nourish it existed, and an ocean flowed over the holy places of Hindustan."

Volcanic activity under the Indian Ocean kept the pressure up, forcing layers of rock to crumple under the Indian margin to create the largest mountain chain on Earth.[1]

So massive are the mountains, so heavy is their concentration of snow, ice, heat, and melting water that they shape Earth's climate. Asia's great rivers are a product of this geological history. They flow south and southeast, and they have shaped the landscape that is visible here: the force of the rivers descending from the mountains eroded rock, creating the gorges and valleys. Over centuries the rivers have carried silt and sediment from the mountains; they have deposited them along Asia's valleys and floodplains to sustain large human populations. Writing in the 1950s, guided by maps and not yet by satellite photographs, geographer Norton Ginsburg described Asia's "mountain core" as the "hub of a colossal wheel, the spokes of which are formed by some of the greatest rivers in the world."[2]

And then your eye comes to rest on what was invisible to the satellite but is now superimposed—evidence of a more recent history lies in the borders that dissect the rivers, their shapes governed by bureaucratic, not environmental, logic. Within the frame of this image alone, the mountains run through southwestern China, Nepal, Bhutan, and northeastern India. The rivers are more unruly; they spill beyond the frame of the photograph. From mountain peaks flow ten great rivers that serve a fifth of humanity—the Tarim, the Amu Darya, the Indus, the Irrawaddy, the Salween, the Mekong, the Yangzi, the Yellow River, and, at the heart of this photograph, the Ganges and the Brahmaputra. The Himalayan rivers run through sixteen countries, nourished by countless tributaries. They traverse the regions we carve up as South, Southeast, East, and Central Asia; they empty out into the Bay of Bengal, the Arabian Sea, the South and East China Seas, and the Aral Sea.

Look at the left of the picture and you can see a more compressed history. The haze of pollution that hangs over North India is a composite "brown cloud" of human-produced sulfates, nitrates, black carbon, and organic carbon. Aerosol concentrations over the Indian subcontinent are the highest in the world, especially in the winter months when there is little rain to wash the skies clean. Individual particles remain in the atmosphere only for a matter of weeks, but cumulatively the cloud lasts for months—what we see here is a fleeting archive of every domestic stove, every truck and auto-rickshaw exhaust pipe, every factory smokestack and crop fire that burned across the Gangetic plain after the end of the monsoon rains that year. But the location of the cloud, and its contributing sources, testify to a longer twentieth-century history of population growth, urban expansion, and uneven economic development through that belt of northwestern India. Over time, a constant succession of transient "brown clouds" may have attenuated rainfall over South Asia over the past half century, transforming the water cycle that binds the clouds, the mountains, and the rivers.[3]

Finally, look at the snow on the mountain peaks visible from outer space. The time horizon this gestures toward is the future. The descent of water is vulnerable, now, to the ascent of carbon. As Earth's surface warms, the Himalayan glaciers are melting; they will melt more rapidly in the decades ahead, with immediate consequences for the flow of Asia's major rivers—and for the planet's climate.

———

ASIA IS HOME TO MORE THAN HALF THE WORLD'S POPULATION, but it contains less freshwater than any continent except Antarctica. A fifth of humanity lives in China, a sixth in India; but China has only 7 percent, and India 4 percent, of the world's freshwater—and within both countries that water is distributed unevenly. The quality as well as the quantity of water is under strain from a multiplicity of new demands and uses. Asia's rivers are choked by pollutants and impounded by large dams. An estimated 80 percent of China's wells contain water unsafe for human consumption; in India, groundwater is poisoned by fluoride and arsenic, or made undrinkable and unhealthy by salinity.[4]

The effects of climate change are already manifest. They compound the water-related risks that Asia's peoples already face. Most predictions hold that the Himalayan rivers will swell as the planet warms and the ice thaws; and then, around the middle of this century, they will start to dry out for part of the year. Existing inequalities will deepen: wet regions will get wetter, and dry regions will get drier. Within that broad pattern, there will be an increase in variability and a rise in extreme weather. The effects of planetary warming have already begun to interact with regional drivers of climate change—changes in land use, aerosol emissions, and "brown clouds"—to multiply uncertainty. Coastal regions in particular face a cascade of threats: heat stress, flooding, rising sea level, and more intense cyclonic storms.[5] Most at risk is the coastal crescent at the southern and eastern edge of the Eurasian landmass, home to the greatest concentration of the world's population. All twenty cities in the world with the largest populations vulnerable to rising sea levels are in Asia.[6] Most threatened, because numbers are compounded by high levels of poverty and inequality, are Mumbai and Kolkata in India, Dhaka in Bangladesh, Jakarta in Indonesia, and Manila in the Philippines.

All the while, statesmen and engineers plot water's final subjugation by technology. Over the next decade, more than four hundred large dams will be built on the Himalayan rivers—by India, China, Nepal, Bhutan, and Pakistan—to feed the region's hunger for electricity and its need for irrigation. New ports and thermal power plants line the coastal arc that runs from India, through Southeast Asia, to China. India and China have embarked on schemes to divert rivers to bring water to their driest lands: costing tens or hundreds of billions of dollars, they are the largest and most expensive construction projects the world has ever seen. At stake in how these plans unfold is the welfare of a significant portion of humanity. At stake is the future shape of Asia, the relations among its nations. Each of these risks, each of these responses, is rooted in ideas, institutions, and choices that earlier generations have made—that is to say, they are shaped by Asia's modern history.

I

To understand why Asia is the part of the world most vulnerable to climate change, why South Asia in particular stands at the front line, we need to turn to the history of water. Across the heartland of Asia—from Pakistan in the west, through India and Southeast Asia, to China in the east—the control of water has underpinned an increase in human population and an expansion in longevity that would have been unimaginable even in the middle of the twentieth century. In a warming world, Asia is distinctive for its sheer scale, and distinctive for the scale of inequality among its peoples. Both are rooted in the quest for water, which is a vital feature of modern Asian history, and one that we have ignored.

The struggle for water in modern history is a global story. We can tell a version of it set in the western United States, or in Germany, or in the Soviet Union, which was an Asian as well as a European power.[7] But nowhere has the search for water shaped or sustained as much human life as in India and China. Their demographic weight is not a fact of nature. It is an outcome of history, a history in which the control of water was pivotal. Today that control is more rigorous than ever, thanks to intensive hydraulic engineering, but the foundations of that control are fragile. Nowhere is the multiplier effect of any destabilization in the material conditions of life greater than it is in Asia. This, too, demands a historical explanation. As rains grow erratic and storms more intense, as rivers change course and wells dry up, the hard-won gains of half a century are vulnerable to reversal. The force of planetary warming combines with the material legacy of earlier quests to control water. Warming seas meet coastal zones that sag under the weight of growing cities, many of them founded as colonial ports in the eighteenth and nineteenth centuries. River deltas are sinking, starved of sediment by large dams upstream that were built in the 1950s and 1960s. We live with the unintended consequences of earlier generations' dreams and fears of water.

The origins of these dreams and fears, the longevity of the policies and infrastructures to which they gave rise, are the subject of this book. *Unruly Waters* tells the story of how the schemes of empire builders, the visions of freedom fighters, the designs of engineers—and the cumulative, dispersed actions of hundreds of millions of people across generations—have transformed Asia's waters over the past two hundred years.

This is not the way we usually understand Asia's modern history. Since the 1990s, identity and freedom have been the dominant themes in historical writing: these have oriented the study of Asia as much as anywhere else.[8] The late 1980s and the 1990s witnessed an upsurge in struggles for democracy in China, Indonesia, the Philippines, and Burma. In trying to explain the weakness or the persistence of authoritarian states, historians looked to political and intellectual history to capture alternative understandings of freedom, especially as earlier clusters of ideas were reinvigorated after the end of the Cold War. In the study of South Asia, the theme of identity has loomed largest. In India in the 1990s, political mobilization along caste lines—and growing recognition of the deep wounds that caste still inflicts on Indian society—clashed with the spectacular rise of a violent and exclusionary Hindu nationalism to focus historians' attention on the cleavages of culture and community that continue to divide South Asia.

These histories shed light on struggles for recognition and justice that are unfinished; they pinpoint inequalities that endure. But there is much that we have missed. Novelist Amitav Ghosh points out the irony that twentieth-century literary fiction proved oblivious to the growing crisis of climate change at the very moment of its escalation—a solipsistic turn at a moment when the material world was in the process of irrevocable transformation.[9] With only a few exceptions, the same charge can be leveled at those of us who write history. My premise here is that the transformation of Asia's environment, and in particular its ecology of water, may be as consequential in modern history as the political and cultural transitions that have compelled our attention—and it is consequential, not least, for its impact on both culture and politics.

Outside the specialized field of environmental history, the disappearance of nature from most broad accounts of historical change has been marked. It is also recent. In the 1970s and 1980s, agrarian history was a vibrant field. In those decades, discussions of water and agriculture in Asia fought to shake off the ghost of the German Marxist sociologist Karl Wittfogel. Wittfogel had argued in the 1950s that the need for centralized control over irrigation lay at the heart of "hydraulic societies" like China, ancient Egypt, and India, predisposing them all to absolutist government, or what Marx had called "oriental despotism."[10] Wittfogel's generalizations crumbled under closer examination. The agrarian histories written in the 1970s and 1980s emphasized the variability of arrangements through which different Asian societies harnessed the power of water. They all insisted on the importance of irrigation, but traced no simple relationship between that hydraulic fact and political forms. Browse through any study of South Asian or East Asian agriculture written in those decades: water is omnipresent. Historians of China were inclined to take a very long view, showing how the control of water shaped Chinese society and civilization over millennia; historians of South Asia were more likely to emphasize discontinuity—and especially the rupture that came with British colonialism, which forced the Indian countryside more fully into the global capitalist economy. Whether on the scale of millennia or of decades, this work exudes a rich sense of landscape. It is alive with a sense of the seasons changing, of the shifting flow of rivers, of the threat that floods or drought posed to human survival.[11] This tradition of historical writing disappeared most conspicuously from the study of South Asia, where the turn to cultural history swept all before it. But in other fields, too, historians decamped to the cities, leaving rural history behind as they turned to urban culture and politics, to intellectual history, to histories of cosmopolitanism and travel and migration. They did so just when a mounting water crisis began to pose an existential threat.

There are two main ways in which my view departs from the perspective of earlier work on the Asian countryside. The first is to see water as more than just a resource. In the pages that follow, the effects of new economic pressures and new technologies on water itself—on the water cycle, on the toxicity of water, on ideas about the value of water—are as important as the effect of water resources on agricultural output, which is what economic historians were primarily concerned with. As Asia's waters were transformed, water was understood in new ways by meteorologists, hydrologists, and oceanographers. Recent scientific research, made possible by advances in imaging technology and statistical capacity, has transformed the possibility of understanding water and climate historically, bringing us to archives we had scarcely thought to look at. The great French historian Marc Bloch believed that human history lived "behind the features of landscape" as much as it lived in "tools and machinery" and in institutions.[12] It lives, too, behind the chemical content of river water samples; behind satellite images of the water that lies underground; behind the composition of the smog that hovers above South Asia every winter, altering its rainfall. It lives in the changing ocean currents and winds.

In Fernand Braudel's three-fold conception of historical time, the first, slowest-moving layer was the time of nature and the seasons: a "history of constant repetition, ever-recurring cycles." His perspective influenced histories of the Indian Ocean, for example, in which the regularly reversing monsoon winds provide a basic material backdrop, enabling long-distance trade and shaping the agricultural cycle.[13] But over the last two hundred years nature has been altered by human intervention to such a profound extent that that stability and "constant repetition" cannot be assumed. By the end of the twentieth century it became possible to ask—as this book will ask—not only how climate has shaped us, but how we have affected the climate.

My second departure stems from a more elastic sense of geography. Like most history-writing until the end of the twentieth century, agrarian history took the nation-state for granted, though often the most meaningful unit of study was the region-within-the-nation: South China or Java, the Bengal or Mekong deltas. To put water at the heart of the narrative is to demand that we adopt a more flexible conception of space. Rivers pay no heed to human frontiers; but political boundaries have had a material effect on their flow. The quest to understand climate has led meteorologists and engineers and geographers to think beyond borders; but they have faced countervailing pressure to fix their plans and dreams in place. Water draws our attention not only to the two-dimensional space between points on a map—as when we trace the crooked line of a river—but also to depth and altitude, which turn out to matter more than historians have realized.

What we end up with is not an alternative to the well-known narrative of modern Asia shaped by empire and capitalism, forged by anticolonial revolution, remade in the second half of the twentieth century by ambitious new states. Rather, water adds another dimension to that familiar story. Asia's waters have long been a gauge for rulers' ambition, a yardstick of technological prowess—and a dump for the waste products of civilization. Water is, in a sense, a "sampling device" for other sorts of change, even as changes in water ecology have had a direct effect on millions of people's lives.[14] We can trace many of Asia's political transitions through the effects they had on water: from the global reach of the British empire in the nineteenth century, to the projects of national reconstruction that the Indian and Chinese states carried out in the twentieth. But the history of water is more than a mirror to human intentions. The history of water shows that nature has never truly been conquered. Water has served as a material constraint on every Promethean plan of growth and plenty. The sheer ferocity of a wet climate—a climate of monsoons and cyclones—remains a source of fear, and no fear is as great as the fear of water's absence, in drought. The cultural history of water is one of reverence as much as hubris. And water has its own chronology—the chronology of the seasons; the episodic chronology of sudden, intense disasters; the imperceptible chronology of cumulative damage, as manifested in the effects of human activity on the oceans.

                                II

Environmental history derives its richness from a close attention to particular landscapes—the most profound works have often been local and regional in scope, ranging from the study of a single village to a city, a forest, or a river. Only at that limited scale can we truly tease out the relationships between nature and human society. But the scale of environmental change has ballooned; its pace has accelerated. Connections between environmental crises have multiplied: the causes of harm and risk in any given locality may lie far away. We need a larger view. In a 2009 article, "The Great Himalayan Watershed," historian of China Kenneth Pomeranz took up the challenge: "For almost half the world's population," he wrote, "water-related dreams and fears intersect in the Himalayas and on the Tibetan plateau."[15] The Himalayan rivers bind the futures of the significant portion of the world's population that depend on them; conflicts over their course, and their use, threaten to ratchet up tensions between bordering states, especially India, Pakistan, and China.

The scale and interconnection of Asia's water crises provide a starting point for *Unruly Waters*. But this is not only a view from the Himalayan peaks; still less is it the omniscient view from a satellite image, for one characteristic of the satellite view is that there are no people in it, even if signals of the human imprint are everywhere apparent. This is a history of Asia's waters with India at its heart—and there are three compelling reasons why India is an illuminating vantage point from which to tell a story that crosses regional and national boundaries.

The first is India's centrality to the history of the British Empire; and empire's centrality, in turn, to the history of climate change. The conquest of most of the world by European powers in the nineteenth century forced a fundamental transformation in the human relationship with the rest of nature. Asian and African lands were drawn more closely into a global capitalist economy. Their absorption was underpinned by imperial gunboats and colonial taxes, but it was driven, too, by new opportunities for enrichment and advancement. India was at the sharp edge of change—exploited more intensively and on a larger scale than almost anywhere else, and pivotal to the further thrust of imperial power into Asia. From European trading companies' earliest expansion into the islands of the Atlantic and the Caribbean in the early modern era, they thrived on the exploitation of "cheap nature" as well as coerced labor.[16] The pace of change stepped up in the nineteenth century. The period from the 1840s to the 1880s witnessed the global triumph of industrial capitalism; in Eric Hobsbawm's words, "an entirely new economic world was added to the old and integrated into it."[17] India's fields and its waters were pushed harder to sustain the colonial state—which depended on agricultural taxes—and to produce the raw materials that fed Europe's industrial machine and its working classes: cotton, jute, indigo, sugar, tea, and coffee. Each of these thirsty crops generated new demands for water.

From India, imperial power and investment spread east and west across the Indian Ocean. British ships, filled with Indian troops, set sail in 1839 to bombard China, to force the Chinese government to allow the sale of Indian opium to Chinese consumers—a traffic that was vital to the East India Company's financial health. A reordering of the entire region between India and China followed. By the last quarter of the nineteenth century, from Burma to Vietnam, Asia's demography changed as migration opened new frontiers of settlement; its ecology altered to accommodate the spread of cash crops for export. Many of Asia's largest coastal cities—Mumbai, Calcutta, Chennai, Dhaka, Hong Kong, Jakarta—began life as colonial ports, built to sustain the global trading networks on which European empires thrived.

Imperial India reached further than the present boundaries of the Indian nation-state, and further, too, than the region we now define as South Asia—present-day India, Pakistan, Bangladesh, Nepal, Bhutan, and Sri Lanka. But British India was also more internally variegated than independent India. Areas under direct British control existed amid a patchwork of other forms of polity, known collectively as "princely states," all of which retained a degree of sovereignty while submitting to overall British domination. Both within the Indian subcontinent and beyond its shores, water constituted the connective tissue of imperial power. In the British imagination, India extended across the vastness of the Indian Ocean, connected to China and Southeast Asia (the "East Indies") through the flow of its rivers and the span of its climate. The ability to imagine India on that scale was, itself, a product of the nineteenth century and its new ways of seeing—maps, censuses, surveys, and photographs. It depended on the compression of space by the railway and the steamship. The contraction of those larger geographies in the twentieth century is a recurrent theme in this book.

In another sense, too, India's experience of imperialism cast a long shadow over the history of Asia's waters. British colonialism was a source of enduring trauma for many Indians, including for the educated elite that led India's nationalist movement in the first half of the twentieth century. Beyond the outright violence that the British government of India deployed, this trauma resided in a sense of profound social and economic destabilization. The effects of British policies combined with drought at various moments in the nineteenth century to create famines that killed millions. At the core of anticolonial thought in and beyond India was a clear imperative: "never again." In China, too, the experience of a "century of humiliation" at the hands of European powers, beginning with the catastrophic Opium Wars, left political leaders with a deep and urgent drive for self-sufficiency and self-reliance. The control of water was central to almost every scheme that arose from this quest for development. Memories of the nineteenth century lie beneath the fervor with which India built 3,500 large dams, and China 22,000, in the decades after independence. The memory of subordination by European empires continues to shape Indian and Chinese foreign policy: it orients their approach to agriculture; it even underpins their responses to climate change.

If India's role in empire is one reason to put it at the heart of this story, the second is India's political history after independence. Alone among Asia's newly independent or postrevolutionary states, India has been a democracy continuously since 1947, for all but three years. Clamorous and vibrant and flawed, Indian democracy has coexisted with glaring social and economic inequalities; the Indian state has often behaved in an authoritarian manner, not averse to exercising the powers it inherited wholesale from the British Raj. In their pursuit of water at any cost, there has been little to distinguish Asian states with different political systems and with varied ideological complexions. Still, the depth and diversity of India's public sphere has been unique.[18] Debates about water in India were never limited to disagreements between experts behind closed doors (as they were in China until the 1980s). They threaded through newspaper columns; they animated social movements; they filled the publications of environmental organizations. Many of these ideas echoed beyond India, and in turn Indian observers marshaled examples and gathered data from around the region. One vehicle for the movement of ideas about water and technology was the cinema. In the second half of the twentieth century India developed a popular film industry that exceeded Hollywood in size and matched it in influence in the postcolonial world: Indian films drew large audiences across Asia, the Middle East, and Africa. To an extent that has no parallel, Indian cinema captured the hopes and fears that fired visions of "development" across the Third World. Water was a recurrent theme.

The third, and perhaps the most fundamental, reason to tell this story looking out from India is a climatic one. The Indian subcontinent is the crucible of the monsoon.[19] And the monsoon is the thread that runs through *Unruly Waters*. In its simplest definition the monsoon is "a seasonal prevailing wind." There are other monsoons, in northern Australia and in North America; none is as pronounced, as marked in its reversal between wet and dry seasons, as the South Asian monsoon. More than 70 percent of total rainfall in South Asia occurs during just three months each year, between June and

Storm clouds, characteristic of the weeks leading up to the burst of the monsoon.
CREDIT: NurPhoto/Getty Images

September. Even within that period, rainfall is not consistent: it is compressed into a total of just one hundred hours of torrential rain across the summer months. Despite a vast expansion in irrigation since 1947, 60 percent of Indian agriculture remains rain-fed, and agriculture employs 60 percent of India's population. Unlike China, unlike most large countries in the world, India's population will continue to be predominantly rural by the mid-twenty-first century. No comparably large number of human beings anywhere in the world is so dependent on such intensely seasonal rainfall. In the first decade of the twentieth century, the finance minister in the imperial government declared that "every budget is a gamble on the rains"; more than a century later, leading environmental activist Sunita Narain reversed the terms but retained the substance of the observation: "India's finance minister *is* the monsoon," she declared.[20]

Climate is woven into the fabric of Indian social, economic, and political thought in a way that it is not (or is no longer) elsewhere. In the late twentieth century that claim would have raised hackles among scholars of South Asia; it might still do so today. A fundamental assumption of modernity was that we had mastered nature. The notion of India in thrall to the monsoon would seem to perpetuate a colonial idea of India's irredeemable backwardness. To emphasize the power of the monsoon would be to portray Indian lives as so many marionettes moved by a climatic puppetmaster. That is how this story would have been understood a generation ago. But to our eyes now, alarmed by the planetary crisis of climate change, a reminder of nature's power has different implications. This is not a story of geography as destiny. It is a story of how the *idea* of geography as destiny provoked, from the mid-nineteenth century on, a whole series of social, political, and technological responses within and beyond India. The monsoon is significant precisely because it has been a unique source of human concern, fear, and adaptive ingenuity. The desire to liberate India from its climate powered hydraulic engineering on an ever-increasing scale, with consequences far beyond India's borders. The struggle to understand the monsoon's dynamics motivated scientific research that remains at the core of our understanding of global climate. Living with the monsoon, India never had the luxury of the climate-blindness that has seeped into many other societies' worldviews. The history of how Indians have understood and coped with the monsoon may have wider lessons at a moment when climate can no longer be ignored, anywhere in the world—in this sense, at least, India is not behind the world but ahead of it. The lessons are not always heartening. As we will see, awareness of the monsoon's enduring power has coexisted with inertia, with negligence, with decisions to put more people in harm's way, and with maneuvers by the wealthy and powerful to insulate themselves from risk.

The South Asian monsoon has effects far beyond South Asia. We know this, at least in part, because of climate research undertaken in India in the twentieth century. Sir Gilbert Walker, a pioneer of global climate science, wrote in 1927 that "the climate of India is of special interest, not merely as that of the greatest tropical region in the British Empire, but also because it seems to have been designed by nature with the object of demonstrating physical processes on a huge scale." That sense of scientific opportunity, combined with the pressing material need to understand the monsoon, inspired a century of study in India. Charles Normand, Walker's successor as head of the Indian weather service, insisted that the monsoon is "an active, not a passive feature in world weather." Subsequent research has confirmed his view—the Asian monsoon is entwined with many aspects of the global climate. It has an important influence on global atmospheric circulation. The future behavior of the South Asian monsoon has implications for the whole world.[21] Arguably no other part of the global climate system affects more people, more directly.

———

AND SO, STARTING OUT FROM AND RETURNING TO INDIA, WE FOLlow the monsoon, the mountain rivers, and ocean currents—straying into Chinese waters, traveling down the Mekong, skirting the coastal arc of Asia, and coming back to the heart of South Asia.

This is a story with many possible beginnings. The recently excavated Liangzhu Ancient City, along the lower Yangzi River delta, reveals the vast scale of hydraulic engineering undertaken along China's coast five thousand years ago.[22] The elaborate step-wells of Rajasthan and Gujarat, and the anicuts (dams) along the rivers of South India, are testament to a long struggle to cope with the monsoon. But our starting point lies in the nineteenth century, when the scale and interconnectedness of Asia's waters first became visible, in tandem with unprecedented pressure to put water to work. The concatenation of political, economic, and environmental transformations this set in train continue to shape modern Asia.

TWO

WATER AND EMPIRE

ALONG THE BANKS OF THE GODAVARI RIVER, CLOSE TO THE MIDSIZED city of Rajahmundry on the eastern coast of India, stands a museum dedicated to Sir Arthur Thomas Cotton. Pointing the way to the museum, a bronze statue of the man watches over a busy bridge across the river; he is mounted on a horse, head cocked, eyes on the horizon.

Arthur Cotton was born in Surrey, one of eleven children. He joined the East India Company's forces in 1819, as second lieutenant in the Madras Engineers. Two years later, he was seconded to the chief engineer of Madras, and developed his lifelong fascination with water. He was an evangelical Christian, a stern and devout man. His career began with a marine survey of the Pamban channel, off the coast of Madras. In the 1840s, Cotton renovated and restored the ancient dam at Kallanai, along the Kaveri River that flows east from the Western Ghats mountain range to the Bay of

Bengal; the Kaveri's fertile delta was, and still remains, the agrarian heartland of Tamil-speaking South India. Cotton's attention moved north of the Kaveri, to the Krishna and Godavari rivers that meet the Indian Ocean in the region of Andhra, farther up India's eastern seaboard.[1] In 1852, Cotton completed a barrage, or dam, over the Godavari River at Dowleswaram, which regulated the flow of the river using large gates. Henry Morris, the district's chronicler, described the barrage as the "noblest feat of engineering skill which has yet been accomplished in British India." It was a "gigantic barrier thrown across the river from island to island, in order to arrest the unprofitable progress of its waters to the sea."[2] The Godavari

Statue of Sir Arthur Cotton at the Arthur Cotton Museum, near Rajahmundry, India. CREDIT: Sunil Amrith

delta gets scarcely a mention in most general histories of British India. It witnessed no major battles or massacres; it was home to few members of India's nationalist intelligentsia; its urban centers were relatively small. But the region epitomizes the transformation of India's waters in the nineteenth century.

The Cotton museum, close to the barrage, conveys both unaffected enthusiasm and palpable neglect. The photographs are faded. Monuments of water technology from the 1850s—pulleys and simple pumps—are dotted around the complex. There is something almost accidental in their placement, as if they had been forgotten there. But the museum is well attended by groups of schoolchildren and by young couples. Most of the explanatory text in the museum is in Telugu. The message is unambiguous: Arthur Cotton saved the Godavari delta. His bold engineering skill turned it from a poor region into an expanse of irrigated fertility. Frescoes on the wall tell the

గోదావరి మండలం 1831, 32 సంవత్సరాలలో అతివృష్ట తుఫానుట వరదలకు గురియైనది.

Fresco inside the Arthur Cotton Museum showing Andhra at the mercy of the elements before Cotton's engineering feats. CREDIT: Sunil Amrith

story: before Cotton, this land was stalked by famine and leached by drought; thanks to his munificence it became the "rice bowl" of India, secure from the fluctuations of climatic fortune. On a local bus the next day, my neighbor turned to me and—prompted by nothing more than the lush landscape around us, and a sense that I was a visitor—told me the story again. "Everything here," he said, his arm sweeping across the horizon, "is here because of Cotton *dora* ('Boss Cotton'); he was a very great person." A Telugu language biography of Cotton was published a few years ago. Farmers' associations are named after him. Every year on his birthday, cultivators gather to garland his statue. In 2009 a small delegation from Andhra, including a former cabinet minister, traveled to England to locate his grave, which they found in a quiet corner of Dorking, Surrey. Such veneration of a colonial Englishman is unusual in contemporary India: curiously at odds with the movement to rename cities and streets and buildings to erase the stain of imperialism. It reflects a sense that water has a value beyond ideology, beyond politics—beyond history.

Boat on the Godavari River. CREDIT: Sunil Amrith

The geography of empire in India was sculpted by wind and water. Until the nineteenth century, the only India that Europeans knew, the only India they were interested in, was the India that was wet. They sailed to India's coasts, swept there by the direction of the monsoon winds; in the eighteenth century, they moved upriver into the Ganges valley, heartland of the successive Indian empires of the Mauryas, the Guptas, the Afghans, and finally the Mughals. By 1800, the English East India Company had defeated its remaining Indian challengers: the Marathas in the west, and Tipu Sultan's kingdom of Mysore in the south. Following the Napoleonic Wars, British power commanded the Indian Ocean. But the British faced the same hydraulic dilemmas of every South Asian empire before them. The sea routes between India and the world were governed by the reversal of the winds. Communication between the coasts and the interior was slow; India's mighty rivers could only be traveled up at certain times of the year; roads were poor. The East India Company's revenues were tied to the cycle of planting and harvesting. Only gradually did the Company incorporate arid zones into its domain—the Deccan and the southeastern

edge of the Peninsula by 1800; and then, by the middle of the nineteenth century, India's northwest frontier.

Over the next half century British engineers and administrators
and investors sought to master nature, as a step toward connecting
India's interior more closely to its coastal ports and from there to the
rest of the world. The quest to understand water in India fused the efforts of adventurers and engineers, mariners and scientists. They were
driven by curiosity and by necessity. Some sought profit and renown.
Others followed their private enthusiasms. Not all of them served the
colonial state. Their work would not have been possible without the
ingenuity of Indian assistants, observers, draftsmen, recorders, porters, and soldiers, whose achievements have been effaced, for the most
part, from the historical record. Women were scarce in this scientific
world, but the few who were involved made contributions of lasting
significance. The science of water in nineteenth-century India traced
the descent of the rivers, the tracks of the storms, and the path of the
rains. Each of these crossed the borders of British India. Knowledge
of each brought awareness of interdependencies and inequalities on
a regional scale. Each provoked new kinds of political intervention.

I

The word "monsoon" appeared in English first in the late sixteenth
century, derived from the Portuguese monção. It comes from the Ara-
bic mawsim (for "season"), which also provides the word for "season,"
mausam, in Urdu and Hindi. In its simplest definition, it is a weather
system of regularly reversing winds, characterized by pronounced wet
and dry seasons. There are many monsoon systems around the world,
but the Asian monsoon is by far the greatest in scale and consequence,
and the Indian subcontinent is its zone of most intensive activity.

South Asia lies at the heart of the monsoon system because of the
geological history that has left the Indian peninsula protruding from
Eurasia into the vastness of the Indian Ocean. India lies at the edge
of the continental landmass that dominates the northern hemisphere,

facing a southern hemisphere that is mostly water. The monsoons have evolved over tens of millions of years. They have left an archive of their natural history on the seabed and on land. Tiny algae, diatoms, and single-celled marine plants called radiolara show that the monsoons first appeared in the Miocene era, soon after the Himalayas irrupted from the collision of the island Indian peninsula and the Eurasian landmass. Traces embedded in tree rings tell us that the Asian summer monsoon has strengthened during warm interglacial periods, as during the Medieval Warm Period up to the fourteenth century, and weakened during periods of planetary cooling, as during the Little Ice Age that lasted from the middle of the sixteenth to the early eighteenth centuries.

As early as 1686, English astronomer Edmund Halley identified the basic driving force of the monsoon as the differential heating of the sea and the land—he saw it as a gigantic sea breeze. In the summer months, land temperatures rise more rapidly than the sea warms. Winds are driven from areas of high pressure over the sea to areas of low pressure over land. "The Air which is less rarified or expanded by heat and consequently more ponderous," Halley wrote, "must have a Motion towards those parts thereof, which are more rarified, and less ponderous, to bring it to an Equilibrium." Halley's understanding missed one crucial dimension, which was understood by George Hadley in the eighteenth century—Earth's rotation affects the winds, causing them to veer right in the northern hemisphere and left south of the equator.[3]

So as the Asian landmass begins to heat up in the spring, the warming air above it rises, and cooler, moist ocean air moves in to take its place. The monsoon winds blow from the southwest, curving and doubling back to grip India, pincerlike, from both the Arabian Sea and the Bay of Bengal. The air sweeping in from the ocean contains vast stores of solar energy in the form of evaporated water, which is released as the vapor condenses as rain: the release of this stored energy sustains the power of the monsoon. The monsoon makes landfall in Kerala and Sri Lanka in late May or early June, reaching the

Bengal delta by the end of the month and moving steadily inward. The arrival, or "burst," of the monsoon is presaged by a period of unsettled weather and frequent thunderstorms. When it comes it can be spectacular. It brings welcome relief after months of building heat; it sustains the land's capacity to feed India.

Torrential rainfall cools the earth's sodden surface as the peaks of temperature and rainfall move steadily inland, finally petering out as they reach the far northwest of India and Pakistan. The Himalayas are a crucial part of the monsoon system. The elevation of the Tibetan Plateau leads it to warm rapidly and so drives the differentials of pressure and temperature that power the monsoon system; but the mountains themselves act as a colossal barrier to the winds, essentially sealing India off from the rest of Asia, and concentrating the monsoon rains to the south of the mountains, along the Gangetic plain.

As the temperature contrast between land and sea begins to even out, the system returns to equilibrium, and another period of transition begins. As winter advances the Asian landmass cools more rapidly than the ocean. The winds now reverse to blow from the northeast, creating dry conditions over much of Asia between November and March. But neither in summer nor in winter is the monsoon uninterrupted. The wet season is characterized by frequent suspensions in rainfall, known as "breaks"; the "dry" winter monsoon brings the bulk of the year's rainfall for a few regions, including the Tamil Nadu coast in southeastern India.[4] The periods of transition, as the winds reverse, are prime time for the devastating cyclones that visit the Bay of Bengal regularly. As we will see throughout *Unruly Waters,* the quest to understand the monsoon, which began in earnest in the second half of the nineteenth century, has been riddled with obstacles. The study of the monsoon remains filled with uncertainty.

IN THE LONG SWEEP OF INDIAN HISTORY THE MONSOON IS BOTH an internal and an external frontier. The monsoon has shaped the limits of cultivation and the distribution of crops. It has facilitated communication between some places and barred it between others. Its ecological niches have created economic unevenness—the stuff of which political power is made. The reach of the monsoon also marks the junction, the ecological nexus, between two very different ideas of India. One is as a settled agrarian empire; the other, as the outward-looking heart of the Indian Ocean world. The pattern of the monsoon draws a rough vertical line down the middle of the Indian subcontinent, dividing the drier west—part of a Eurasian "arid zone" stretching across Central Asia and as far as the Sahara—from the wet, marshy east, which stretches beyond, to Southeast Asia, to form a region that twentieth-century geographers called "monsoon Asia."[5] The bulk of India's population has always lived to the east of that line.

The line dividing wet and dry zigzags across the subcontinent. The arid zone reaches down from Rajasthan in the northwest to the Deccan plateau at the heart of central India. The Deccan lies in the "rain shadow" created by the hulking Western Ghats mountain range—the rain clouds that sweep in from the Arabian Sea collide with the high mountains and disgorge their contents, leaving little for the plains beyond. From there arid bands snake down to the very far southeast of India, interspersed with more fertile coastal or riverine belts. The frontier corresponds to the ancient division—still visible today, though now modified by technology—between the major staple crops that have fed Indians for centuries: rice in the monsoon zone, and wheat or millet in the drier region.

The great rivers of South Asia modify the patchwork of wet and dry. They interact with the monsoon in a hydraulic cycle of colossal proportions. Because of their fertility, their ability to sustain life and to produce a surplus, the Gangetic plains have been the heart of every Indian empire. Their rich alluvial soils have supported a large population for centuries. The river system watered crops and provided an artery of transportation and trade, if never on the same scale or with the same reach as the Yangzi River system in China and its ancient complex of canals.[6]

Given their immense power both to sustain and to destroy life, India's rivers have been among the most revered on Earth. The river Ganges—often styled as "mother," or "Ma Ganga"—is the archetypical sacred river, spiritual source of all of India's rivers, writes scholar of Hinduism Diana Eck. In some sense, all of India's other rivers are microcosms of the Ganges. For millennia, the Ganges has been a site of pilgrimage, most especially at the point of its confluence with the Yamuna River at Prayag. For many Hindus, *moksha*—liberation from the cycle of rebirth—has been believed to come from bathing in the waters of the Ganges, or being cremated on it banks. The purity of the waters of the Ganges (*gangajal*) has long been accepted and valued by people across India. Hindu scriptures contain many versions of the origin myth of the Ganges, known as the *avatarana,* or descent to Earth. In the version of the story in the *Ramayana* and the *Mahabharata,* the unruly Ganges tumbles from heaven, tamed as it flows through the serpentine locks of Shiva's hair before it spills onto the plains of India. In all of these stories, the Ganges epitomizes liquid *shakthi,* the energy that sustains the universe. The Ganges is not alone, it stands at the apex of a land of sacred waters. In many regions of India, rivers have been personified; their flow helps people to imagine how distant places are connected to one another. In many spiritual traditions in South Asia, the rivers have been thought to channel the power of all the water in the world, from the clouds to the oceans.[7]

THE LINE BETWEEN PLENITUDE AND SCARCITY MIRRORED THE trajectory of the rain-bearing winds and followed the paths of the rivers. Over centuries India's rulers built irrigation canals, storage tanks, channels, and dams. These shifting arrangements bore little resemblance to Wittfogel's ideal type of a "hydraulic society." Regional leaders and imperial administrations spearheaded construction projects, but so too did local lineages, temple complexes, and landowners. In the pre-modern period the most widespread infrastructures of irrigation were found in Sri Lanka and in central and southern India. The biggest of them were elaborate hydraulic systems, individual waterworks linked in a larger web. And some of the dams, like the sixteenth-century Daroji reservoir in arid northern Karnataka, were large even by modern standards. Spurts of hydraulic ambition alternated with stasis, construction with disrepair. The power that arose from the control of water spread unevenly, liable to seizure or decay.[8]

Water was never far from the minds of the mounted conquerors from the highlands of Inner Asia who stormed their way to the Gangetic plain in the second millennium to forge a new political power in India. The heart of their power lay at the frontier between the monsoon and the arid lands. They harnessed the benefits of both. Established in 1206, the Delhi Sultanate was the first Persian-Islamic state in South Asia.[9] Though it collapsed in the second half of the 1300s, riven by internal division and threatened by fresh invasions from the northwest, the reach of the sultanate's power into the heartland of the Indian subcontinent was a prelude to the Mughals' even greater empire.

The Mughals were a Turko-Mongol dynasty with roots in present-day Uzbekistan. They unified much of the Indian subcontinent during the two centuries when they were at the height of their powers. Zahir-ud-din Muhammad Babar (1483–1530), known as Babur, was the first Mughal emperor. He claimed descent from Timur (Tamerlane), the Turkic conqueror, and on his mother's side from Chingghis Khan. Driven from Samarkand, Babur established a new kingdom in Kabul, Afghanistan. From there he launched an assault on the Indian subcontinent, where he established the Mughal Empire in 1526. From the age of twelve, he kept a diary from which he later composed the *Babur Nama,* one of the earliest autobiographies in the Islamic tradition. He was a meticulous observer. He was driven by naked ambition. He was not averse to brutality. He was a lover of nature. The *Babur Nama* is filled with references to water. Babur's primary interest was in water as both ornament and practical necessity in constructing the gardens that he loved. In the Mughal tradition of landscape architecture, gardens played both a symbolic and an aesthetic role: they were places of beauty and sensual pleasure. Their proportions embodied the principles of order and harmony.

Babur's interest in water went beyond the requirements of his exquisite gardens. He commented on the entire system of irrigation at work as he advanced into North India. The cultivation of gardens and the sustenance of agriculture were related endeavors.[10] "The greater part of the Hindustani country is flat," he observed of the Yamuna valley. "Many though its towns and cultivated lands are, it nowhere has running waters"—by "running waters," Babur meant the canals well known in the Central Asian lands of his birth. Rather, "rivers and, in some places, stagnant waters" in wells or tanks, irrigated the Indian plains. He saw that "autumn crops grow by the downpour of rain themselves," but that "some vegetables" had to be "watered constantly." Babur observed cultivators at work. He was struck particularly by the method that the British would later dub the Persian Wheel. "In Lahore, Dibalpur and those parts people water by means of a wheel," he wrote:

> They make two circles of ropes long enough to suit the depth of the well, fix strips of wood between them, and on these fasten pitchers. The ropes with the wood and attached pitchers are put over the well-wheel. At one end of the wheel axle a second wheel is fixed, and close to it another on an upright axle. This last wheel the bullock turns. Its teeth catch in the teeth of the second, and thus the wheel with the pitchers is turned. A trough is set where the water empties from the pitchers and from this the water is conveyed everywhere.

Further down the Yamuna valley, toward Agra, he noticed that "people water with a bucket"—leather buckets lifted by yoked oxen—which he described as "a laborious and filthy way."[11]

The Mughal realm expanded ceaselessly between 1560 and 1605, and again between 1630 and 1690. Its territories stretched from Gujarat in the west to Bengal in the east, and far into South India.[12] The

Mughals mobilized long-distance trading networks that followed the caravan routes to the western edge of central Asia and beyond. They stored wealth in precious metals. Once the Mughals conquered the Gangetic plains, they filled the state's coffers from productive, densely settled agrarian lands and large populations. The Mughals inaugurated a rigorous system of land taxation. The administration relied on an interlocking system of larger landowners (*zamindars*), who served as tax collectors, and, below them, subordinate holders of rights to the land. The state used land grants to reward its loyal officials and to co-opt local elites into the system.[13]

Mughal military and financial prowess came from the mounted martial traditions of dry lands. As many as one in five men served in local military forces, often as seasonal labor in a climate where agriculture was uncertain and horses were plentiful.[14] But India's unruly waters had a bearing on Mughal military strategy, as they would later constrain the options of the British. As they approached the Bengal delta, Mughal armies struggled beyond Rajmahal—their horses were ineffective in the humid conditions; they had to use boats. The *Akbar Nama,* Abul Fazl's account of the Emperor Akbar's reign (1556–1603), describes the challenge of climate. In 1574, as Akbar's forces captured the city of Patna—near the site of the ancient Ganges River port of Pataliputra—they "chose the river route, in this season full of turbulence, and with constant rain and tempest."[15]

Aside from military campaigns, the instability of India's rivers were a source of woe to local inhabitants. Historian Irfan Habib's *An Atlas of the Mughal Empire*—a painstaking work of recovery—pieced together, from the minutiae of Persian sources and European travel accounts, how often, how abruptly, rivers changed course. This turmoil came from the huge loads of silt the rivers carried down from the Himalayas—when the rivers were in full force they threw up sandbanks and islands as obstacles around which the waters found a new path; silt deposits raised riverbeds and pushed the rivers into new channels. At other times, sudden changes in course were driven by violent earthquakes. The only response was flight—

abandoned settlements lined the banks of vanished rivers, as when the
Ganges "deserted" the once-flourishing town of Kanauj in the early
sixteenth century. People had no choice but to move with the migrating waters. Beginning in the early seventeenth century, the Ganges
began to shift eastward. A major earthquake in 1762 and another
in 1769–1770 jolted the river away from its channel, forcing it into
contact with new tributaries: the Tista and the Jamuna, the Jelangi
and the Mathabhanga, the Kirtinasa and the Naya Bhangini. Even
the names that rivers bore could be a testament to their instability:
"Naya" means new, suggesting memories of an old Bhangini. The
coastline shifted with the rivers: Habib reconstructs the coast of Gujarat through "decaying ports": places that had once been on the coast,
that now lay silted up.[16]

As their realm expanded to reach India's coasts, the Mughals incorporated within their realm port towns that faced the eastern and
western Indian Ocean. Long before the arrival of Europeans, merchants in India had trading links that spanned the Indian Ocean rim.
Indian textiles filled marketplaces across Southeast Asia and China,
the Mediterranean and West Africa. Many regions of India—Gujarat,
Bengal, and the Coromandel coast—thrived on long-distance trade.
Textiles brought forth medicinal products, spices, local crafts, and
large quantities of precious metals. On one estimate the Indian economy absorbed 20 percent of the world's supply of silver between
1600 and 1800.[17] Throughout Southeast Asia's era of commercial
expansion in the sixteenth century, Indian traders from the coasts of
Gujarat, Madras, and Bengal shipped cloth to Pegu and Tennasserim
in Burma, to the thriving port of Melaka on the Malay peninsula,
and to the Indonesian islands of Sumatra and Java.[18]

The Portuguese apothecary Tomé Pires observed in Melaka in
1512 that ships from Bengal brought "five white cloths, seven kinds
of *sinabafos,* three kinds of *chautares, beatilhas, beirames* and other
rich materials. They will bring as many as twenty kinds." In their
holds came "very rich bed-canopies, with cut-cloth work in all colors and very beautiful," and "wall hangings like tapestry." Pires con-

cluded that "Bengali cloth fetches a high price in Melaka, because it is a merchandise all over the east"—from Melaka these textiles would make their way to markets across the Indonesian archipelago. In return, Indian traders exported from Melaka "camphor and pepper— an abundance of these two—cloves, mace, nutmeg, sandalwood, silk, seed-pearls a large quantity, copper, tin, lead, quicksilver, large green porcelain ware from the Liukiu [Japan's Ryukyu islands], opium from Aden . . . white and green damasks, *enrolados* from China, caps of scarlet-in grain and carpets; krises and swords from Java are also appreciated."[19] The variety of Indian cloth gave rise to a lexicon that seeped into languages of trade everywhere: longcloth and salemporis, moris and gingham, dungarees and guinea cloth and kaingulong. Indian weavers targeted diverse markets. Their weaves, patterns, colors, and designs were all adapted to local tastes.[20]

The Indian Ocean's trading world reached deep into the interior. It intersected with circuits of commercial exchange that went overland to Central Asia. In India, as elsewhere, the sixteenth and seventeenth centuries saw the deepening commercialization of the countryside. From 1500, South India saw the rise of a merchant class whose diversified business included overseas trade, collecting and keeping a share of local tax revenues, and financing local rulers' military ambitions. Global demand for Indian cotton set up a chain of transactions that linked the cotton-growing countryside to the port towns. Cultivators became more dependent on credit from urban merchants to finance the next year's crop; gold and silver currency came into more widespread use. Agrarian commercialization fed the Mughal state's treasury. The state's demand for cash taxes, historian Victor Lieberman observes, "acted like a giant pump, sucking foodstuffs from the countryside into towns and cites."[21]

This was the world that European trading companies entered; initially, they were players among many others. The monsoon winds brought the first Portuguese ships to India at the end of the fifteenth century. Arriving at Malindi on the eastern coast of Africa, Vasco da Gama sought advice from Indian traders there. They counseled him on the pattern of the winds that would take him across the Indian Ocean to Calicut. A local pilot guided his voyage. Chartered companies from Europe's western edge took to the sea because the landed power of the three great Islamic empires—the Ottoman, the Safavid, and the Mughal—blocked their path across central Eurasia. So they found an alternative route to the profitable cotton textiles of India, the spices of the Indonesian archipelago, and the ceramic manufactured goods of China. Their ships rounded the Cape of Good Hope and crossed the Indian Ocean. They arrived on the coast of western India to find a region already open to the world, tied by commerce to the littoral of the Indian Ocean right up to the Mediterranean. Coastal India already, in the sixteenth century, faced both ways: connected to the far reaches of the Indian Ocean, but also to mountainous Central Asia.[22]

Successive invaders of South Asia had come on horse from the northwest. Now, parvenu claimants to power approached by sea. Conquest was neither easy nor initially attractive. Monopoly was what the Europeans sought. In staking a claim to exclusivity they militarized the Indian Ocean's sea lanes in a new way. Roiled by the Dutch revolt against the Spanish crown at home, the Portuguese soon met competition from the Dutch and English East India Companies in the Indian Ocean. European power was concentrated in a scattering of "factories" along the Indian coast. Each served as a dormitory, a trading post, and a warehouse all in one. European power rested on relationships delicately negotiated with local rulers. Competition among Europeans fueled the companies' expansion—armed conflict at home and rivalry on the high seas heightened commercial contests in India.

By the eighteenth century European chartered companies had mobilized relationships and resources that spanned the globe.[23] They paid for their purchases in Asia with silver from the mines of Potosí (in present-day Bolivia). They transported Indian cottons to the coast of West Africa, and exchanged them for enslaved human beings. Europeans possessed many advantages as they inserted themselves into the fissiparous politics of regional kingdoms that had risen to fill the political void of a declining Mughal Empire. As well as precious metals, they could offer their local allies the security promised by the sophisticated weaponry at their disposal. The companies raised a lot of capital, quickly, aided by their structure as joint-stock firms. They intervened in local succession disputes. They did deals with bankers and *zamindars*. British, French, and Dutch companies became enmeshed at the frontier between the coast and the interior.[24]

The transformative moment came in the second half of the eighteenth century. British maritime power pushed inland. Not long after the Battle of Plassey in 1757—which stabilized the British "bridgehead" in Bengal—the East India Company received the *diwani,* or the right to the land revenues of Bengal, the most fertile province of the Mughal Empire. Bengal's agrarian wealth funded a violent cycle of English expansion. The Company's army rose to become one of the largest military forces in the world. The English were the first truly to harness both of the monsoon frontiers, welding maritime and landed wealth. Between 1757 and 1857, British control expanded up the Ganges valley from Bengal; the most rapid period of expansion came in the 1790s and 1800s, as British confidence and ambition were fueled by the worldwide war against France. The Company subdued the proudly independent southern Indian kingdom of Mysore in 1799, the Gangetic kingdom of Awadh and the southern domain of Arcot in 1801, the Dutch-dominated island of Sri Lanka by 1815, the Maratha lands in western India by 1818, and the coast of Burma by 1826. The late Christopher Bayly, whose early work on the merchants and markets of North India is unsurpassed, observed that at all times the Ganges valley remained "the main axis of Britain's Asian Empire." The valley's commerce "pointed northward to the high regions of Central Asia," while "huge quantities of cotton, opium, and indigo bound for China and Europe" flowed downriver to Calcutta, along with hides, oilseeds, and the saltpeter used to make gunpowder. Shipments of rice, opium, and tobacco went upstream toward India's northwest frontier.[25]

As British power forced its way into the interior, into southern and
western India, the distribution of water created different possibilities
for agricultural, fiscal, and therefore political expansion. For scientists
and travelers, mastering the map of India's water became a matter of
curiosity; for revenue administrators it was a matter of urgency.

II

One of the earliest maps of the British domains in India was the
work of James Rennell (1742–1830), surveyor-general of Bengal.
Given the centrality of the Ganges to British power in India, Rennell
set out to map the Ganges, from its source to the Bengal delta. He
described the descent of the river from the "vast mountains of Thi-
bet" to the Indian plains, where it "serves the capacity of a *military
way* through the country," before reaching the ocean in "a labyrinth
of rivers and creeks." Rennell was impressed by the power of India's
rivers: "next to earthquakes," he wrote, "perhaps the floods of the
tropical rivers produce the quickest alterations in the face of our
globe." He described the "extensive islands" of silt (known locally
as *chars*) formed "during an interval far short of that of a man's
life." Rennell saw that "it is no new thing for the rivers in India to
change their course." He described how the confluence of the Kosi
and the Ganges had migrated forty-five miles in a short span of time.
The mighty Brahmaputra, India's second-largest river, had "varied its
course still more." Rennell described the Brahmaputra as unknown,
forbidding, and turbulent. Where the Brahmaputra met the Ganges, in the Bengal delta, he described "a body of running fresh water,
hardly to be equalled in the old hemisphere." Its governing influence
was the monsoon. The characteristic pre-monsoonal storms, known
as northwesters, were "the most formidable enemies that are met with
in this inland navigation." Of enemies there were plenty. Traveling upriver, he wrote, a regular "budgerow" (*bajra*)—large boats with cabins
that covered their length, most common along the Ganges—"hardly
exceeds 8 miles a day, at ordinary times."[26]

Map of the Ganges and Brahmaputra as they meet in the Bengal delta, from James Rennell, *A Bengal Atlas: Containing Maps of the Theatre of War and Commerce on That Side of Hindostan* (London, 1781). CREDIT: Courtesy of the Map Collection, Harvard College Libraries

Rennell was not alone in his fascination with, even fear of, India's climate. As they established gardens as a site of botanical and commercial experiment, Company officials in India took a greater interest in the weather—none more so than William Roxburgh (1751–1815). Roxburgh studied anatomy and surgery at Edinburgh University at a time of intellectual ferment; he left Edinburgh in 1772 to join the East India Company's ship *Houghton* as surgeon's mate on a voyage to India. The following year, he signed on for another voyage, which took him via Saint Helena and the Cape to Madras.[27] Upon his arrival at Fort Saint George in 1776, Roxburgh began a meteorological diary. He equipped himself with a portable barometer "made by RAMSDEN" and an indoor thermometer supplied by Nairne and Blunt—scientific instruments, like texts and theories, moved along imperial shipping lines. His outdoor thermometer he placed "under a small, shady tree." His observations, three each day, devised a scale for describing the winds: "gentle, brisk, stormy, and what we call a tufoon in India." The rain gauge he had initially proved worthless; he assured his correspondents that he had installed a better model on the roof of his house on the hospital grounds.[28] Based initially in Madras, Roxburgh moved south down the coast to the small port of Nagore, long connected to Southeast Asia by Tamil Muslim merchants; from there, he settled in Samulcottah (Samalkot) on the Godavari delta, midway along India's eastern coast, until finally in 1793 he became director of Calcutta's botanical garden. Roxburgh followed many pursuits. He made a small fortune as a private trader. He kept an "experimental botanical plantation" where he grew indigo and pepper, breadfruit and sugarcane; he collaborated with botanist Johann Gerhard König, who was stationed at the Danish settlement of Tranquebar on the Madras coast.[29]

Roxburgh was a keen observer of life around him. He took an interest in how the monsoon's rhythms shaped farming on the land. "The rains generally set in, in June," he wrote in his description of the growing season of the Godavari delta, "towards the end of that month, the coarse or early Paddy, is sown, and in July the better sorts, or great crop." He described how "our rains continue from the time they set in, June, 'till about the middle of November; July and August, are generally our wettest months: in October and November the weather is more stormy, being the period we call the Monsoon." In his usage, "monsoon" was the period of change, as the winds switched from southwest to northeast. "The cultivator has to depend on the rains," he said again; "the more favorable they are, the better is the crop." Higher, more arid lands, "as in every other part of India," were given over to "dry grain."[30] He undertook a detailed study of hardy crops that thrived in dry conditions, locally and around the world. He ordered samples from across the empire; he planted them in his experimental garden.[31]

Roxburgh became an astute observer of South India's climate, both in its regularity and its extremes. He could not himself escape its risks. In 1787, a severe cyclone struck the Godavari delta; it destroyed Roxburgh's home, his herbarium, his library, and most of his personal wealth. His family escaped narrowly with their lives. He observed at close quarters the prolonged drought that brought famine to the region in the late 1780s and early 1790s. In 1791 Roxburgh wrote to his friend, celebrated English naturalist Joseph Banks, that "the famine of these provinces begins to rage with double violence, owing to a failure of our usual rains." Two years later, Roxburgh's friend Andrew Ross declared that "the dreadful effects of the famine here have . . . far exceeded any description from us." He saw that "in many places where populous villages formerly stood, there is at present neither vestige of man or beast."[32] Roxburgh looked for patterns in the data he had collected—he sought to understand the cycles of the seasons, and variations from year to year.[33] During his years on the Coromandel coast, Roxburgh had collected an extent of meteorological data on the Madras coast that one historian describes as "unrivalled elsewhere until the 1820s except among indigenous Chinese observers."[34] The early initiatives of Roxburgh and his colleagues formed a foundation on which India's modern meteorology was built.

He began to wonder, like so many others of his generation and education, whether India's nature might benefit from "improvement."[35]

He wanted to harness the water that "passes annually unemployed into the sea."[36] He was "astonished" to find not the "least trace of any work, ancient or modern, for retaining, or conveying the water to fertilize their Paddy Lands" in the region; the result was that "the cultivators here depend entirely on rains, when they fail, a famine is, and must ever be, the consequence." Roxburgh observed, and sketched, the Godavari delta; he imagined it transformed. "In consequence of the favourable level and descent of lands," he wrote, "we clearly see the infinite benefit that must arise from the waters of large rivers when a method of making them subject to the will of man is affected." The solution he saw was to use natural basins to store large quantities of water as the Godavari descended from the hills.[37]

III

However we understand India's economic transformation in the nineteenth century, water is at the heart of it. The flow of water—the flow of India's rivers, their seasonality, their propensity to change course—constrained how India's producers could respond to new market opportunities and new compulsions. Britain's industrialization had benefited from an elaborate network of canals; India's economic development, by contrast, was limited by the difficulty and expense of water transportation. China had a far more extensive network of canals than India at the time, but unlike in Britain, energy sources were far from the waterways.[38] The availability of water enabled a changing landscape of cash crop production, for some; water's absence tested others' capacity for bare subsistence. Water was instrumental to making the Indian soil produce more of the commodities the world demanded.

In the 1830s and 1840s, India's British rulers still faced constraints that would have been familiar to their predecessors. Transport was slow—and dangerous. "No part of the inland navigation of India is so dreaded or dangerous," wrote the botanist Joseph Hooker in 1848, "as the Ganges at its junction with the Cosi"; in the rainy season the

Cosi "pours so vast a quantity of detritus into the bed of the Ganges that long islets are heaped up and swept away"; boats "are caught in whirlpools formed without a moment's warning."[39] The monsoon governed not only the harvest—and threatened the possibility of harvest failure—but also threatened the health of Europeans. Cholera, malaria, and other ailments led many British officials in India to an early grave. Water was still a source of both awe and foreboding for British residents in India. A medical topography of Calcutta published in 1837 stated that "without taking into view the expanse of the Bay [of Bengal], the coup d'oeil of a good map of Bengal will at once show how bountiful nature has been to that country, by means of her majestic rivers with innumerable tributaries." But these waters were at the same time the source of "aqueous exhalations"—a product of the "commerce of land and water" in a monsoon climate—that menaced life. The author James Martin saw a clear "connexion of the rainy season with disease," and suggested that "among Europeans, the diseases of the rainy season assume a character of diminished vital action." Throughout the nineteenth century, fears persisted of whether Europeans could survive tropical climates.[40]

To make India productive, to integrate it more fully with the global capitalist economy that was in formation, to exploit more effectively its natural resources to feed Britain's industrialization, British engineers and investors and administrators looked to master the unevenness of water, its extreme seasonality in India; and they sought to conquer space. Both of these quests unfolded between the 1830s and 1870.

HALF A CENTURY AFTER ROXBURGH'S TIME, THE GODAVARI DELTA was still "entirely without any general system of irrigation, draining, embankments or communications."[41] This was the verdict of Arthur Thomas Cotton (1803–1899), the museum to whose memory opened this chapter. Like Roxburgh's before him, Cotton's problem was the distribution of rainfall across the landscape. His task: "counteracting the irregularity of natural supplies of water." "One year a portion of the whole crop . . . is destroyed by the overflowing of the rivers," Cotton observed, "in another, the crop is destroyed by a failure of the rains over three-fourths of the district." He was convinced that "not an acre . . . need be dependent at all" upon the rains if a comprehensive system of irrigation were introduced. He insisted that the Godavari delta needed not a piecemeal restoration of existing irrigation works, but rather "works of a general nature." Perennial irrigation; an improvement in the "roads and bridges" of the region; a restoration of the port of Kalinga ("Coringa") so that it could fulfill its potential as "incomparably the best port" between Hooghly and Trincomalee—such investments in infrastructure would free the district from its uneven and capricious rainfall.

Cotton made a fervent case for government intervention. India was unlike Britain, he argued; the rules governing public expenditure could not be considered akin to the principles of household economy. The problem was that "there is almost literally no capital to enable landowners to make improvement." An outlay of three hundred or four hundred thousand rupees each year by the state "would put life and activity into the whole district"—in time, revenue would flow into the treasury far exceeding what the state might spend. Possessed by evangelical self-confidence—nothing less than a sense of destiny—Cotton went further. He condemned what he saw as his countrymen's "proneness . . . to lower ourselves to the level of natives" instead of "diligently applying the means which God has placed in our hands to benefit the countries He has given us charge of."[42] Cotton found the support and the money for his grand scheme. In 1852, he completed his barrage at Dowleswaram. But his dreams were bigger. Cotton imagined a network of canals that would, one day, bring the Himalayan rivers to the southern tip of the peninsula. He also saw that the rivers had unrealized potential for navigation. In 1867, Cotton dreamed of a link between the Brahmaputra River—its upper reaches were still at that time unknown to British explorers—and the Yangzi.

"The throwing open of all India to all China, the access of a country containing 200 millions of people to the produce of a country occupied by 400 millions," he wrote, would be "a work of such magnitude as that nothing approaching it has ever been seen in the world."[43]

In the British imagination as well as in administration, Peninsular India was quite distinct from the "heartland" of Gangetic India. Separated by half a century, William Roxburgh and Arthur Cotton in turn sought to mold a riverine landscape that attached the dry interior of the Deccan plain to the coast of the Bay of Bengal. They sought both to harness and to overcome the political inheritance that distinguished South India from the north. In the former, political power was contested within a system of small states that arose to fill the void of the troubled Mughal Empire; the hydraulic landscape was dispersed in thousands of tanks, wells, dams, and weirs, many of them now lay in a state of disrepair after decades of warfare—not least the warfare that accompanied English expansion. But Cotton's counterparts along the Ganges were no less anxious to see what could be done to "improve" nature: to repair or replace the hydraulic remnants that scattered the valley. They faced different challenges, they chose different solutions, but they shared many assumptions with their counterparts in the south. Just two years after Cotton's barrage was complete, a project still more monumental opened its floodgates: the Ganges Canal, at the time (and still today) the largest in the world.

The Ganges Canal was the creation of Proby Cautley—Arthur Cotton's contemporary, classmate, and eventually his bitter rival. Cautley arrived in India in 1819 as an artilleryman. A few years after his arrival, the first Anglo-Burma war in 1824 drew many of the East India Company's engineers across the Bay of Bengal; their absence created new openings in India for those without formal training. Like so many Company officers, Cautley was an autodidact. He learned his craft through practice and observation. Working in different ecological settings, Cautley and Cotton embraced different hydraulic approaches. By the 1860s, they fought their battles in a bitter and public war of pamphlets. Cotton accused Cautley of making fundamental mistakes in the design of the Ganges Canal; at stake was not only prestige, but also a debate over the ownership and financial management of India's hydraulic works.[44] Along the Ganges, as everywhere else in India, the infrastructure of water control long preceded British rule. But in the nineteenth century British engineers turned the Ganges valley into one of the most "thoroughly engineered" landscapes in the world.[45]

The Gangetic plain's hydraulic transformation began with the Company's effort to restore the old Yamuna Canal's supply of water to Delhi. The waterworks dated back to pre-Mughal times: Delhi's water infrastructure owes much to the rule of Sultan Iltutmish in the thirteenth century. He ordered the construction of an elaborate web of tanks and step wells. The Mughals brought them to a new level of sophistication. They built a complex of ornate gardens along the banks of the Yamuna River, laid out around the tombs of Mughal leaders. They watered their new capital at Shahjahanabad from a canal and an interlocking system of smaller canals and drains. Emperor Akbar ordered the renovation of the West Yamuna Canal—first built by the ruler Firoz Shah—for irrigation, and extended it to Delhi.[46] Akbar's Canal Act of 1568 declared the canal's aims to be "to supply the wants of the poor," to "leave permanent marks of the greatness of my Empire by digging canals," and to ensure that "the revenues of the Empire will be increased."[47] The British found the canal gone to ruin, yet traces of its sophisticated engineering remained. In 1820, British engineers restored the water supply to Delhi through the West Yamuna Canal. They followed quite consciously in the footsteps of Mughal architects.

With this success in hand, local administrators turned to the restoration of the eastern branch of the Yamuna Canal. Second in command of this project was young Proby Cautley, who had no prior experience of hydraulic engineering. Cautley was open, perhaps unusually open, to learning from local practices: he suggested adapting local well-building techniques to provide a stronger foundation for bridges than usual European methods could sustain in the soils of the Gangetic plain.[48] As he took charge of the canal project, Cautley ordered the construction of rest houses every ten or twenty miles along the path—in keeping with the old Mughal tradition of *caravansarais* along the Grand Trunk Road. Besides water, Cautley's interests encompassed archaeology, paleontology, and botany. In 1831, while supervising the digging of a well as part of the canal project, he discovered evidence of an ancient settlement at Belka. With even more enthusiasm, Cautley and his colleague Hugh Falconer began collecting fossils of mammals and birds and fish, eventually shipping to the British Museum in London a collection that took up 214 crates. The history of science in nineteenth-century India often saw the blurring of lines between disciplines.

By the middle of the 1830s, though, Cautley was first and foremost a water engineer. In 1835, he became the Company's superintendent of canals. His predecessor in that role, John Colvin, had left him with an idea: to build a canal to bring the waters of the Ganges to the arid Doab (the name means "between two rivers") that lay between the Ganges and the Yamuna. Early investigations concluded that the canal would be too expensive—and probably an engineering challenge too far. The calculus of costs and benefits, so central to the Company's mode of administrative thought, changed in 1837 when a major famine devastated the drought-prone Doab. By 1840 plans were in place to build what would become the Ganges Canal.[49]

The centerpiece of the canal complex was a headworks at Haridwar, where the Ganges meets the plains. Its most complex feat of design was the Solani aqueduct, which ran sixteen miles below Haridwar—civil engineer G. W. MacGeorge, author of an 1894 treatise on the infrastructure of British India, called it the "most interesting and remarkable modern structure in India."[50] The technical challenges were formidable. The project created a hybrid landscape as an artificial "river" crisscrossed Himalayan streams that in the summer became torrents. "To carry the great canal—itself a small river—across such a country," one British engineer observed, "to see it pass silently on, uninterrupted and uninjured by these torrents" was "a triumph of art and engineering ability."[51] Above all, it was a feat of labor. The works were labor-intensive; machines played little role in the initial stages. The canal was the work of thousands who molded and fired bricks, their kilns fed by timber from local forests.[52] Earthworkers (*bildars*) dug the canal. Hundreds of men were deployed in transporting materials. Much of the work was organized by local contractors, who recruited workers from across the region. We know almost none of their names. Historian Jan Lucassen, in studying a strike by the brickworkers in 1848–1849, has uncovered a few of their stories. When their employers tried to cut wages, brickworkers first deserted the site, and later set fire to a number of encampments.[53]

At the time of its opening in 1854, the Ganges Canal was more than seven hundred miles long. A pamphlet, "A Short Account of the Ganges Canal," was distributed in English, Hindi, and Urdu at the opening. It declared that "the great motive by which the British government was led to sanction the Ganges Canal" was "to secure to its people, in the country between the rivers Ganges and Jumna, an immunity from the pains and losses that famine brings with it."[54] The famine of 1837 and 1838 was still fresh in the spectators' memories. For Company administrators, those memories involved the loss during the famine of land revenues, and relief expenditures amounting to over 5 million pounds—financial loss was the spur to action, however sincere the humanitarian considerations might have been.

A year after the canal opened, the *North American Review,* a Boston literary journal, published an account of the Ganges Canal and its opening ceremonies. It evoked the "double sanctity" that the "mysterious river of the farthest East" now possessed—the Ganges had long been revered and worshipped as a divine river, a place of pilgrimage for people from the distant corners of India; now it was newly (or doubly) blessed by the bounty of technology. The canal was hailed as "the largest of its kind in the world, adapted for navigation as well as for irrigation"; it was "designed not less for the benefit of a remote future than of the present age." The inauguration of the canal drew large crowds. "From the most distant parts of India pilgrims came up this year," the American correspondent wrote, "when the revered Ganges was about to leave her ancient and hallowed channel for one formed for her by the hands of strangers." Quoting from a "private account" that had come into the journalist's possession, he described how the aqueduct's embankments were "lined by our own work-people, to the number of more than thirty-five thousand men," in "long lines of stout forms." The military presence was strong, for here as in every development of infrastructure in India, military imperatives were paramount. "The infantry were on the tops of the aqueduct parapets" while "the artillery were stationed on a high piece of ground." The crowd gathered to celebrate the new canal was estimated at not fewer than five hundred thousand people.

The canal was a monument to imperial power, a symbol of English conquest over India's land and water. In his opening speech, Lieutenant Governor John Colvin—Cautley's predecessor, and the originator of the idea for the Ganges Canal—declared that "we have an answer . . . to the old reproach, that the British have left no permanent mark upon the soil of India to attest the power, the wealth, and the munificence of their nation." But the canal also marked a symbolic step up in the justification of British rule on humanitarian grounds. In the eyes of the American journalist, it was "difficult to conceive of a more impressive service" than the opening prayers consecrating the complex. Seen through the observer's evangelical imagination, the entire Ganges Canal complex was the work of a "few hundred Christians in the heart of a foreign country, surrounded by many thousand heathens"—a "work of civilization . . . for the benefit of these unenlightened multitudes." He declared a new "era of intelligent and liberal government" that "regards and cherishes the interests of the governed." He conceded that, for all its benefits, the advance of British rule in India had been attended by "the bitter consequences of evil" and "past misgovernment." But the tide had turned: "The night in which false religion, tyranny, and war have enveloped India," he wrote, "is giving place to the day of Christianity, good government, and peace."[55]

Not long after the completion of the Ganges Canal, the Indian Rebellion of 1857 brought an end to the East India Company's rule.

A mutiny within the army spiraled into widespread social protest that
spread across North India; the old Mughal emperor, Bahadur Shah,
was the rebels' symbolic leader. The rebellion was suppressed with
spectacular violence. The British government took control of India
from the East India Company. The colonial state intervened more extensively in the countryside. It used the law to reconfigure property
rights and to reshape relations between landlords and tenants, men
and women, Hindus and Muslims, dominant and subordinate castes.
It used force to settle mobile people, and the force of punitive contracts to mobilize labor for the plantations of Southeast Asia. In 1869,
the government's approach to the land found clear expression in Lord
Mayo's dictum that "every measure for the improvement of the land
enhances the value of the property of the State"; especially so, he
added, because "the duties which in England are performed by a good
landlord fall in India, in a great measure, upon the government."[56]

IN THE ODES OF THE EMPIRE'S PRAISE SINGERS, THE ACHIEVEMENTS
of British engineers in nineteenth-century India stood without parallel. But what truly was new about the hydraulic fever of the second
half of the nineteenth century? For Cotton, it was the capacity to
design the world anew through "works of a general nature." But the
ancient tank irrigation of southern India was just as ambitious, just
as systemic. Landscapes of water have always been shaped by human
intervention. Water historian Terje Tvedt warns us against the conceit
that the "conquest of nature" is a modern phenomenon.[57] But there
can be no question that the scale of the works designed and built in
the nineteenth century were without precedent. Steam power broke
the physical limits of earlier modes of construction—even though, as
we have seen, old methods were used extensively as better adapted to
local ecology.

The British justification of large-scale public works, too, had new
dimensions. In precolonial India, though not as markedly as in China,

the control of water lent legitimacy to local rulers. Irrigation works bolstered local resilience to drought, and ensured states' coffers remained full. Maximizing revenue remained the be-all and end-all of British rule in India, from start to finish; behind every investment in infrastructure lay the aim of extraction. Like many local rulers before them the British government of India used irrigation works to signal their benevolence, to demonstrate their power, to satisfy their own vanity. But some British architects of water went further than this. Driven by an aggressive evangelical Christianity, engineers like Arthur Cotton saw their mission as going far beyond the sustenance of revenue for the state. Irrigation, alongside other technologies, would usher in the social and moral transformation of rural India: midwife to an ever-expanding universe of commerce and trade. The moral argument for infrastructure in India laid deep roots in the nineteenth century—it would not be long before it was turned against British rule.

In earlier times the benefits of hydraulic infrastructure resided entirely at the local, or at most the regional, level. In this sense, perhaps the most far-reaching change that British imperial engineering brought was its spatial expanse—in the imagination of British engineers and administrators and investors, the irrigation of a particular district of the Godavari delta or the Gangetic plain would have repercussions across the globe, as more and more of the products of India's soil found their buyers in the markets of London and Liverpool, Hamburg and New York.

IV

The irregular availability of water was one challenge; the conquest of space was another. To take the augmented products of India's irrigated lands to market, India's great rivers had to be made navigable. In the first three decades of the nineteenth century, little had changed since cartographer James Rennell's description in the 1790s. The inland waterways of Bengal sustained "a system of regional trade that served a population of some 60 millions." An array of specialist vessels plied the waterways of the Bengal delta: salt boats, the boats used by wood-choppers in the Sundarbans, the small craft for the traffic in betel leaf, and the distinctive port lighters that serviced European merchantmen, loading and unloading their cargoes. Higher up the hierarchy in this catalog of rivercraft were the *bajra* preferred by European employees of the Company, with a large sail on a single mast. Most luxurious was the pinnace, reserved for higher officials and the wealthiest Indian merchants. The river teemed with life, animated by the labor of boatmen with a panoply of specialized skills. A distinctive Anglo-Indian lexicon emerged to describe work on the river—the product of British translations and mistranslations, transcriptions and mispronunciations of local words.[58] *Serangs* and *tindals* were boatswains; *manjhees* and *seaconnies* were steersmen; *dandees,* expert oarsmen, *lascars,* sailors. Each group knew the river intimately.[59]

But fluctuations in the Ganges River's flow between the wet and the dry season, the heavy loads of silt that it carries, forging and undoing sandbanks and shoals along its course—these all made it treacherous for larger vessels. Sandbanks deceived the most seasoned boatmen. Insurance firms charged the same premium on freight going from Calcutta up the Ganges to Allahabad as they did to London. If anything, Company rule had slowed traffic along the river, because of its punitive taxation. This was the view of Charles Trevelyan, who traveled down the Ganges and the Yamuna in 1830 to report on the oppressions inflicted by countless customs posts, acting in the Company's name, along the riverbanks. "These streams, intersecting, as they do, the whole of the Bengal Provinces from one end to the other and terminating in the Sea port of Calcutta," Trevelyan wrote, "must be the great channels and high roads of trade of the country." But they fell short of that potential. Trevelyan found that the number of customs inspections was enough "not only to embarrass the navigation of the Jumna, but to close it entirely for nearly half its course from the hills." He noted how few of the "great staples" of salt, cotton, ghee, and asafoetida traveled by river—instead, merchants resorted to "tedious and expensive land carriage." Only one cargo of salt arrived in Agra from Delhi in the year 1830 even though vast quantities were traded yearly from Delhi "for the consumption of our Eastern Provinces." In his sloping copperplate hand, Trevelyan was damning: "It may appear extraordinary," he wrote, "that the Officers who are charged with the collection of the customs should possess so imperfect an idea" of the inspections and exactions conducted in their name. Those who suffered most were the "poorer class of merchants and traders who can ill afford to pay." Trevelyan noted that "speed is the Life of Trade." Like so many other Company officers, he was a trader himself: he spoke from experience.[60] Trevelyan's report played a key role in the East India Company's abolition of internal customs duties.

By the time Trevelyan made his voyage downriver, the Ganges was the site of some of the earliest experiments with steam technology in India. British administrators and businessmen looked eagerly to the expansion of trade and navigation upriver. The first steam engine to arrive in Calcutta, in 1817 or 1818, was there to clean the river Hooghly. The eight-horsepower engine from Birmingham powered revolving buckets to clear the river of the silt it carried from the hills. A few years later, the steamboat *Diana,* launched in July 1823 by the Calcutta shipbuilding firm Messrs Kyd & Co, drew large crowds to witness its maiden voyage. As a commercial proposition it failed. War spurred technical improvement. In 1825 the company launched a military expedition against the Burmese kingdom after tension along India's expanding northeastern frontier. The old dredger was converted into a warship; unprofitable *Diana* was pressed into service to carry medical supplies and the wounded between India and Arakan, on the eastern littoral of the Bay of Bengal. It was deployed up the Irrawaddy River, where local people labeled the ship the "fire devil."[61]

By the 1830s, steamboat agents had set up shop at every point along the river—most of them as a sideline to their primary occupations. Many profited from the arrival of steam. J. P. Leslie was by day a pleader at the Allahabad High Court; he made himself the government's agent at the port and charged commission for overseeing the loading and unloading of cargo. Carr, Tagore, & Co., managing agents, secured the contract to supply coal to the government's Steam Department from their mines in Burdwan, in eastern Bengal. The Ganges was a microcosm of India's economic transformation. Steamboats began to carry to Calcutta the revenues on which the East India Company's administration depended—the proceeds of the annual harvest. Steamboats were filled with "boxes loaded with five thousand rupee coins," each one "roped, ticketed, and sealed with lead and wax" and guarded by one or more soldiers. Upriver, private money traveled from Calcutta merchants to Patna, Benaras, and Allahabad—advances on the crops eagerly awaited by traders in London and Liverpool and New York. The Ganges was a conduit for India's economic integration with the world. The bulk of the cargo traveling up from Calcutta consisted of the material paraphernalia of British imperialism in India—arms, medical supplies, printing presses, seals for opium agents, compasses and theodolites for the staff of the Indian Survey, the mammoth project to map and survey every inch of British territory in India. Consignments of three commodities dominated steam traffic along the Ganges: cotton, indigo, and opium. Each was of global importance. Long transported to Calcutta on country boats, indigo was small enough to be stashed away—it was a favored way for company officers to smuggle their ill-gotten gains out of the country. From 1836, consignments began to travel downriver by steamboats.[62]

By the end of the 1830s, ironclad steamers traversed the 780 miles between Calcutta and Allahabad in three weeks, but beyond Allahabad they found their passage blocked.[63] The Ganges continued to challenge the power of steam. Even under experienced pilots, steamboats ran aground. They foundered on the river's treacherous shoals and sandbanks. Silt blocked the path of large vessels, confining them to the most easily navigable stretches of the river. More nimble vessels lacked the power to push upriver. And steamboats were expensive. The most valuable commodities formed the bulk of the cargo on steamboats along the Ganges. But steam freight was too costly for most merchants. Bulk goods—rice, sugar, saltpeter, linseed, hemp, and hides—continued to travel on country boats, or overland.[64] Far from supplanting earlier uses of the river, the steamboat took its place in a varied economy of energy and transportation: the oldest and the newest technologies coexisted and competed with one other. Often the river itself—its currents, its seasonality, its contours—set the boundaries of what was possible or financially viable.

In the end it was not road but rail that emerged as the biggest competitor to steam vessels on the river. Because of British investors' embrace of the railroad, steam transportation along the Ganges never really flourished.

THE WORLD OVER, THE COLLAPSE OF SPACE AND TIME BY THE RAILroad, the steamship, and the telegraph underpinned the transition to industrial capitalism. Writing of the same period in the nineteenth century, environmental historian William Cronon describes how Chicago's tentacles reshaped the entire landscape of the American Midwest. Cronon writes of the "railroads' liberation from geography"—their ability to operate "quite independently of the climatic factors that had bedeviled earlier forms of transportation."[65] But how much did this apply in India? Could rail operate "independently" of the monsoon, where river transport could not?

India's railway dreams were born in the 1830s; by the 1840s, these visions had become a sort of "mania." The construction of India's railway network began in the 1850s and reached its zenith in the last quarter of the nineteenth century, financed by private investors at public risk, their returns guaranteed by the state. By this time railway lines snaked across Europe and North and South America. Now speculators eyed India with anticipation. After false starts and burst bubbles, construction began in the 1850s.[66] In 1853 the Marquess of Dalhousie, governor-general of India, inaugurated the construction of India's railways: investors were guaranteed a rate of return of 5 percent. As ever, military needs loomed largest in the government's calculations.

As Dalhousie told Parliament in a statement that inaugurated India's railway age, a countrywide rail network "would enable the Government to bring the main bulk of its military strength to bear upon any given point in as many days as it would now require months." He envisaged the "commercial and social advantages" of the railway as "beyond all present calculation." Beyond calculation, too, were the "extent and value of the interchange which may be established with people beyond our present frontier."[67]

Within three decades, civil engineer George W. MacGeorge estimated that the railway had reduced India to "one-twentieth of its former dimensions." Trains could cover up to six hundred kilometers a day. Bullock carts could manage, at best, twenty to thirty kilometers a day; river boats could cover sixty-five kilometers a day going downstream, but even fewer than a bullock cart when battling upriver. The railway grid consolidated the state's control over Indian territory. India's cotton and indigo, jute and opium moved faster to the ports of Bombay and Calcutta for export. The railway implanted the colonial state upon the recently conquered lands of the northwest.[68]

Among those who watched with interest was Karl Marx, who saw the railway as a necessary tool to break down feudalism and social division. Rail collapsed distance. Rail integrated markets. Marx quoted from a British observation just a few years earlier, in 1848, that "when grain was selling from 6/- to 8/- a quarter at Khandesh, it was sold at 64/- to 70/- at Poona, where people were dying in the streets"; the sole reason was that "the clay-roads were impracticable." In Marx's view, one of the railway's potential benefits lay in what it could do to the hydraulics of the land: it could "easily be made to subserve agricultural purposes," he thought, "by the formation of tanks" along the railway embankments, "and by the conveyance of water along different lines." A few years later, in 1860, railway engineer Edwin Merrall published a riposte to Sir Arthur Cotton's condemnation of expensive railway construction in India; it was a defense of the value of railways, faced with Cotton's alternative—investing in India's waterways. For Merrall, too, water was central. Merrall insisted that the railway could overcome climatic variation, operating "at all seasons of the year," whereas rivers swelled during the monsoon and dwindled in the dry season, leaving few months of the year when they could be traversed safely. India regularly suffered from famine, "from the failure of the periodical rains"; but "such scarcity," he argued, "is not general, but partial and local," and could "very easily be met by an increased supply of food from other and more fortunate districts." The railway's greatest promise was to connect the driest parts of India with those "which never want water."[69]

It is difficult to write about India's railways without a flurry of astonishingly large numbers. Over the second half of the nineteenth century, the Indian railways expanded to encompass twenty-four thousand miles of track, and India possessed the fourth-largest railway network in the world: by far the largest in Asia. India's railways demanded a "prodigious consumption of mineral or vegetable food, in the shape of coal, coke, or wood," vast quantities of iron and steel, and the complex manipulation of water to supply the engines. Because so many of the materials were imported, Indian railway expansion was a fillip to British industry.[70]

But what did this mean for villagers in a district facing drought or flood? Some observers worried from the outset that, contrary to the dominant view, transportation was no panacea for social and economic inequality. Writing in 1851, C. H. Lushington, railway commissioner, saw that the lands of the Ganges valley were "sublet in very small portions"; they were worked by the "poor and needy" who were "without capital who live from hand to mouth." They lacked the stocks of grain that would "make it worth their while" to seek distant markets. He worried that railway lines would cut through small holdings, dividing them further; he feared the "serious and tangible injuries" that would come from the way the railway lines interfered with drainage by building over natural floodplains. A quarter of a century later, many of his fears would prove prescient.[71]

The railways reached deep into the interior. They carried the force of the state and the pull of global markets to even the smallest villages.

Railway bridge over
the Godavari River.
CREDIT: Sunil Amrith

An Indian economist writing in the mid-twentieth century described
the railways as sparking a "revolution in the economic pattern of the
country": the "age-old walls of localized economy," he wrote, "were
collapsing." Recent analysts reach a similar conclusion based on the
modeling of district-level data. Dave Donaldson estimates that the
arrival of the railroad in any given district raised real income by 16
percent, thanks to "previously unexploited gains from trade due to
comparative advantage." He echoes a sense widely shared in the late
nineteenth century: "Districts that had been largely closed economies
opened up as they were penetrated by railroads." His data show that
the dependence of local grain prices and even local mortality rates on
local rainfall vanished at last with the arrival of the railways.[72] But these
numbers tell us little about how resources were shared within each village, district, or household. Markets grew more integrated, but many
without land or capital were not well positioned to benefit from this

expansion. And what effect did these transformations have on those—the lower castes, women and children, those ill or disabled—who had to look outside the market, to benefits in kind or to always-tenuous customary entitlements, for their security and survival?

However dazzling its scale, the rail network met the needs of a colonial export economy, driving produce to the river mouth ports. It was designed to take India's jute and cotton and tea and coal to where merchants needed them; it was designed to transport labor to the plantations and mines and factories. The massive expansion in mobility both within and beyond India in this period coexisted with deepening pockets of social and geographical immobility. Swaths of India remained off the railway map. The inequality between regions that benefited from the change and those left behind grew starker. Where there were no rail connections, roads also tended to be poor, and waterways poorly maintained. The National Sample Survey of India showed that, even a century after 1870, no fewer than 72 percent of journeys in many parts of rural India were undertaken on foot. Always, the newest and the oldest technologies depended on one another: as historian David Arnold observed, "The railways relied on country carts to bring raw cotton and other cash crops to the railheads or distribute grain to needy villages in times of famine."[73]

But in this age of evangelical faith in the "civilizing" effects of capitalism, of which British rule was the "Providential" vehicle in India, the railway seemed to augur an "extraordinary awakening": a "wonderfully rapid subversion of previous habits of life and thought" in India. The engineer George MacGeorge's encomium concluded that "one of the most rigid and exclusive caste systems in the world" had been "penetrated on every side by the power of steam." Many Indian observers shared his enthusiasm. Madhav Rao, chief minister of the princely states of Travancore, Indore, and then Baroda, wrote of the "glorious change the railway has made," as "populations which had been isolated for unmeasured ages, now easily mingle in civilized confusion"—it offered no less than the prospect of India becoming "a homogeneous nation." Not only did the railway change people's behavior, it transformed the landscape: trains ran through the rainy season and the dry, they crossed bridges across rivers and mountains, connected humid and arid regions. As they cut a track through the Western Ghats, railway engineers had turned "the stupendous natural inequalities of the precipitous hills into a series of uniform inclined surfaces," and "the whole rugged and inhospitable region has been smoothed down."[74]

Behind the self-regarding, if genuine, heroism of the engineers' accounts lies the forgotten heroism of those Indian workers who built the steel lines, the canals, the bridges. Countless among them paid with their lives. However fervently the railway engineers believed they could import their methods from England—just as they imported steam engines, coal, locomotives, tracks, sleepers, and even prefabricated bridges—what emerged in practice was a hybrid approach to building infrastructure.[75] India's ecology could not easily be "smoothed down." Whether through railway lines or irrigation canals, reshaping the Indian landscape was a colossal feat of work.

It was India's hydrology that challenged every scheme. Building the rail line from Howrah to Burdwan in Bengal, engineers faced an "inland sea" of water channels that required, in response, "viaducts, bridges, culverts, and flood openings" on a scale that no engineering project in the nineteenth century had ever attempted. Building bridges and aqueducts over the Himalayan rivers demanded ingenuity and a great deal of improvisation. Designs had to take into account the "immense volumes of water periodically brought down" by the great rivers through "seasons of flood," and the "erratic and unstable character of their channels." The rivers "scoured" the piers and abutments of bridges: torrents of water scraped away their foundations. So the engineers harnessed the knowledge of the people who knew the land most intimately. Even with the arrival of steam dredges and sand pumps, the terrible labor of divers and hand-diggers was vital; the piers of India's railway bridges often reached one hundred feet below the water's surface.[76] Other skills came into play. Work was overseen by men who had experience as seafarers, as *lascars,* in the British merchant marine. Their mastery of the winds, the tides, the currents; their mastery of a language of command—these were put to new use to reshape India's inland seas, far from the ocean air.

Few believed as fervently as Rudyard Kipling in Britain's imperial mission. But in his short story "The Bridge Builders"—based on his experience watching the Kaisar-i-Hind bridge being built across the Sutlej River, a tributary of the Indus in the northwest—what emerges most strongly is a sense of fragility before nature, and the deep dependence of British engineers on local expertise. The driving force in the story is the character of Peroo, a *lascar* from Kachch, "familiar with every port between Rockhampton and London." His mastery of the sailing ship had found a new outlet:

> There was no one like Peroo, serang, to lash, and guy, and hold, to control the donkey-engines, to hoist a fallen locomotive craftily out of the borrow-pit into which it had tumbled; to strip, and dive, if need be, to see how the concrete blocks round the piers stood the scouring of Mother Gunga, or to adventure upstream on a monsoon night and report on the state of embankment-facings.

As a storm threatens the bridge he has designed, chief engineer Findlayson falls into an opium-induced hallucination, haunted by the question of whether his bridge would survive the onslaught of the water: he asks himself, "What man knew Mother Gunga's arithmetic?"[77]

In the story the bridge survives. But the fears were well founded. India's ecology of water threatened not only the stability of bridges but the lives of the thousands of men who built them. Railway workers faced punishing conditions; infectious diseases were a constant threat. New infrastructure diverted watercourses, altered drainage channels, modified the water cycle; new risks were not far behind, posed above all by malaria. However far the power of steam had advanced, the monsoon rivers retained the capacity to surprise. In 1868, Sibganj, a "great grain market" along the Ganges, was destroyed: "A northward movement of the river in 1868 swept away the bank on which the market stood." Traders moved on; they set up at Karik, six miles to the northeast.[78]

V

If India's empire builders needed a reminder of their fragility, it came from the ferocity of the monsoon climate.

In October 1864, a "cyclone of unparalleled fury" struck Calcutta and the coastal districts of Bengal. The "rivers raged and tossed like a sea" and left the city "in ruins." "Far as the eye can see," a British correspondent wrote, "there is unbroken waste and gloom."[79] The cyclone originated in the Bay of Bengal, to the west of the northern Andaman Islands, on October 2. That morning, from the deck of the *Conflict,* sailors observed that "the stars had a sickly appearance." The sailors saw that the sun "rose blood red." The cyclone had built up in the southwest, its effects felt in Ceylon and Port Blair a few days earlier; it gathered force as it approached the Andamans. From the Andamans the cyclone swept up the Bay of Bengal, traveling at ten miles an hour toward the mouth of the Hooghly River. As the cyclone approached the coast of Bengal, the steamer *Martaban* was at anchor in the Saugor Roads. By the morning of October 5, the wind-whipped vessel drifted with its "jibboom gone and likewise the fore-royal and top-gallant masts." By the afternoon the gusts had eased, the captain wrote, "leaving us a total wreck." The crew realized they had been "dragged 17 miles across the banks at low tide."[80]

Another vessel, the *Ally,* foundered. It had departed from Calcutta on October 4, carrying 335 migrants bound for Mauritius. They were indentured workers, hundreds among the hundreds of thousands of indentured laborers from India bound to labor on the sugar planta-tions that met the British Empire's taste for sweetness. The ship was overturned by a gale. Only twenty-two of the emigrants and seven of the ship's crew survived.[81]

However visibly Calcutta was affected, the storm was worse in rural Bengal. It swept through coastal districts and moved inland to the northeast, finally fizzling out over Assam on October 7. Few people survived to bear witness. The storm surge generated "great sea waves . . . which, on reaching shallow waters, were piled up to a height greatly exceeding that of the highest spring tides, when they broke over the low lying lands at the mouths of the Hooghly and Godavery." A lighthouse keeper wrote to Calcutta in despair: "I cannot accurately state what the loss of life has been by the Cyclone and inundation, but I am afraid the fatal malady has carried off more." Disease killed many more than the initial flood. "Every tank, pond and well," he wrote, "is stagnant with decaying matter." More than fifty thousand people died. Flooding drove millions from their homes.[82]

This account of the storm, penned in 1867, comes from Henry Francis Blanford. He was born in London in 1834. His father, William Blanford, owned a workshop manufacturing gilt moldings—one of innumerable small manufacturers propelling Britain's industrialization. In 1851, Henry joined London's Royal School of Mines, and from there traveled to the Bergakademie in Freiberg to continue his study of mining. Along with his brother William Thomas Blanford, Henry joined the Geological Survey of India in 1855. Their first assignment was to explore the Talchir coalfield in Orissa, in eastern India—vital in this new era when India's hunger for coal was spurred by the railways. The Blanfords' inquiries in Orissa established some of the key groundwork for the later discovery of Gondwana, the supercontinent that fused the southern hemisphere's landmasses together with the Indian subcontinent and the Arabian Peninsula. Gondwana broke up into eastern and western segments—separating Africa, South America, and Australia—while the Indian subcontinent's northward drift and collision with the Eurasian continent, around 50 million years ago, created the Himalayas.

The following year, 1856, Henry took over as curator of the new Museum of Geology in Calcutta, and supervised the official Geological Survey of India, charged with exploring India's geology with a view to exploiting its mineral resources. He spent the rest of the 1850s in southern India, investigating the stratigraphy and paleontology of rock

formations between Tiruchirapalli and Pondicherry. After a sojourn
back in Europe to recover from ill health brought on "by the exposure
incidental to geological surveying in India," Blanford returned in 1862
to teach physics and chemistry at Presidency College, Calcutta.

Around that time, Blanford became involved with the Asiatic Society. Founded by famed Orientalist and linguist Sir William Jones
in 1784, for "enquiring into the History, Civil and Natural, the An-
tiquities, Arts, Sciences and Literature of Asia," the Asiatic Society of
Calcutta emerged as the most influential among a network of learned
societies in the colonial world. Its journal was a storehouse of cultural, linguistic, and scientific research. The inclusion of meteorology
among its subjects was due in large part to the work of a retired
ship's captain and president of the Marine Courts of Calcutta, Henry
Piddington (1797–1858). Inspired by the work of Colonel Henry
Reid—pioneer of American meteorology and author of *An Attempt
to Develop the Law of Storms*—Piddington's interest in the characteristic storms of the Indian Ocean was deeply practical. His aim is
clear from the title of his 1848 treatise, *The Sailor's Horn-Book for
the Law of Storms,* which he dedicated to "mariners of all classes
in all parts of the world." In his catalog of different types of storms,
Piddington proposed a new word, "cyclone," to describe those driven
by "circular or highly curved winds." He derived his term "from the
Greek *kukloma* (which signifies amongst other things the coil of a
snake)." The new science of cyclones demanded attention from sailors, he wrote, "for it is . . . a question of life and death, of safety or
ruin." He described a "storm wave," of the kind that struck Bengal in
1864, as a "mass of water . . . driven bodily along with the storm or
before it"; crashing upon bays and estuaries, they caused "dreadful
inundations." Piddington published in the Asiatic Society's journal a
series of ships' logs, from which he derived his work on the forces
driving the Bay of Bengal's cyclones.[83] Blanford began lecturing at
Presidency College, Calcutta, a few years after Piddington's death; he
studied Piddington's writings and developed an interest in the science
of storms. Given his prominence in Calcutta's world of science, given

his expressed interest in the weather, Blanford was an obvious candidate to lead the society's inquiry into the great cyclone of 1864. James Gastrell, Blanford's collaborator in compiling the report on the great cyclone, was the Asiatic Society's treasurer, and he also worked as the deputy surveyor-general for the government of India. What began as a report for the Asiatic Society became an official inquiry.

To reconstruct the path of the storm, Gastrell and Blanford pored over the logbooks of ten ships dispersed across the Bay of Bengal through the storm. The storm's chroniclers tallied the ships' barometric pressure readings with sailors' descriptions of sky and sea; they plotted these against the ships' likely positions to track the storm's path. The record of the *Moneka* gives a sense of the terse precision of the ship's log, heavy with foreboding:

> From midnight to noon, light and variable winds from north-west to west, with cloudy weather; sea more composed, but south south-west swell as lively as ever. No rain this day. Barometer 29.74. Thermometer 82°. From noon until midnight, light and unsteady winds from west by north, with cloudy weather; sky looking very black and lowering to the north and north north-east, with a high rolling sea from the same quarter. Sea rose very quickly; observed lightning in the north north-west. Barometer inclined to fall. Midnight, gently increasing wind from west, with gloomy appearances to north north-east. Sea still very heavy from that quarter. Ship pitching, bows under.

Gastrell and Blanford traced the cyclical motion of the winds by comparing the logs of the *Conflict* and the *Golden Horn*. The two vessels began the afternoon of the storm one hundred miles apart; by midnight, they were at most twenty or thirty miles apart, having been blown toward each other by the rotating cyclone. Sailors' most common response to the power of the storm was awe. The power of steam was no match for the cyclone. The *Alexandra* was a steam tug, at the mouth of the Hooghly River when the storm hit; traveling into a headwind, its engines "were set going with seven revolutions, at full power," making no progress. The "frightful roar of the hurricane" drowned out even the din of the steam engine. The winds overcame the ship: a "moaning sound," a "sudden blast from the northwest," and suddenly the ship was lying on its side.[84]

Vital though ships' readings were, their barometers were not always standardized, and their readings were difficult to compare with one another. The ships' records had to be read against measurements from land observatories. Gastrell and Blanford had access to the records of sixteen land stations, from Agra and Benares on the Gangetic plain to Kandy in the mountains of Ceylon and Port Blair on the Andaman Islands; farthest south was the station in Singapore. They were few in number and widely spaced. Their records had to be supplemented by the private journals kept by individuals such as one Mr. Barnes of Kandy. He noted on October 1, 1864: "raw and very damp, low scud *nimbi* covering the greater part of the sky, (dense *cirro-cumuli* beyond), and moving generally from west south-west, the wind veering from west to south and back."[85]

To follow the storm after it made landfall, the investigators relied on eyewitnesses: lighthouse keepers, railway stationmasters, European missionaries and district officials, captains of riverboats, government engineers. Gastrell and Blanford were tireless in their work of archiving the storm. Accounts of the devastation were heartbreaking. The storm's detectives searched these reports for telling details; they were intrigued by accounts that described the changing color of the sky, the shifting quality of the light, as the cyclone advanced. It was the best record they had of the formation and movement of clouds. Observers saw clouds of "dark lead" and indigo; they saw blackened afternoon skies lit up with "balls of fire," and nocturnal landscapes glowing with eerie light.[86]

Storm investigation was a form of narrative, and nature was the protagonist. Gastrell and Blanford depicted a battle of forces. A few days before the storm's arrival on the coast of Bengal, "the northerly current retreated before its stronger opponent, now forcing its way up the east of the Bay." That current was in turn "opposed" by the Yamadoung Mountains of Arakan on Burma's western coast—an obstacle the winds circumvented as they "curved round" them. The story of the storm took form akin to a travel account: it had a point of origin, an itinerary, and a destination—and it left a trail of destruction in its wake. The investigators depicted the storm's "tracks" using wood-block print, in two dimensions on a map. But their picture of the storm—bringing together pressure readings, wind speeds, latitude, and longitude—was fully three-dimensional. They were as interested in its vertical as its horizontal dimensions: the dance of rotating winds, the churning ocean currents, the contours of landscape.[87]

Gastrell and Blanford were limited to retrospective reconstruction. What they hoped for was instantaneous information. Their goal was to track future storms as they unfolded, through a network of monitoring stations linked by telegraph. They emphasized the "great importance" of meteorological telegraph stations "along both coasts" of India; ideally, they wanted an archipelago of observatories encompassing Ceylon and the Burma coast, "if possible as far down as Port Blair on the east side of the Bay." From the very detail of their inquiry into the storm of 1864, they sought an expanded sense of correlation and consequence across space and time, using "indications from distant stations."[88] What were the telltale signs, in the Andamans or in Ceylon, that presaged trouble in Bengal a few days later? How could some form of warning be delivered in time? Their view of the world was one that linked land, sea, and atmosphere. The Indian Ocean, on this view, was a weather factory: the source of India's climate.

———

WITH THE ADVANCEMENT OF STORM SCIENCE CAME A DIFFERENT way of thinking about space—and about India's place in the world. A new understanding of the monsoon emerged from the fusion of maritime and terrestrial observation. In the same decades, British explorers and scientists began to study the Himalayas. From there, botanist Joseph Hooker observed the monsoon from the other side—from the

mountain peaks. In Sikkim, Hooker witnessed a watery realm reaching from the ocean to the atmosphere; the sky was a mirror to the sea.
"The ocean-like appearance of this southern view," he wrote, "is even
more conspicuous in the heavens than on land, the clouds arranging
themselves after a singularly sea-scape fashion." "Upon what a gigantic scale does nature here operate," Hooker exclaimed. He described
a climatic system where "vapours raised from an ocean whose nearest
route is 600 kilometres distant are safely transported without the loss
of one drop of water to support the rank luxuriance of this far distant
region." And then, "the waste waters are returned by the rivers to the
oceans, and again exhaled, exported, recollected and returned."[89]

The enduring power of the monsoon to cause distress and upset
political calculation, within and far beyond India, would become am-
ply clear in the 1870s.

THREE

THIS PARCHED LAND

ETWEEN 1876 AND 1879 THE DECCAN PLATEAU IN THE SOUTH and parts of northwestern India suffered famine as intense as any ever recorded. Twenty years later, in 1896 and 1897, drought ravaged millions of lives again, this time across a large expanse of central India. Before they had recovered, another serious famine struck those same regions in 1899 and 1900. Crops withered. Cattle perished. Tanks ran dry. Employment vanished. Those with the least power in society—the landless, the aged and infirm, women and children—were the first to find that they could earn no money with which to purchase the food that made it to market, its price swollen by scarcity and rumor. People moved to the cities, where some of them survived on private charity; hundreds of thousands moved to British famine camps, where they received meager rations and a cash wage for strenuous labor building roads, digging ditches, breaking stones. Rarely in the voluminous reportage on the famines do we read the actual name of a person who died. They succumbed to starvation; weakened by hunger, they fell to cholera, to plague, to the catchall "fevers" that medical officers inscribed as their "cause of death."

"The rains failed." The phrase recurs in almost every account of these catastrophes, always intransitive. What this meant is that the rains failed to behave as they were expected to—they failed to fall when, as much, or where they usually did. The rains failed to obey the patterns upon which human societies had organized their material lives. The suffering of those years reached as far as China, Java, Egypt, and Brazil's northeast. In China five northern provinces—Shandong, Zhili, Shanxi, Henan, and Shaanxi—bore the brunt of the suffering. Between 9.5 and 13 million people died in China, most of them from diseases that spread hand-in-glove with starvation. We know now that failures of rainfall in the late nineteenth century were caused by El Niño Southern Oscillation (ENSO) events of exceptional intensity. El Niño is a quasiperiodic rise in sea surface temperatures in the equatorial Pacific, with effects on global atmospheric circulation. Local fishers off the coast of Peru had identified the phenomenon as early as the seventeenth century, and had called it El Niño (Christ child), since it tended to appear near Christmas time. Those who tried to make sense of it in the nineteenth century had only a growing sense that the droughts were global in reach—somehow connected.[1]

More searching questions followed. The rains failed—did individuals, societies, and governments fail, too? In an era when technology promised to collapse both time and space—as so many enthusiastic observers of rail and steam foresaw—need drought always turn to starvation? In an era when the British Empire proclaimed its superiority and its benevolence, did the colonial authorities act with foresight and justice? Could the famines have been prevented? These questions animated supporters and critics of colonial policy, economists, and meteorologists; they haunted administrators who carried the guilt for starvation in their districts. Drought and famine sparked many discussions about the future of water, for it was water's absence that had spelled disaster. Water unleashed new claims upon states by journalists and humanitarians and engineers; water unleashed new claims by states upon their subjects. Here, in the realm of ideas—in competing visions of the future, in the articulation of fears about nature and hopes of betterment—lay the lasting legacies of the nineteenth century's nightmares. We live with them still.

I

The first portent of trouble in southwestern India came with the drought that set in over the princely state of Mysore in 1875. The following year, rainfall during the southwest monsoon was much lower than usual over the whole Deccan plateau. By October, as local supplies were exhausted, the first murmurings of famine began to be heard in districts across Madras and Bombay. In the western Indian countryside of Bombay Presidency, the alarm was raised early on by workers of the Poona Sarvajanik Sabha. Founded in 1870, the association was, in the words of its constitution, a "mediating body" between the state and "the people." The Sabha was a bold experiment in representative politics—each member had to produce a *mukhtiarnama* (a power of attorney) signed by at least fifty people, authorizing him to speak on their behalf. Dominated by the landed and the wealthy, exclusively male in membership, the Sabha flourished under the leadership of Mahadev Govind Ranade, a judge and social reformer who carried a reputation as an orator when he moved from Bombay to Poona in 1871. The Sabha pioneered a new tradition of social investigation in India. The British kept close watch, sensing that the Sabha, in the words of one official, "threatens to grow into an *imperium in imperio,*" noting that "popular representation is a sharp weapon, and a very perilous one to play with."[2] The Sabha's growth coincided with the worst famine to hit the region in living memory—from the start, the Sabha played an important role in drawing attention to the suffering.

In a series of letters to the government of Bombay, written in the last few months of 1876, the Sabha gave an account of the famine's spread: one of the letters insisted that it contained "details accessible only to those who, like the agents and correspondents deputed by the Sabha, live among, and form part of, the people overtaken by this calamity." The Sabha's workers mirrored the "tours" undertaken by British officials through their districts, but they presented a view closer to the ground. The dispatches are brief; village by village, they chart a looming catastrophe. One of the entries reads:

> Pangaum, Oct 11—No rain except the first showers. . . . Drinking water scarce, as in the hot season. The tank will last for 3 months. No new water in the wells. . . . Neighbouring villages in a worse condition. The only relief work is to be commenced at Mohol. Great distress is expected of the respectable and poor people. Grains should be imported and sold gratis.[3]

The northeast monsoon brought no respite that winter, extending the drought to southeastern India. One contemporary observer noted that prices "sprang at a bound" to levels previously unknown. In a brutal reversal of what had been thought would be the impact of the railways, newly laid rail lines raised prices as grain was "hurriedly withdrawn by rail and sea from the more remote districts," channeled to urban markets where speculators fed on fears of shortages to reap higher prices. In the summer of 1877, the monsoon was slow to begin and then patchy, followed by a deluge late in the season which destroyed many of the limited crops that had managed to take root. Water's absence, followed by a short burst of excess, brought cultivators to ruin across Madras Presidency, their reserves depleted by the previous year's crop failures. That summer the drought spread north; central India and parts of the northwest saw their lowest rainfall ever on record. They had few reserves to fall back on because so much grain had been exported from India, drawn by high prices on the London market. What began as a localized drought in Mysore became a catastrophic famine. The winter rains of 1877 brought some relief, but only in the summer of 1878 did "normal" rainfall produce a good harvest. The drought coincided with the most severe El Niño event in 150 years; its effect was global.[4]

Everywhere drought sparked a rapid rise in food prices accompanied by an abrupt loss of agricultural employment. Landless laborers in the worst-affected districts were the first to feel the effects. What followed was a collapse in the livelihoods and incomes of the most vulnerable sections of the population. Many people undertook long journeys in search of relief. By the end of 1876, starvation began to kill those who were most vulnerable. Illness thrived where immunity had been weakened by widespread starvation. Cholera and dysentery accompanied the movement of people, and social disruption contributed to their spread. The return of the rains in 1877, and then more fully in 1878, saw another spike in death: most likely as a result of malaria, which thrived in the sudden change in the ecology of water after a long dry spell.[5]

IN THE MIDST OF INDIA'S DISASTER, IN MAY 1877, RICHARD STRAchey presented a lecture in London on the "Physical Causes of Indian Famines." Strachey was from an aristocratic family deeply connected with empire. From the 1840s, as a member of the Bengal Engineers, he took an interest in irrigation. Meteorology was among his most enduring interests, and from 1867 to 1871 he served as director-general of irrigation. He would go on to lead the government of India's Famine Enquiry Commission of 1880. He began his lecture by depicting a constant struggle between the forces of life and death: "Among the most active of forces are the conditions of local climate, and notably those of atmospheric heat and moisture." Many contemporary observers saw climate as an active force in the world. In its late-nineteenth-century English usage, "agency"—from the medieval Latin *agentia*—was used as often to refer to natural as to human actions.

India was overcome by the "agency of drought," preyed on by "agencies of destruction." These "devastating forces of tempest, drought, flood and disease" rendered human life fragile in their wake.[6]

As the drought gripped the Deccan region, administrators and journalists and missionaries followed its path. Drought appears in their accounts as an unwelcome visitor, leaving telltale marks upon the land: "The whole country is bare and brown," one letter described; "tanks, which at this season of the year ought to be wide sheets of water, are now nothing but vast expanses of dry mud."[7] Drought left its fingerprint on market prices. The most widely traveled famine observer was Sir Richard Temple, sent to tour the affected districts by the imperial government in Calcutta. He had one overriding goal—to spend as little money on famine relief as possible. Critics charged that he barely dismounted from his carriage: he swept through the land and saw only what confirmed his prejudices. But his own pen left us a vivid description of monsoon failure. Temple recounts the progress of drought as if it were on a journey. "On the right or southern bank of the Toongabhadra river," he reported, "the drought developed all its most destructive agencies, and showed its greatest force all along the frontier." Drought "visited" the city of Madras, and then "rested for some time on the districts of South Arcot, Tanjore, and Trichinopoly, and threatened them with evil." It had the force of a marauding army as it "extended itself with havoc throughout the southern peninsula, laying waste the districts of Madura and Tinnevelly, right down to the sea-shore near Cape Comorin." Worse was to come. By the middle of 1877, "all hope of the south-west monsoon was given up," the Madras government wrote to the British Indian capital in Calcutta. The government of southern India pleaded for resources. Endowed with a cruel propensity to tease, the clouds tantalized only to disappoint: "Very heavy showers would fall with a dash from blackened skies over a small area, whilst all around the skies continued as iron."[8]

Drought was an "agency" of famine; its prime characteristics were violence and caprice. In many eyes, a Christian God brought or withheld rain: "I now see but little chance" of rain, wrote Mr. Price, collector of Cuddapah district, "except by a special dispensation of Providence."[9] Some invoked Hindu cosmology: the rains rested in the power of "Indra and Vayu, the Watery Atmosphere and the Wind," who were "still the prime dispensers of weal or woe to the Indian races."[10] Others turned to the language of science to describe the physical drivers of climate: "The true cause of all movements of the atmosphere which we describe as wind is wholly mechanical, being difference of pressure at neighboring places," Strachey declared.[11] Mechanical or divine, to see climate as an active force was also to attribute a certain inevitability to drought. Famine appeared to be a natural characteristic of India, no less than its landscape. At best, governments and communities could adapt to the certainty of periodic monsoon failure. In the shadows of this resignation lay a British reluctance to acknowledge that now, to an extent unimaginable fifty years earlier, the means were at hand to mitigate the impact of famine quite drastically. That, after all, was the boosters' claim for the railways. But faced with the scale of the crisis, faced with demands to spend more money, it was easier to insist that famine was, and ever would be, India's climatic fate.

IF THE DROUGHT WAS A FORCE OF NATURE, IT ALSO PRESENTED AN all-too-human crisis. Prolonged and exceptional in its severity, it threw into sharp relief the fractures of society. It exposed the fragile infrastructures of economic life. It pressed upon the limitations of the physical infrastructure—the word only came into widespread use in English in the early twentieth century, from the French—that had impressed so many observers of India's landscape. In the late nineteenth century, human dependence on water—rains and rivers, wells and streams—began to be posed as a moral and political challenge.

In India, as in many parts of the world, the weather reflected moral concerns. Throughout Christian Europe, extreme weather and geological events—floods, droughts, eruptions, earthquakes—were seen as manifestations of divine judgment. China had a deep tradition of "moral meteorology," as historian Mark Elvin describes: "Rainfall and sunshine were thought to be seasonal or unseasonal, appropriate or excessive, according to whether human behavior was moral or immoral"; the conduct of the Emperor mattered most of all. "The people of the empire bring floods, droughts, and famines on themselves," the Yongzheng emperor declared in a decree of 1731. Because extremes of climate were unevenly distributed, a moral geography of rain and drought could be discerned—areas that suffered most, on this view, were those where standards of public behavior or administration had slipped. North America, too, had its version of "moral meteorology." In the 1870s and 1880s, ideas about rainfall and virtue underpinned conflicting views of how land should be allocated in the Great Plains of the United States. The idea arose that "rain follows the plow": industrious white settlers would transform the land, and their labor would in turn bring rain. Settlers, historian Richard White writes, saw themselves as "the agents of climate change." Aridity was a form of cultural or spiritual malaise.[12]

A form of "moral meteorology" pervaded the writings of missionaries in India who observed the great drought—but it acquired a radical edge, as they suggested it was a judgment not on the morality of Indian society but rather on that of the British government. Florence Nightingale wrote from India in 1878 that "the land of India is not especially subject to famine"; she insisted that "the cultivators of the soil are industrious; the native races compare favorably with other races in capacity to take care of themselves." The true nature of the problem was simple: "We," the British, "do not care for the people of India."[13] "Famine in India is no invincible foe," another observer wrote—"in a climate whose great danger is drought, and where Nature therefore teaches the necessity of precaution, we reap only a legitimate punishment when we suffer the penalties of imprudent neglect."[14] Here was a reversal of the language of Malthus: the "precaution" was lacking, the "imprudence" manifest on the part of the British rulers of India and not its people. An American writer in the *New York Times* went so far as to describe a "state of society in India whose only parallel in recent times was to be found in American slavery."[15] In a bold Tamil work depicting the great famine in verse, Villiyappa Pillai, court poet of the small kingdom of Sivagangai, turned the conventional view on its head. His bitingly satirical poem, published at the end of the nineteenth century, depicts the lord Sundarweswara (Siva) confessing to the starving people of the area that he was helpless in the face of their suffering—he directs them instead just to write a letter to the local *zamindar*.[16] The "agencies of destruction" were not climatic—they were human.

Who personified, or what embodied, the transformation sweeping India and leaving millions vulnerable to a failure of the rains? A clear culprit emerged from critical accounts of the famines of the 1870s: "the new class of capitalists." They were described by W. G. Pedder, an official who had worked in Gujarat and Bombay, as "men possessed by no ennobling ideas of public duty, cowards by caste and confession, citizens in no sense beyond that of benefiting society by selfish accumulation." Pedder insisted: "The dearth was one of money and labour rather than of food; the cultivators were without the resources their own fields should have furnished, the labourers could not obtain work and wages." The core of Indian cultivators' vulnerability, he observed, was their "constant relations with a mercantile class" who combined "the functions of general shopkeepers, dealers in agricultural produce, bankers, and money-lenders," known as *banias,* or *saukars.* Already indebted to the local moneylender, cultivators found themselves trapped without reserves if the harvest should fail. Pedder juxtaposed the wily "unscrupulousness" of the lenders with the "ignorance and timidity of the peasants."[17]

A lack of capital was the root cause of cultivators' vulnerability. The railway commissioner Lushington had seen this as early as the 1850s, when he cautioned against undue optimism about the railways. Andrew Wedderburn, the collector of Coimbatore district in Madras—a humane and direct voice railing against official inaction during the famine—saw that "some villages have sold their brass vessels, their ornaments (even including their wives' 'talis'), their field implements, the thatch of the roofs, the frames of their doors and windows." In Bombay, G. L. Hynes, the master of the mint, saw something similar: "Silver ornaments and melted country silver discs are pouring in," he wrote, to the tune of 9 lakhs (900,000) of rupees each month.[18]

The leaders of the Poona Sarvajanik Sabha also had the malign influence of the moneylender firmly in their sights. In the first issue of their journal, in 1878, they published their analysis of the famine. They argued that India's cultivators "are found to be too poor, too hopeless of retaining their independence, too inextricably involved in debt to be able to undertake agricultural improvements." The famine had served to "throw the Ryots [cultivators] more and more into the hands of the Sowcars [moneylenders], and leave them little ground to hope a change for the better."[19] Families fortunate enough to have sufficient resources to survive three seasons of drought now found "the savings of years utterly exhausted." They, too, had no choice but to turn to the moneylenders—and, at once, "from free men they have been degraded into slaves."[20]

———

WHILE HUMANITARIANS, SOCIAL REFORMERS, AND EVEN SOME CO-lonial officials attributed India's vulnerability to famine to the grip of social and economic inequality, other observers drew a direct connection between nature's "agencies of destruction" and human actions. In the eighteenth century, a group of European naturalists known as "desiccationists," many of whom had traveled and worked in tropical lands, argued that cutting down trees caused drought. Deserts, on their view, were but ruined forests.[21] This view gained prominence through the work of Alexander von Humboldt, who wrote in 1819 that "by the felling of trees that cover the tops and the sides of mountains, men in every climate prepare at once two calamities for future generations; the want of fuel and a scarcity of water."[22] Desiccationist views were common in British India by the middle of the nineteenth century; very often they were used to condemn local pastoralism and to restrict the use of forests by India's tribal peoples, known today as *adivasis*. Justified by arguments for conservation, the colonial state encroached upon India's forests, claiming more and more forest land, as well as uncultivated "wastelands" for itself; this limited the use rights of local people in a punitive way.

The strongest iteration of the desiccationist view was penned by an anonymous correspondent to *Macmillan's Magazine* in 1877. He called himself "Philindus"—an echo, probably deliberate, of the liberal publication *Friend of India*. His dispatch reached a wide audience, discussed not only in England but as far away as Japan.[23] Philindus begins his piece by recalling an episode at "the most sacred sanctuary of South Indian vapidity: I mean the bar of the Madras Club," where he heard two men, "Jones and Brown of the Civil Service," denigrating Arthur Cotton's achievements as driven by vanity and wasteful expenditure. After a strong defense of Cotton's genius, echoing Cotton's belief that water was the key to India's security and prosperity, Philindus set out his main claim: the "disastrous action of drought in Southern India" owed much to the "enormous extent" to which "the jungles of the Carnatic, and of the Peninsula generally, have been cut down during the past century." He invoked the authority of the American geographer George Perkins Marsh, whose influential 1865 book contained the first global account of environmental transformation by human hands. Marsh had seen that "as forests are cut down, the springs which flow from them, and consequently the water-courses which are fed by these, diminish in number, continuity and volume." Philindus approved: he was quite certain that "the most crying of the evils" afflicting South India was "the increasing desiccation of the country from the reckless destruction of its trees and forests."[24] In the first of their "Famine Narratives," published for India's growing reading public, the Sabha's reformers invoked a similar sense of drought being caused by human intervention: "Owing to denudation of forests and the absorption of waste lands under the mischievous system of a wrongly conceived revenue settlement," they argued, "the occasional rain that falls is never retained by the soil."[25]

At the peak of the great famine, the Government of India's Forest Act of 1878 brought India's forests under public ownership. Was this the avant-garde of global conservation, or a land grab that dispossessed India's most marginal people? It was both.[26] The attempt to limit the pace at which India's forests were being cut down began with a material concern about rapid depletion of a profitable resource, vital not least to India's railways. By the 1870s, the ecological argument pushed in the same direction, spurred by concerns about the long-term climatic harm that deforestation brought. By the same token, the Forest Act compounded the creeping attack on the rights of India's *adivasis*. Wrote Valentine Ball, an Irish geologist and anthropologist who had spent many years in central India: "The reservation of forest tracts which prohibits the inhabitants from taking a blade of grass from within the boundaries" had the effect of leaving people "cut off from . . . food sources throughout wide areas."[27] Forests were a refuge particularly in lean times: a source of tubers, fruits, and other foodstuffs that staved off starvation, even if they were no defense against hunger. The encroachment of the state upon India's forests threatened the lifestyles as well as the livelihoods of *adivasi* communities.

II

The famine provoked a searching look at Indian society; most of all, it turned the spotlight on the state. The British government's inaction, its seeming indifference to Indian lives, catalyzed criticism from within and from outside.

The most immediate charge was that the government of India had neglected the most basic precautions in advance of the drought, driven by a damaging parsimony. It had neglected southern India's web of tanks; by the 1870s, many of them lay in ruins. One British observer lamented that "the former rulers of India, if not so great or so powerful, yet had more of that simple craft and homely benevolence which show themselves in storing the rain and diverting the torrent to the first necessities of man."[28] In these humble tasks, the British colonial state failed. Their failure was evident when the rains at last came, late in the summer of 1877—so much rain that "the famine year of 1877 will appear in the Madras meteorological records as the year of the heaviest rainfall in a long period." The maintenance of tanks had fallen victim to the state's drive to cut spending, but many saw that "when economy is required, [tanks] are the very last thing that should be tampered with." The neglect of India's infrastructure of water allowed "a flow towards the sea of a precious fluid which represents in passing away unused a sacrifice of human lives."[29] Florence Nightingale observed that the Madras rains of 1877 "were lost, because the tanks were left unfinished in the autumn of 1876; the order having been issued for the stoppage of all public works" as a measure to cut government spending. The result: "Millions of tons of precious water so ran to waste," and millions of people starved for want of water.[30] From this point on, the "waste" of India's water came to be a battle cry for humanitarians, engineers, and critics of colonial policy.

A greater indictment of colonial rule arose from what the famine revealed about the conditions of life in India—it was as if the catastrophe lifted a veil covering the everyday workings of Indian society and economy. The reason that so many people in India were so vulnerable to the failure of the rains, many argued, was that British domination had impoverished them, eroded their defenses. "A people must be poverty-stricken beforehand to be thus absolutely cut down by want of food," an American observer of the famine wrote.[31] The idea that British misrule had undermined Indian economic independence was not novel; Adam Smith had been damning in his assessment in *The Wealth of Nations*. By the 1860s, the notion of a "drain of wealth" from India was most closely associated with the writings of Dadabhai Naoroji, a Parsi merchant and scholar who would go on to be elected the first Indian member of the British Parliament. Naoroji penned a devastating verdict on the economic effects of British rule in India. The famines of the 1870s sharpened his scalpel. The vast bulk of India's people were "living from hand to mouth," he wrote, so much so that "the very touch of famine carries away hundreds of thousands." Despite this, he argued, the Indian peasantry bore a "crushing" burden of land tax—India essentially paid for its own colonization through the "home charges" remitted each year to Britain. "Every single ounce of rice . . . taken from the 'scanty subsistence'" of India's people, Naoroji charged, "is to them so much starvation." Heedless of India's suffering, British rule, in contravention of its own stated principles, "moves in a wrong, unnatural, and suicidal groove."[32]

Naoroji's analysis was all the more powerful for being couched in the language of loyalty to empire: his fury was measured, every sentence backed by the colonial government's own statistics. Like so many political economists at the time, Naoroji was fond of fluid metaphors—he wrote of a "drain" of wealth from India, of "flows" in the wrong direction. India's closeness to the brink of disaster and the fragility of its people's subsistence arose from its acute dependence on water. In the view of so many critics of British policy in the 1870s, to mitigate that dependence—to secure life against the "very touch of famine"—ought to be the overriding concern of government.

———

As the disaster unfolded, the most pressing charge against the British government was that it had failed to provide adequate relief to starving people: it was too slow, too callous, too divided within itself, too concerned with economy, or simply too incompetent. The deliberate withholding of relief by the state, underpinned by an unrelenting faith in free markets and notwithstanding plentiful stocks of food in other regions of India—this remains at the core of historians' later attempts to provide a moral reckoning of India's nineteenth-century famines.[33]

William Digby (1849–1904) wrote the most detailed account of the Madras famine, in two exhaustive volumes that appeared in 1878.

Digby was a journalist and campaigner. From a humble start at his local paper, the *Isle of Ely and Wisbech Advertiser,* Digby edited the *Ceylon Observer* and, from 1877, the *Madras Times*.[34] Digby's account is a tragedy in slow motion: a story of warnings ignored, precautions abandoned, decency abrogated. A recurrent theme in his account of the famine is the conflict between local officials, many of them humane and observant, and an imperial government hidebound by ideology. As the first reports of starvation deaths began to filter through to the government in the last months of 1876, every village magistrate in North Arcot district received a warning from the collector that they would be "held responsible for the safety of individuals whose deaths may have been occasioned by starvation." But responsibility was precisely what the imperial government wished to avoid. Late in 1876, as it seemed more likely that large-scale famine relief would be necessary, the Madras government tried to bolster its granary by purchasing food in secret, acting through Messrs Arbuthnot & Co. The Madras government was at pains to do so anonymously so as not to interfere with the market, but its actions drew the ire of the imperial government, which ordered an immediate halt to the practice: "The supreme authorities objected to interference with trade," Digby observed laconically.

Famines had recurred periodically in India through the nineteenth century. British responses were ad hoc: they depended on the proclivities of the officials in charge at the time; they drew on local precedents and memories of earlier dearth; they were shaped by the security or fragility of colonial control over the affected areas. The most recent crisis before the disasters of 1876–1878, the Bihar famine of 1873–1874, was anomalous: on that occasion British intervention in the face of food shortages was unusually energetic and effective. Rather than imposing controls on the grain trade, the government imported directly 480,000 tons of rice from Burma, which by that time was emerging as India's new rice frontier. With regular and plentiful rainfall in the Irrawaddy delta, Burma seemed immune from the fluctuations of climatic fortune that bedeviled Indian agriculture. More locally, the state purchased grain incognito, acting through agency houses to stockpile food without creating panic in the market. Local people were employed on relief works for a cash wage. By contrast with common British practice, eligibility tests for relief were neither exacting nor punitive: the administration trusted the knowledge of village leaders and local-level officials. Even by the standards of, say, the late twentieth century, British intervention in the Bihar famine was a success. The official in charge of relief was Richard Temple.[35]

Far from celebrating the policy's success in Bihar, public reaction in England was harsh. *The Economist* condemned the large expenditure on famine relief, concerned that it had led Indians to believe that it was "the duty of the Government to keep them alive."[36] Worse was to follow. A tract titled *The Black Pamphlet of Calcutta*, circulated in 1876, turned its fire on Temple in particular, describing his policies as "an economic catastrophe, a culmination of unthrift and unreason."[37] It was published anonymously but its author was soon revealed as Charles O'Donnell, an Irishman who served in the colonial government of India. In these critics' eyes, the famine was a fiction, because for all the money spent, hardly a single death from starvation was recorded; they refused to see that this may rather have confirmed the policy's wisdom.

Temple was stung by humiliation. An ambitious man, he was determined to learn from the experience, and more determined to see that it did not hold back his advancement. When Temple was appointed the viceroy's envoy to the famine districts in 1877, there were murmurings that he might be too generous. He set out to prove them wrong. As Digby pointed out, "Sir Richard was commissioned to the distressed districts to economise, and it was known . . . that he would exercise economy." Temple fought to trim the scale of relief that the Madras government was providing to its starving subjects— in what became notorious as the "Temple Wage," he cut wages on relief works to (or even below) the level of basic subsistence. A newspaper in Ceylon pointed out with dark humor that the scale of rations provided in the famine relief camps of Madras was significantly lower than those enjoyed by a prisoner, Juan Appu, "recently convicted of knocking out the brains of a near relative."[38] In a missive "from the affected districts," a missionary correspondent took aim at Temple's "extra-economical theory of managing a famine"—perhaps there is a double meaning here, suggesting both that Temple's policy was excessively stingy, and that it was "extra-economical" in the sense of being motivated by ideology more than by economy. The letter concluded: "The duty of a great Government is not only to prevent its subjects from dying of starvation but to save life."[39]

Others around the world drew different lessons from India's famine. In Shanghai, as news of the horror of North China's famine reached the port city's elite, a feature in *Shenbao*—one of China's first modern newspapers—praised the British response to India's famine to underline the paper's charge that the Chinese state had failed to relieve its suffering subjects. The reference to India was mostly rhetorical, a way to call attention to the Chinese state's weakness. It was not based on a deep understanding of the Indian famine; rather, *Shenbao*'s account of the Indian famine drew largely on the accounts of Shanghai's British press, which reflected the ideological orthodoxy in support of Temple's parsimony.[40]

—

BY THE TIME THE SUMMER MONSOON ARRIVED IN 1878, SWATHS OF India lay in ruins. At a most cautious estimate, 5 million people had died from starvation or from disease. When the British government appointed a commission of inquiry to investigate the catastrophe, it was all set to be a whitewash. The viceroy, Lord Lytton, sought vindication in the face of widespread criticism at home and abroad; the finance minister, John Strachey, pushed for the appointment of his brother, Richard, as chair of the committee.[41] Despite its origins, despite a heavy dose of self-justification, the Famine Enquiry Commission's 1880 report was an "intellectual and administrative masterpiece," in the words of one of the most astute observers of hunger and famine in late-twentieth-century India.[42] Under Strachey's leadership, the commission included (as "English" member) the Irishman James Caird, Madras official H. E. Sullivan, and C. Rangacharlu and Mahadeo Wasadeo Barve, officials of the princely states of Mysore and Kohlapur. They toured the country. They accumulated hundreds of hours of testimony. One theme dominated their report—water.

The famine commissioners' diagnosis of the root cause of India's famines was unequivocal: "All Indian famines," they wrote, were "caused by drought." The commissioners saw that India's task was to find means for the "protection of the people of India from the effects of the uncertainty of the seasons." The seasonality and unpredictability of the monsoon was at the heart of the matter. The report's opening sketch of India's geography takes the form of a narrative map of water, an account, by now familiar, of the frontiers of wet and dry lands. Strachey's long-standing interest in meteorology was one reason why the committee's report went beyond justifying imperial policy: Strachey saw, here, an occasion for the elevation of meteorology to a position of greater prominence in India's future. To Strachey's dismay, there was vigorous debate within the committee. The final report was accompanied by a dissenting note by Caird and Sullivan. They disagreed with the majority view that the government of India was right not to intervene in the grain trade. Caird and Sullivan condemned the reliance on harsh "tests" to determine eligibility for relief; above all, they advocated for the establishment of public granaries in remote districts, where private trade was unlikely to reach. India's suffering, they saw, came from a "want of timely preparation to meet a calamity, which though irregular in its interval, is periodical and inevitable."[43]

If there was disagreement on how far the state should intervene in markets, on the "inevitability" of drought all agreed. Turning to the question of whether droughts may better be predicted, the famine commission dismissed an idea that was fashionable at the time— the theory that sunspots, dark and cool patches on the sun's surface caused by magnetic flux, followed eleven-year cycles correlated with droughts across large parts of Earth. The theory was championed by figures including the economist W. Stanley Jevons, famed logician and proponent of the "marginalist revolution" in economics, and William Wilson Hunter, editor of the *Gazetteer of India*.[44] The commissioners, however, concluded that sunspots "cannot be said to be in any sufficient degree established, still less to be generally accepted"—they had been "contested on various grounds, such as that the evidence is directedly opposed to them."[45] Instead, the commissioners lauded the patient work of meteorological observation. "As at present no power exists of foreseeing the atmospheric changes effective in producing the rain-fall, or of determining beforehand its probable amount in any season," the commissioners concluded, "the necessity becomes greater for watching with close attention the daily progress of each season as it passes, for ascertaining with accuracy and promptitude the actual quantity of rain in all parts of the country." They observed that "within the last few years a very satisfactory system of meteorological observations has been established all over British India"—there were more than one hundred rainfall observation stations across the country by 1880, and these tracked the progress and development of the monsoon across the country. The famine commissioners insisted that "it is of primary importance" that this infrastructure "shall be maintained in complete efficiency."[46]

The most important institutional innovation of the commission lay in the Famine Codes, which sought to break the link between drought and famine. They consisted of prescriptions for local government officials faced with scarcity. Their ultimate tool was the provision of large-scale public works in famine-affected areas to generate employment income, and to attract food supplies to the area by boosting purchasing power. The codes were, by definition, an emergency measure, but they also stimulated thinking about how to reduce India's dependence on the vagaries of the monsoon. "Among the means that may be adopted for giving India direct protection from famine arising from drought," the famine commission concluded, "the first place must unquestionably be assigned to works of irrigation."[47] The famine commission's overriding conclusion was that India's water resources needed to be managed better.

III

Twenty years after the suffering of the 1870s, India experienced two more major famines that marked what historian Ira Klein has aptly called a "grim crescendo of death."[48]

The drought of 1896 was felt first in the "black soil" region of Bundelkhand in central India, in the nineteenth century a region at the frontier of cotton production for export. By the end of the year, the summer rains had fallen short; the suffering spread across central India, reaching up toward Punjab, down to Madras, and east to upper Burma. The return of the rains in 1897 then unleashed a lethal epidemic of malaria. The famine coincided with an epidemic of bubonic plague that arrived in Bombay in 1896. The epidemic would persist for a decade, thriving on the large-scale migration sparked by famine, feeding on a population weakened by hunger, spreading along the railway lines to rural areas. In the eastern region of Chota Nagpur, creeping deforestation and colonial forestry laws had imperiled local *adivasi* communities, which had no local source of subsistence left, even as they fell through the cracks of the minimal safety net provided by colonial relief works. Mass starvation ensued. Only three years later, the same regions of central India faced another failure of the summer monsoon. Rainfall in 1899 was the lowest ever recorded in India. The drought covered an expanse of a million square kilometers of territory—central India was again worst hit—affecting tens of millions of people. In Bombay, the famine of 1899 and 1900 was the worst of the nineteenth century.[49]

By the 1890s, photographic technology had become cheaper and more portable than it had been two decades earlier. Haunting images of starving people in India circulated through missionary and humanitarian networks around the world. Fund-raising drives garnered millions of pounds in donations. Missionaries, writers, and photographers traveled to India. Among them was the American George Lambert, representing the Home and Foreign Relief Commission that drew its members from Ohio, Pennsylvania, Indiana, Michigan, Illinois, Nebraska, and Kansas. The famines of the 1890s helped to bring about a new global humanitarian sensibility on the part of middle-class publics in Britain, the United States, and Europe—a sense of identification with the suffering of distant strangers. But the same imagery that brought forth donations often reinforced the idea of a helpless India at the mercy of the elements, diverting attention from the political and economic subordination to British interests that left such a large number of Indians vulnerable to the monsoon. The question that had raised itself in the 1870s remained: was famine a "natural" disaster, or a political one?[50]

The backdrop to so many of the photographs, and to so many famine travelogues by European and American missionaries and journalists, was the sheer dryness of the land. In 1899, the *Times of India*'s correspondent in the princely state of Kathiawar, in Gujarat, suggested that even a photograph might not be enough to convey the absence of water:

> Were I an artist of the impressionist school and did I wish to represent the scene, I should dash in yellowish grey, a long diminishing streak, which would be the road throwing up the heat that made the distance shimmering and indistinct; a great splash of reddy-brown on either side would indicate the land where the crops should be; and above all a liberal dash of blue from the horizon to the top of my canvas would be the sky. I do not think I ever hated blue before; but I do now.[51]

Vaughan Nash (1861–1932), a British journalist and correspondent for the *Manchester Guardian,* described "tracts of dismal sun-cracked desert," and "brown wilderness spreading to right and left"—there was "no water in the wells, no water in the rivers," and the people he met at famine camps had their "lips and throats too parched for speech," so much that "the silence is unbroken." The famine camps offered a bare minimum. At a famine camp outside Poona, he wondered "whether the people can subsist on this penal allowance without ripening for cholera and other famine diseases." "There must be something wrong with India," he concluded, "when one finds a collapse like this."[52]

The "collapse" was not as severe as it had been in the 1870s. The Famine Codes had taken effect in 1883—each province had its own, modeled on the template proposed by the 1880 famine commission. Once a district officially declared that it had crossed from "scarcity" into "famine," the codes' machinery started up: public works were initiated to provide employment to boost local incomes, combined with relief for those not able to work. However, local governments worked under enduring, sometimes intolerable, pressure to economize; there was a clear incentive for district officials not to declare famines, and many waited until it was too late in 1897 and again in 1899. Nash pointed out that India's Famine Codes were "excellent on paper," but in reality local governments were "short of administrators, short of doctors, short of medical assistants, short of material." The infrastructure of relief was creaky, but it was extensive. At the peak of the famine of 1896–1897, 6.5 million people were receiving public relief. The monsoon failures of the 1890s were more severe than those of the 1870s—1899 saw the greatest shortfall—but mortality was lower. Nevertheless, at the very least a million people died.

The more concerted response to famine in the 1890s cannot be ascribed to imperial benevolence, though there is no doubting the good intentions of many local officials. Rather the colonial state was newly aware of pressure from Indian civil society—from journalists, lawyers, industrialists, and activists who came together in a growing number of associations: professional associations, caste associations, religious associations, reformist associations. They met in study clubs and book rooms, in university halls and public parks; they expressed their views in an expanding universe of print, in multiple Indian languages as well as in English. If the British were quick to see the malign influence of "agitators" behind every criticism of their rule, they could not avoid growing public scrutiny of their actions—or their failure to act. The Poona Sarvajanik Sabha, which had played such an active role in documenting the 1876 famine, adopted a more confrontational approach under the leadership of Bal Gangadhar Tilak, who took over in 1890. When before they petitioned government, now the Sabha convened large public meetings, speaking directly to *kisans* (farmers) about their rights to relief. The Sabha's local informants produced reports that contradicted official statistics and fueled criticism. Unable to trust government to intervene, Indian civic leaders took famine relief into their own hands, often working with charities overseas. The private charity of wealthy families had always played a vital role in providing food to the hungry. Arguably it mattered more, for most of Indian history, than government policy; very often it was religiously motivated. But indigenous charity organized itself on a larger scale in the 1890s, mirroring as well as challenging state infrastructures. The two most prominent Hindu reformist movements of the age—the Arya Samaj and the Ramakrishna Mission—undertook large-scale charity work for the first time during the famines of the 1890s.[53] The famines of the late nineteenth century spurred the development of pan-Indian political anger and activism.

IV

More than a century after the great famines, the question of responsibility still haunts us. For Mike Davis, author of a path-breaking global history of the famines, the answer is present even in his title, *Late Victorian Holocausts*. "Imperial policies towards starving 'subjects' were often the exact moral equivalent of bombs dropped from 18,000 feet," Davis writes; the millions who died during the late Victorian famines were "murdered . . . by the theological application of the sacred principles of Smith, Bentham and Mill." Yes, the rains failed on a colossal scale; but what turned drought into disaster was imperial policy: in the long term, by undermining the resilience of rural communities in the process of dragging India into modern capitalism; in the short term, by denying relief to starving people because of an unflinching refusal to interfere with "free" markets. Far from alleviating famine, the railways encouraged speculation, sucked food out of regions where it was needed most, and hastened the spread of epidemics.[54]

Without absolving colonial high officials of callousness, other writers paint a more ambivalent picture. They point to how life and livelihood in rural India had long been acutely dependent on rainfall; as long as India's infrastructure remained patchy, this would continue to be the case. "Famines were frequent and devastating" in colonial India, writes geographer Sanjoy Chakravorty, "but were they more frequent or more devastating than famines in pre-colonial regimes? Very doubtful."[55] Economists continue to believe that "railroads dramatically mitigated the scope for famine in India" and made Indian lives "less risky," but they point out that it took until the early twentieth century for these effects to be widely felt.[56]

In his global history of the nineteenth century, Jürgen Osterhammel draws a contrast between the Indian and Chinese famines of the 1870s. Osterhammel calls the Indian famine a "crisis of modernization," which is to say, it was a crisis brought on by the uneven impact of global markets on the Indian countryside. The Chinese famine, by contrast, he called more a "crisis of production than a crisis of distribution." The affected parts of North China were already under strain; they inhabited an "ecologically precarious niche, where for centuries state intervention had been able to ward off the worst consequences of disastrous weather conditions." Still reeling from the massive rebellions that rocked China in the 1850s and 1860s—the Taiping Rebellion the largest of them—and under pressure from European encroachment, the Qing state was much less capable than it was in the past of responding to the crisis. Osterhammel is cautious in his assessment of the historical consequences of the great famines of the 1870s. They brought no real change, he argues. In China there was "no really significant increase in political or social protest"; in India, British rule "held firm."[57]

But beyond the immense suffering they caused, there is a sense in which the famines were profoundly consequential for the future. The catastrophes of the late nineteenth century left many people—Indian economists and British administrators, water engineers and humanitarian reformers—with an acute anxiety about climate and water. To borrow a phrase from an earlier work of Davis's—a book about California and not India—climate was at the heart of a new "ecology of fear."[58]

THE AQUEOUS ATMOSPHERE

N OCTOBER 1876, JUST AS ALARM SPREAD ACROSS MADRAS OVER THE failure of the rains, India's eastern seaboard was struck by the worst cyclones ever recorded. Two storms followed in rapid succession: the first hit Vishakapatnam on the Orissa coast; the second inundated the Meghna delta in eastern Bengal. The loss of life was incalculable—the greatest toll was taken by the storm surge that accompanied the cyclone as it hit eastern Bengal. The commissioner of Dacca division, surveying the devastation, wrote of one locality that "not a single hut and hardly a post was left standing"; it was, he said, "too soon to attempt to compute with anything like accuracy the loss of life which has occurred." In district after district, local people estimated that 40 or 50 percent of the local inhabitants had died. In another village on his journey the commissioner listed the victims not by their names but by their positions: "Moonsif, rural sub-registrar, native doctor, post-master, court sub-inspector, abkaree darogah, two abkaree burkundauzes, seven constables, a mohurir of the moonsif's court, and a post-office peon."[1]

John Eliot, meteorologist of Bengal, set out to archive the storm. He relied on the usual combination of ships' logs and eyewitness accounts, supplemented now by records from the many land-based observatories that had been established over the preceding decade. Eliot conveyed the ferocity of the storm as it built over the Bay of Bengal: "This piled-up mass of water advanced under the pressure of the acting forces towards the head of the Bay" at twenty miles an hour. He estimated that the storm contained latent energy from the evaporation of water over the Bay of Bengal "equal to the continuous working power of 800,000 steam-engines of 1,000 horse-power."[2] Eliot proceeded to narrate an epic battle of forces between the storm wave rushing in from the ocean, and the Himalayan rivers—the combined power of the Ganges and the Brahmaputra—seeking an outlet to the sea. "These two vast and accumulating masses of water opposed each other over the shallows of the estuary," he wrote; their "struggle and contention for mastery" brought death and destruction to millions of people. Eventually, the "larger and more powerful mass of water forming the storm-wave" overcame the river waters. It deluged the islands at the mouth of the Meghna, one of three rivers that constitutes the Ganges delta; the islands were themselves "formed chiefly from the *detritus* of the Himalayas deposited over the area in which the tidal and river waters wage incessant warfare."[3]

This was a vision of India's climate shaped by water in every dimension: the descent of water from the Himalayan rivers and the ascent of water vapor from the Bay of Bengal and the winds stirring the ocean surface and transporting clouds to shore. On what scale should the climate of India be understood? The problem, Eliot noted, was that "so little is known of the action and independent motion of the aqueous vapour in the atmosphere, and of its relations to the atmosphere of dry air."[4] Tracking alongside the story of the disastrous famines, this chapter charts the quest to understand the monsoon that unfolded in the last quarter of the nineteenth century. Famine spurred the development of Indian meteorology. As knowledge of the monsoons grew, so, too, did awareness that India's climate was shaped by distant influences. As India's boundaries hardened, it mattered more to understand the rivers that crossed them. New knowledge of water raised new questions about India's place in Asia—and uncomfortable questions about how far science could conquer nature.

I

Meteorology was an international science by the 1870s: the telegraph had allowed the world's weather to be tracked with unprecedented immediacy. In 1873, the United States and a number of European countries agreed to form the International Meteorological Organisation (IMO). Like many international associations at the time, the IMO was a voluntary initiative, founded on an aspiration for greater cooperation between national weather services in the sharing of information. Like many international associations at the time, its concerns were dominated by those of industrialized, imperial powers. Britain claimed predominance in international meteorology because it represented a vast empire of climatic variation.[5]

Among the priorities of the new international meteorology was to devise a common and standardized language in which to describe the weather anywhere in the world. Especially daunting was the challenge of finding words to describe clouds in all of their variety, in their mutability and evanescence, in all of their profoundly local manifestations. The tools of Linnaean classification struggled to capture the texture of the skies. The basic cloud types that we still use—the puffy white *cumulus;* the gray blanket of *stratus;* the wispy *cirrus;* the dark rain cloud, *nimbus*—date from the early nineteenth century, in the parallel but independent work of Luke Howard in England and of the French statistician Jean-Baptiste Lamarck. The midcentury advent of photography was a fillip to cloud watchers, giving them the tool to capture clouds in a fleeting instant. Under the leadership of Swedish meteorologist Hugo Hildebrandsson, director of the Uppsala obser-

Meteorologists in the nineteenth century studied the genesis of storm systems in the Indian Ocean. CREDIT: Illustration by Matilde Grimaldi

vatory, the IMO published its first international cloud atlas in 1892. The atlas aimed to standardize the cloud observations that professional and amateur meteorologists and cloud-watchers were compiling the world over. Hildebrandsson and colleagues illustrated their atlas primarily with photographs, each illustrating a typical instance of a particular cloud form, even if each individual cloud that had been photographed would have changed shape moments later. As historian of science Lorraine Daston observes, long after the effort to standardize observations, knowledge of clouds and weather remained a profoundly local affair.[6]

Formal classification could not always capture the nuance of clouds in different climates. In agrarian societies (and so across much of Asia) the mutability of clouds had an immediate bearing on people's fortunes and the sky was a series of signs to be read, or warnings to be heeded. Every Indian language contains a rich lexicon to describe clouds, capturing their relation to the seasons and to the landscape.

In Tamil, *mazhaichaaral* invokes the drizzle from clouds that gather atop hills; *aadi karu* are the dark clouds that gather in the month of *aadi*, promising a good harvest to come. Notwithstanding the promise of meteorological advances, British officials in India often turned to local, or what they called "folk" knowledge, when they wanted to understand how the weather shaped the harvest. Historian Shahid Amin found in the local archive of Gorakhpur district, in the northern Indian state of Uttar Pradesh, a handwritten account from 1870, penned by a British district officer, compiling local aphorisms about the seasons. The sayings take the form of instructions to farmers issued by each season, personified by its name in the Sanskrit calendar. The hot summer season of "Jeth" (in May and June) says: "Be undaunted by the heat of the season. Make ready your threshing flood; work hard and gather the produce before the rains set in." The winter season of "Magh" warns farmers to "leave your cane mill and drive the water full into your fields. If God be pleased to give you rain you will be truly blessed." Other regional traditions across India had their own stores of wisdom about the clouds and the rains. In the words of a Tamil proverb, "If clouds withhold their gifts and grant no rain, the treasures fail across the ocean's wide domain"—local wisdom conveyed an awareness of the connectedness of the weather across large areas. Cultivators searched the skies for signs of foreboding. "Oh farmer!" another proverb pleads, "get out of the field with the young seedlings in your hand, should you see the first crescent moon in [the month of] *Arpisi*." Right up to the 1930s, alongside the development of meteorology, local governments in British India collected and published proverbs about climate and weather and cultivation.[7]

But the monsoon failures of the 1870s were so total, so devastating, that new answers were sought from the new science of climate.

II

Faced with the total failure of the rains in 1876 and 1877, India's meteorologists sought an explanation. Leading the quest was Henry

Blanford, the geologist-turned-meteorologist who had risen to prominence with his study of the Calcutta cyclone of 1864, and who was now director of the Indian Meteorological Office. In his regular report on India's climate and rainfall for the year 1876—written with factual detachment in the midst of disaster—Blanford ascribed the drought to the "remarkable and unseasonable persistence of dry northwest winds"—winds he had studied a few years earlier.[8] Blanford observed two abnormal forces at work: the first was exceptionally high pressure across northern and western India; the second was a sharper than normal temperature contrast between northwestern and eastern India. He concluded that "some cooling influence more potent than usual was at work, probably in the Punjab and on the northern mountain zone."[9] The following year, again, Blanford reported that "the land winds have been so persistent in the upper provinces and on the plateau south of the Ganges, as to cause an almost complete failure of the summer rains in that region."[10] As he sought to understand what had happened, what Blanford needed above all was data.

By the time of the famine commission report in 1880, India had more than one hundred meteorological observation centers. In the decade that followed, Madras, for instance, maintained eighteen observatories under the directorship of Elizabeth Isis Pogson. She was the daughter of Norman Pogson, an astronomer who was director of the Madras Observatory for decades. Isis was taken on in 1873 in the role of "computer," earning the salary of a "cook or a coachman." The family grew up in poverty, and Isis was forced to work, as well as looking after her siblings upon her mother's death. By the 1880s, as meteorology developed as a branch of science separate from astronomy, she was placed in charge of Madras's network of monitoring stations.

Pogson was zealous, inspecting regularly as many of the rain monitoring stations as she could. Her reports exposed the shaky edifice upon which India's weather data were based. Weather observatories tended to be built on hospital grounds, under the responsibility of the local medical officer; some were more enthusiastic than others about this addition to their duties. At Cochin, Pogson found that the local station needed better fencing, "to prevent stray cattle straying into the shed"; she personally arranged for supplies from Oakes and Company of Madras. She battled vandals as much as cattle. In Cuddapah, "the grass minimum thermometer had only been in use for six days when it was . . . found broken outside the hospital compound, evidently done out of sheer mischief." From Kurnool, she had to report that the data were "perfectly useless" because of the positioning of the apparatus. There, the local postmaster had to double as the meteorological assistant, and he struggled with the job. "He was very willing and anxious to learn," Pogson wrote, "but . . . could not possibly undertake to record" the data "as his combined duties as Postal and Telegraph Master were too much."[11]

The traces she has left in the archive are filtered through the technical language of meteorology and contained within columns and tables of official forms. As far as I know she left no personal papers—we can only speculate about how Isis Pogson experienced being a rare woman within the scientific apparatus of British India. As a young science struggling for legitimacy meteorology was likely more open, a little freer from prevailing hierarchies, than more established fields. Meteorology was among the "field sciences" that, as historian Kapil Raj shows, were more open to local knowledge than the laboratory sciences. But the obstacles Pogson faced getting her due recognition as a pioneer of global meteorology are telling. In 1886, she was nominated for membership of the Royal Astronomical Society; she was turned down after the council decided that the use of the masculine pronoun throughout the Society's charter meant that women could not be admitted as fellows. She finally became a member only in 1920, after she had returned to England, leaving India and meteorology behind.[12]

The Indian staff of the meteorological department, too, found a degree of openness they would not have encountered elsewhere in the bureaucracy of the Raj, though this was always weighed down by the knowledge that they could never rise beyond a subordinate position. Much was left in their hands by an institution that was young, understaffed, and underfunded. The most senior Indian meteorologist under Blanford, Lala Ruchi Ram Sahni, wrote a memoir in the 1930s, by which time he had risen to prominence as a patriot and a social reformer in Punjab; his recollections give us a rare insight into everyday life in the cockpit of monsoon science. Ruchi Ram recalled the global reach of Indian meteorology even at that early stage; Blanford would invite him home regularly to sit down and discuss the latest research findings "made in Russia, America, or somewhere else." Blanford's emphasis was always "on the interdependence of the weather in different parts of the world." Ruchi Ram recalled that this "made a deep impression on me in its widest implications."

But Ruchi Ram concluded his account of the meteorological department on a more personal note. He suggested that "if all Englishmen were like Mr Blanford, the social and political relations between the two races [British and Indian] . . . would have been quite different from what, unfortunately, we find today." He absolved Blanford of any sense of racial arrogance. Ruchi Ram's most powerful memory was "trifling," but revealing. Most British officials in those days, he recalled, would bark orders at their subordinates, keeping them waiting, and standing—but not Blanford. Blanford, Ruchi Ram wrote, "never once shouted to me from his chair, or even sent for me through the *chaprasi*." Instead, "the old man would get up from his seat, and opening the door that separated our rooms, would say gently, 'Lala Ruchi Ram.'"[13]

Looking forward, it is surprising to note how many Indian intellectuals spent time working in the meteorological department in the early twentieth century. They included Chintamani Ghosh, founder of the influential nationalist periodical *The Modern Review*, and Prasanta Chandra Mahalanobis, a master statistician who would play a leading role in making economic policy in independent India. In part, this may have been down to the way meteorology posed a daunting challenge of statistical analysis, which attracted many of India's brightest minds; in part, it may have come from a sense that understanding the monsoon was of vital importance to India's future. And perhaps the

Meteorological Department also left them freer from restrictions and pressures. On that count, Ruchi Ram learned one important lesson from Blanford: "He would ask me not to do this or that work myself, but got it done by one of the clerks so as to find [me] more time for self-study."[14] But that is to get ahead of the story.

———

AS RECORDS OF THE MONSOON ACCUMULATED, METEOROLOGISTS looked to capture its "normal" characteristics. In the late 1870s, Henry Blanford described the monsoon as a self-contained system; a climatological force that shaped and demarcated the Indian subcontinent: "India, together with the circumadjacent seas, is, in the main, a secluded and independent area of atmospheric action."[15] In Blanford's vision, the monsoons were an active force—he saw that "the goal of the monsoon, the place of low barometer to which its course is directed, is constantly changing."[16] Blanford viewed the monsoon as driven by the "primary contrast of land and water." He confirmed what was by then well known, that the driving force of the monsoon was the difference in solar heat received by the land surface of India at different points in the year, and its contrast with the relative heating or cooling of the Indian Ocean. But Blanford was able, now, to introduce new complexities to the science of the monsoon. He showed that the monsoon was not, as many had believed, "one current flowing alternately to and from Central Asia," but rather that it was formed from the intersection of "several currents, each having its own land centre." He described an alternate opening and closing of the Indian subcontinent to wider atmospheric forces. In the periods of transition, as the winds reversed—between March and May, and again in October and November—Blanford observed that "the interchange of air currents between land and sea is, in a great measure, restricted to India and its two seas." But once the southwest and northeast monsoons had set in, they connected the Indian "wind system," as Blanford called it, with "those of the Sunda Islands and

Australia, and, at one season, the trade winds of the South Indian Ocean."[17]

Driven by clear laws, Blanford believed that the monsoons were strongly predictable. "Order and regularity are as prominent characteristics of our atmospheric phenomena," he wrote, "as are apparent caprice and uncertainty those of their European counterparts."[18] Beneath this broad predictability, however, the monsoon was characterized by its unevenness. Within any given monsoon season, Blanford observed, rains were not "persistent and unvarying"; rainfall was subject to "prolonged periods of suspension" as well as "regular interruptions known as 'breaks.'" Even more striking was the monsoon's spatial unevenness: meteorologists found "a great diversity of rainfall" in different parts of India. "No country in the world," Blanford insisted, "furnishes such contrasts." Even as the broad contours of the monsoon seemed amenable to prediction, uncertainty was a defining climatic feature in many parts of the country, and "those provinces which have the lowest rainfall are also those in which it is most precarious."[19]

Meteorological anxiety about the unevenness of India's rainfall shaped perceptions of the land itself. In the most lyrical passage in his guide to India's weather, Blanford contrasted tropical with temperate landscapes:

Instead of feeding perennial springs, and nourishing an absorbent cushion of green herbage, the greater part flows off the surface and fills the dry beds of drains and watercourses with temporary torrents. In uncultivated tracts, where jungle fires have destroyed the withered grass and bushy undergrowth, and have laid bare the soil and hardened its surface, this action is greatly enhanced; . . . not only is water lost for any useful purpose, but by producing floods, becomes an agent of destruction. Under any circumstances, the character of the rainfall is hardly compatible with economical storage and expenditure in any high degree; and much more, therefore, than in temperate regions is it incumbent on us to safeguard such provident arrangements as nature has furnished for the purpose.[20]

The rhythms governing the distribution of rainfall proved an enduring mystery. Indian meteorologists pored over the correlations of monsoon failure across different parts of the country. Blanford observed the "curious relations in the way in which certain provinces are prone to vary alike," suffering drought or excessive rainfall simultaneously, "while others vary in the opposite direction."[21] The 1880 famine commission observed that on five occasions over a century, severe droughts on the Indian Peninsula were followed, a year later, by drought on the plains of North India. The causal mechanisms at work eluded meteorology until well into the twentieth century. Much uncertainty still remains.

As clues mounted, Blanford looked back at his annual reports for 1876 and 1877, and he grasped the significance of two tentative observations he had made at the time. The first was that the years of the great drought had also seen unusually heavy snowfall over the Himalayas, later than usual in the winter. Blanford investigated this puzzle over the years that followed. By 1884, he was convinced that the "extent and thickness of the Himalayan snows exercise a great and prolonged influence on the climatic conditions and weather of the plains of North-Western India." He suggested that keeping a close watch on Himalayan snowfall might hold the key to predicting the strength of the summer's monsoon to follow. But he was also quick to acknowledge that the forces at work might be far larger. Between 1876 and 1878, he wrote, "excessive pressure was shown to affect so extensive a region, that it would be unreasonable to attribute it to the condition of any tract so limited as a portion of the Himalayan chain," vast though the mountains were. Blanford's calculations showed that high pressure had prevailed across "extra-tropical Asia . . . and in Australia."[22]

Weather scientists across the British empire sought to pool their expertise and their information. Isis Pogson's detailed account of her library's holdings in Madras gives us a glimpse of these global connections. It included the proceedings of the First International Meteorological Congress in Vienna and reports from observatories in

Batavia and Singapore and Manila.[23] The development of monsoon science probably owed more to imperial and interimperial networks within Asia and Oceania than to wider international ones. As Blanford pursued his intuition about the great drought, he relied heavily on "private correspondence" with district officials in the Himalayas, and with meteorologists across the British Empire.[24] He wrote to his counterparts at other stations across the Indian and Pacific oceans asking them to furnish him with data on atmospheric pressure from 1876 to 1878. Charles Todd, chief meteorologist of South Australia, was quick to respond with records from South Australia and the Northern Territories. Todd and Blanford both saw that their data correlated. By 1888, Todd concluded that "there can be little or no doubt that severe droughts occur as a rule simultaneously" over India and Australia. Information filtered in to Blanford from island observatories, too, which had long been central to British and French ecological investigations: Mauritius, Reunion, the Seychelles, and Ceylon.[25] It was clear, by the 1880s, that the scale of influences on India's climate reached far beyond India's shores.

The famine commission of 1880 had expressed hope that the development of meteorology may provide some advance warning of monsoon failure. In 1881, Blanford was asked to come up with concrete proposals to implement the famine commission's recommendations for the development of India's meteorological infrastructure. His priority was the establishment of more monitoring stations. But he also looked to the more systematic collection of data from ships: a strengthening of the earliest maritime roots of meteorology. Information from ships was "urgently required" not only to track storms, but also "to throw light on the causes of the variations of the southwest monsoon rainfall." From 1881, data was collected systematically from every ship entering the port of Calcutta.[26]

In 1882, Blanford began to produce his first, tentative monsoon forecasts. A long-range monsoon forecast was a fundamentally different enterprise from the storm warnings that had dominated the concerns of Indian meteorologists. Especially with the aid of the telegraph, the approach of storms was now immediately visible. Cyclones were dramatic; their impact was urgent. Forecasting a year's monsoon, by contrast, required a more fundamental understanding of climate and climatic variation, founded on the slow analysis of a wide range of parameters on longer timescales. Despite his own awareness that India's climate was subject to oceanic or even planetary influences—a phenomenon that we now know as "teleconnection"—Blanford chose to base his forecasts on one primary indicator: snowfall in the Himalayas. From 1885, the Indian Meteorological Office's annual monsoon forecasts were published in the *Gazette of India*—and for the first few years, they proved accurate, at least as a broad-brush indication of whether monsoon rainfall was likely to be normal, excessive, or deficient.

III

The investigation of the oceanic and planetary influences on India's climate became an enduring concern for John Eliot (1839–1908), Henry Blanford's successor as director of the Indian Meteorological Office. The son of a schoolmaster and a graduate of Cambridge, Eliot began his career in India lecturing at the engineering college in Roorkee, which Proby Cautley had established at the head of the Ganges Canal; he moved on to Muir College in Allahabad, where he also served as director of the local meteorological observatory. In 1874, he took up a position as professor of physical science at Presidency College, Calcutta, where Blanford had also taught—and Eliot took over Blanford's role as meteorological reporter to the government of Bengal. In 1886, Eliot again succeeded Blanford, now as the meteorological reporter to the government of India—effectively the head of India's meteorological service—and held that position until 1903. Tall and heavy-set and prone to bouts of illness, Eliot was also an "accomplished musician" on the piano and organ.[27] In contrast with Blanford, Eliot was known, by his Bengali staff, as "the native hater."

Blanford's most senior Indian officer, Lala Ruchi Ram Sahni, decided
he would rather quit his job and move to the Punjab Education Department than work for the irascible and prejudiced Eliot.[28]

Like Blanford, Eliot first served in Bengal. Like Blanford, his early
work was on the cyclones that threatened the Indian coast. Even as
prolonged drought stalked the land, sudden tropical storms continued
to pose a recurrent threat, as Eliot had seen during the cyclones of
1876. Eliot pursued simultaneously the two strands of Indian meteorology that Blanford handed on to him: the study of extreme weather
events, and the quest to forecast each year's monsoon. The first proved
easier than the second to achieve.

Eliot's greatest influence on the field came from his understanding
of cyclones. A few years after taking over as chief meteorologist of
British India, Eliot published his *Handbook of Cyclonic Storms in
the Bay of Bengal*.[29] The nautical roots of monsoon science remained
evident: Eliot's book was, above all, a practical guide for seafarers.
It gained readers across Asia. Among those who learned from Eliot's
book—calling it both "masterful" and instructive"—was Father José
Algué (1856–1930), a Spanish Jesuit meteorologist who led the Manila Observatory and stayed on after the 1898 American conquest
of the Philippines to head the weather bureau.[30] From a network
of observatories across the western Pacific, local weather watchers
grappled with the power and the unpredictability of tropical storms
known as "typhoons" in the Chinese-speaking world: storms identi-
cal in nature to the cyclones of the Indian Ocean and the hurricanes
of the Atlantic. The Japanese government invested in a centralized
system of weather observation in keeping with its modernizing thrust
after the Meiji Restoration of 1868. Elsewhere private bodies took
the initiative—particularly the Jesuits, who founded a series of meteorological observatories across and beyond the Spanish empire: at
the Real Colegio de Belém in Havana (in 1857), at Ateneo Municipal
de Manila (in 1865), and at Zikawei (Xujiahui) on the outskirts of
Shanghai (in 1872). From 1869, the British-run Chinese Maritime
Customs Service developed a network of meteorological observation.

As in India, the practical value of storm forecasts for mariners provided the spark. Robert Hart, director of the Chinese customs, wrote of his hope of "throwing light on natural laws and . . . bringing within the reach of scientific men facts and figures from a quarter of the globe which, rich in phenomena, has heretofore yielded so few data for systematic generalisation."[31] These observatories exchanged information and developed a network of observation across the Pacific coast of Asia; but this remained walled off from the Indian Ocean, even as data began to reveal the connectedness of Asia's climate across its whole expanse. Algué was a towering figure, and he found much in Eliot's work on the Bay of Bengal to echo his own studies on the Philippine archipelago.[32]

In his 1904 treatise on *The Cyclones of the Far East,* a revised English version of a Spanish text written in 1897 (penned under the "roar of the cannon and the rumors of war" that "rob the mind of the calmness which is so necessary in a work of this kind"), Algué insisted that "there is no tropical storm which is developed or felt in the sea or on the coasts of China which has not exercised some influence upon this Archipelago." The Manila Observatory built weather monitoring stations across the Philippines, staffed by Filipino volunteers and a growing cadre of trained local technicians. Algué aimed to educate the public about the tropical storms that posed a recurrent threat to their lives and livelihoods. But just like Blanford and Eliot in India, Algué imagined a broader climatic region. The telegraph allowed for the transmission of instantaneous weather information. The Manila Observatory could now warn the China coast of approaching storms; meteorologists could "watch" storms developing in the Pacific. A French journalist wrote admiringly of the "completeness with which the Asiatic continent, from Cape St. James to the mouth of the Amur river, is safeguarded against surprises thanks to the meteorological services of Japan and the Philippines."[33] "Owing to the opening up of the Far East in recent years," Algué wrote, he had revised his work on the Philippines to give it "a greater compass."[34] That "compass" reached beyond the South China Sea and toward the Bay of Bengal.

Part of Algué's book was devoted to an account of "two very remarkable storms" described by Eliot. In a feat of meteorological detection, Algué matched up Eliot's accounts of the Port Blair cyclone that hit the Andaman Islands in November 1891, and the Chittagong cyclone of October 1897, with his own records in Manila. Of the storm of 1891, Eliot had simply written that there was an "absence of information" on the cyclone's origins in the South China Sea. Algué found a small item in the *Bulletin* of the Manila Observatory for October 1891 that might provide the missing context: "Very probably the typhoon which was felt on the 30th and 31st in Singora and other cities," Manila's meteorologists wrote, "then traversed the Peninsula of Malaca after running through the Gulf of Siam, to obtain new strength in the Bay of Bengal." Ships' logs allowed observers in the Philippines to track the path of the storm down the South China Sea, until they lost sight of it in the Gulf of Siam—which is precisely where Eliot began his account. Eliot picked up the storm in Siam, where on November 1, 1891, a storm wave flooded Chaiya, and "387 religious buildings and 4,238 other buildings were more or less completely destroyed." It moved out over the Andaman Sea causing devastation in Port Blair, and fizzled out over the east coast of India.

Algué reconstructed the Chittagong cyclone of 1897 with equal precision. There, too, Eliot began his account of the storm on the Malay Peninsula, but Algué traced it back to the seas around the Philippine archipelago. He combed through accounts from Jesuit observers in the Philippines; he tracked the storm's path through the logs of the German steamer *Sachsen,* heading from Singapore to Hong Kong, and the British ship *Faichiow,* traveling from Bangkok to Hong Kong.[35]

Algué's account reveals how little communication there was at the time between weather observatories in the South China Sea and those in the Indian Ocean. Each body of water seemed a closed system, each with its own characteristic storms—the typhoon seas and the cyclone seas. The expansion of telegraphic communication allowed for a new sense of scale to emerge, a new way of envisaging weather in time and space. Algué's map of the two "remarkable storms" presents a differ-

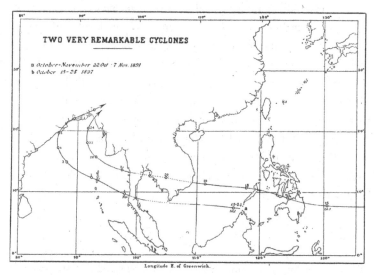

"Two Remarkable Cyclones": a map from Algué's study of typhoons and cyclones. CREDIT: Rev. José Algué, *The Cyclones of the Far East* (Manila: Bureau of Public Printing, 1904)

ent Asia: an Asia of storm tracks that traversed sea and land, crossing imperial borders; a coastal rim from the Philippines in the east to India in the west that shared risks to an extent previously unimagined.

But where Eliot succeeded in illuminating the nature and the threat of cyclones, his long-range forecasts of the South Asian monsoon fared less well. The number of weather monitoring stations in India grew from 135 in 1887 to 230 in 1901. By the turn of the twentieth century, the meteorological office issued five daily weather reports— one for India as a whole, and one for each major region (including one for the Bay of Bengal). Eliot's most significant innovation was to introduce what he called an "extra-Indian" dimension to his forecasts. He incorporated into his forecasts data from the southern Indian Ocean; he was particularly convinced of a correlation between pressure in Mauritius and monsoon rainfall in India. Eliot's forecasts grew increasingly elaborate through the 1890s as they "extended to thirty printed foolscap pages."[36]

However, Eliot's forecasts failed to warn of the climatic disasters that arrived in India in 1896–1897, and again in 1899–1900. Both, later research would show, were strong El Niño years; both droughts reached beyond India to affect China, Southeast Asia, and Australia. In 1899, Eliot's forecast predicted that monsoon rainfall would be "on the average of the whole area . . . slightly above the normal." As it proved that year the shortfall from "the normal" was worse than ever before. But even had Eliot predicted the drought accurately, it is unlikely that the British colonial government would have had the willingness or the drive to intervene on the scale that would have been necessary to avert starvation.

IV

What did it imply to think of Asia as an integrated climatic system? According to the evolving understanding of storms and monsoons, Asia appeared as an expanse of depth and altitude put in motion by the circulation of air. It was a land- and seascape defined by nature rather than by empires, its boundaries dictated by the winds and the mountains. But as soon as this picture of climate was translated into two-dimensional maps, the weight of political boundaries became evident. The first *Climatological Atlas of India,* compiled by John Eliot after his retirement from India, began with a map of winds and pressure across an interlinked oceanic and continental system. It showed how the climate of India was shaped by the transfer of heat and energy between the Eurasian continent and the vastness of the Indian Ocean. This was in keeping with Eliot's own, evolving understanding of India's monsoon. But the flurry of maps in the atlas—monthly maps of temperature and humidity and rainfall, cloud cover and wind direction and wind speed—confined themselves to the territorial expanse of British India. In map after map, the territory of British India was shaded a different color from the surrounding mass in order to stand out; even the arrows showing wind speed are limited to the subcontinent, as though the winds were self-contained. Only the map of storm tracks stretches out toward the Bay of Bengal, as if the ocean were but an external source of weather as it affects the land.[37]

Climate science was forced into contact with geopolitics. Ideas about India's climate echoed, and informed, broader debates about India's place in the world. The networks of storm warnings along the coastal crescent from India to China mirrored Asia's maritime geography. The names of the stations that broadcast telegraph reports were the names of the great ports; the tracks of the tropical cyclones they monitored were the tracks of busy shipping lanes. Research on the longer-term regularities of India's *climate,* as opposed to episodic weather, pointed in a different direction. India's climatology emphasized its distinctiveness, even its isolation. As Blanford put it, the monsoon system rendered India "a secluded and independent area of atmospheric action." Ideas about climate coincided with new understandings of both geology and geopolitics. Blanford had begun life as a geologist, and in the late nineteenth century others in that field delved deeper into India's natural history. They argued that India was a breakaway fragment of the lost supercontinent of Gondwana that had collided with Eurasia in what was, on a geological timescale, the recent past. In the realm of geopolitics, at the same time, British strategists were increasingly worried by threats to their dominance that came not from sea, but from land, through the mountains of Central Asia—the threat from Russia above all. These arguments about India—each of them depicted visually in the form of maps—came together to produce a "subcontinental" as opposed to what had been essentially a maritime view of India. The use of the term "the Indian subcontinent" dates from the early twentieth century. The Himalayas were crucial to this vision of India. They came more clearly into view in the last two decades of the nineteenth century: their role in India's climate, their place as the source of India's rivers, their strategic importance to India's security.

The official compendium of India's history and geography, the *Imperial Gazetteer*—edited by William Wilson Hunter, a keen meteorologist as well as an ethnographer, and published in eight volumes in 1881—insisted that "by India we now imply not merely the wide continent which stretches southward from the Himalayas to Cape Comorin, but also the vast entourage of mountainous plateaux and lofty ranges." India, Hunter insisted, "can no longer be considered apart from that wide hinterland of uplands." This was a political as much as a geographical imperative. "India," he wrote, "must be held to include those outlying territories over which the Indian adminis-tration extends its control, even to the eastern and southern limits of Persia, Russia, Tibet, and China." It was overland, across the moun-tain passes from the northwest, that every invading force—save the British—had arrived in India. India's imperial rulers continued to fear a resurgent threat from their rivals among landed Eurasian empires. But Hunter's concern was with the future as much as with a repeti-tion of the past. In "a future of railways developments," with a "rush of motor traffic," it was possible that "the land approaches to India" would "rival those of the sea"—"then will some of these again be-come the highways of the eastern world." By that time, "we shall take out tickets in London for Herat, and change at Kandahar for Kabul or Karachi." Among these frontiers, the least explored but one "poten-tially destined to play an important part in Indian history" was "the great highway of the Brahmaputra valley from the plateau of Tibet to the plains of Assam."[38]

Hunter drew two conclusions. The first was that "the mate-rial wealth of India largely depends" on its "capacity for the stor-age of that water supply which carried fertility to its broad plains." The other was that British India's security depended on "guarding the gateways and portals of the hills," preventing "those landward irruptions" that had reshaped Indian history on many occasions in the past. These propositions would endure; they would outlast the British Empire that Hunter and his contemporaries wanted above all to preserve. And here is the contradiction at their heart: envisioning India through its rivers encouraged a far-reaching consideration of its connections with distant places, of "world highways" (or potential "highways") that situated India in relation to the flow of goods, people, money, and water to and from China, Tibet, and the expanse of Central Asia. By contrast, to see the Himalayas as a natural barrier, one always under threat of breach, was to advocate for a vision of India as a bounded place, an "amphitheater" sealed off from the rest of Asia. Natural frontiers became synonymous with the security of the realm. Hunter's concern with the economic value of water to Indian agriculture reinforced this bounded view. Water was a resource to be stored, possessed, harnessed, and put to work: the essence of India's "material wealth."

Hunter's was a view moving up to the mountains, and away from the ocean. The last part of the nineteenth century saw a final push of Himalayan exploration, which had begun a century earlier. When Trelawney Saunders drew his map of India's mountains and river basins in 1870, he noted that Tibet was still *terra incognita;* Arthur Cotton argued the same when he dreamed of a canal link between India and China. Locating the source of Asia's great rivers was the final frontier in the spate of expeditions undertaken by European explorers in the nineteenth century. It was not until his expedition of 1905–1908 that the Swede Sven Hedin finally discovered the source of both the Indus and the Brahmaputra on the Tibetan Plateau. Describing his first sight of the Brahmaputra, Hedin fell into rapture. "Above the dark-grey ridge rises a world of mountains which seems to belong to the heavens rather than the earth," he wrote of the northern Himalayas, "between them and the dark grey crest, comparatively near to us, yawns an abyss, a huge fissure on the earth's crust, the valley of the Brahmaputra or the Tsangpo." He described the water: "Bluish-green and almost perfectly transparent, it flows slowly and noiselessly in a single bed to the east, while here and there fishes are seen rising."[39]

A decade after Frederick Jackson Turner's famous address to the American Historical Association, on the "closing" of the American frontier, the British imperial geographer and strategist Halford Mackinder made the point on a much larger scale. In 1904, just the year before Hedin's expedition, Mackinder argued that "geographical exploration is nearly over." There were no "blank spaces" left on the map of the world. In Asia, he witnessed "the last moves of the game first played by the horsemen of Yermak the Cossack and the Shipmen of Vasco da Gama." Mackinder foresaw that the heartland of Eurasia would, again, become the pivot of global power. "A generation ago steam and the Suez Canal appeared to have increased the mobility of sea-power relatively to land-power," he declared, but now transcontinental railways were "transmuting the conditions of land-power."[40] In this light the mountainous frontier of the Himalayas still appeared remote and forbidding. But it was now clear that they might contain vast water resources—resources to be captured for the development and security of the plains.

And here we have the paradox that deepened in these years: water and climate were boundless. Their boundlessness became clearer with every advance in the technologies of measurement. Yet, as the next chapter will show, they came under ever tighter but more fragmented territorial control. We turn now to the fevered quest for water that gripped India, and most of Asia, in the early decades of the twentieth century.

FIVE

THE STRUGGLE FOR WATER

N *INVISIBLE CITIES*, HIS TALE OF AN IMAGINED ENCOUNTER BETWEEN the Italian explorer Marco Polo and the Mongol emperor Kublai Khan, Italian novelist Italo Calvino describes the fictional city of Isaura like this:

> Isaura, city of the thousand wells, is said to rise over a deep, sub-terranean lake. On all sides, wherever the inhabitants dig vertical holes in the ground, they succeed in drawing up water, as far as the city extends, and no farther. Its green border repeats the dark outline of the buried lake; an invisible landscape conditions the visible one; everything that moves in the sunlight is driven by the lapping wave enclosed beneath the rock's calcerous sky.[1]

Calvino's exquisite depiction of "Isaura" distills the themes of this chapter, which is concerned with the search for new sources of water in India. At the turn of the twentieth century, India still reeled from the famines of the 1870s and the 1890s. Advances in meteorological science had pointed to the awesome power of the monsoon climate, and sketched its continental span. Indian economists and British administrators, water engineers and industrialists joined in a search for water to secure India from climatic vulnerability. They looked for water underground. They built dams to harness the power of rivers. They explored new ways to store rainwater. Across India, water formed "invisible landscapes" that shaped "the visible one." In trying to exploit water more intensively, administrators and engineers came to see it as a bounded resource. Control was their aim, but it proved an elusive goal. For in those same decades, climate science pointed in the opposite direction: it threw up the enormity of the scale on which nature worked. The study of the monsoon became ever more spectacularly global. Understanding that the climate of any given part of Asia was subject to remote oceanic and atmospheric influences still too complex fully to comprehend, meteorology called into question the confidence with which many believed nature could be conquered.

I

In almost any account of modern Indian history, the last two decades of the nineteenth century appear as a moment of political awakening. The formation of the Indian National Congress in 1885 represents, in most textbooks, the beginning of organized nationalism, albeit a tentative nationalism still dominated by an urban and professional elite, still outwardly loyal to the empire. The famines of the last quarter of the nineteenth century catalyzed criticism of colonial rule. They made brutally evident the insecurity of life endured by most of India's people; they called into question the effectiveness of the colonial government; above all, they laid bare whose interests the British had in mind when they governed India. "Britain has appropriated thousands of millions of India's wealth for building up and maintaining her vast

British Indian Empire . . . [and] has thereby reduced the bulk of the Indian population to extreme poverty, destitution, and degradation," said Dadabhai Naoroji, an early proponent of economic nationalism in India, and the first Indian to be elected to the British Parliament, in a speech at the Plumstead Radical Club in London in 1900. Naoroji argued that it was Britain's "bounden duty in common justice and humanity to pay from her own exchequer the costs of all famines and diseases caused by such impoverishment."[2] The experience of drought and famine gave rise to new ways of thinking about state and economy, nature and climate. The material consequences of those ways of seeing continue to shape our world.

By the 1890s, many Indian economists saw the monsoon as a limiting condition on India's future development. Mahadev Govind Ranade—social reformer and High Court judge and early leader of the reformist Poona Sarvajanik Sabha—declared in 1890 that in many parts of the country "the last margin has been reached, and millions die or starve when a single monsoon fails." Ranade believed that only a complete transformation of land and water could protect India from future disasters. He lamented the "ruralizing" of India—the destruction of local industries as a result of Lancashire textiles flooding the market. Ranade felt that only an industrial future, with some measure of protection for local industries from foreign competition, would free India from its vulnerability, which arose, he believed, from the country's acute dependence on agriculture. He wrote admiringly of the Cultivation System in the Dutch East Indies, a coercive Dutch colonial policy that forced cultivators to set aside a portion of their lands for the production of export crops. Whether or not Ranade was aware of the system's harshness, which provoked protest and resistance in Java, he saw it as a welcome alternative to the British worship of free trade and even as a boost to Indonesian industry.[3]

Romesh Chander Dutt—Bengali economist and poet, civil servant, and translator of Hindu epics—went further than Ranade in pinpointing responsibility for India's famines. In 1901 Dutt began his

open letter to the viceroy, Lord Curzon, with the "melancholy" observation that 15 million people had died in famines in India within
Dutt's own lifetime. He dismissed British observers' fixation on the
predatory moneylender as the chief cause of distress to Indian cultivators in times of drought. "The money-lender is the result, not
the cause, of the poverty of the cultivators," Dutt wrote, suggesting
that Indian farmers fell back on credit only because their earnings
were insufficient to cover the heavy burden of taxation. Turning to
Malthusian fears about India's population growth, Dutt pointed out
that India's population grew more slowly in the nineteenth century
than England's. The "immediate cause of famines in almost every
instance is the failure of the rains," he said, and the threat of famine
would persist "until we have a more extensive system of irrigation."
But he distinguished the "immediate" cause of famine from the root
cause—the reason why a failure of the rains should bring so many
to the brink of starvation. That root cause was the "chronic poverty
of the cultivators, caused by the over-assessment of the soil." Developing the argument that Naoroji had made earlier, Dutt argued that
the weight of the land tax—used to finance the "most expensive foreign government on earth," bloated by its imperial adventures beyond
India's shores—was the main cause of India's poverty. In 1903 Dutt
published his masterpiece, an economic history of India in two volumes. Written with flair and backed by reams of statistics, Dutt's book
argued that Britain had "drained" India of its wealth and contributed
to the devastation of its industries. In his catalog of misguided colonial policies, however, Dutt expressed his great admiration for Arthur
Cotton, whom he thought to possess "a reputation higher than that
of any other engineer who has ever worked in India." Dutt agreed
with Cotton that India's salvation lay in irrigation. The problem, Dutt
discerned, was that colonial administrators had ignored Cotton; they
"returned again and again to the narrower view, based on the immediate financial return of works constructed." The British government
of India was so concerned with avoiding unnecessary expense that it
was trapped in a short-term view of India's economic development,

unwilling to invest in expensive infrastructure that might, in time, have paid off.[4]

———

AT THE BEGINNING OF THE TWENTIETH CENTURY, THE INDIAN Irrigation Commission undertook a grand tour of India's water. Famine was a raw and recent memory. Colonial officials and Indian critics alike believed that irrigation was essential to protect India from a recurrence of the horror. The viceroy, Lord Curzon, appointed Scotsman Colin Scott-Moncrieff to lead the investigation. Scott-Moncrieff was famous for directing the hydraulic transformation of the Nile in the 1880s through a restoration of the Nile Barrage, but he had started his career with the Bengal Engineers and had served as Burma's chief engineer in the 1870s. Twenty years later, he returned to India. His brief was as simple as it was daunting—to determine how far irrigation could protect India from the vagaries of its climate.[5]

The British colonial government of India has been described as "ethnographic": it collected obsessively information about every aspect of life within its domain; it classified India's people by caste and faith; it published reports of "moral and material progress."[6] In the first three decades of the twentieth century, four commissions of inquiry—on irrigation, agriculture, banking, and labor—cemented the state's knowledge of its territory and its population. This vast enterprise of information-gathering generated authoritative tomes, divided by province, backed with appendices. To read them now is to leap into the minutiae of economic life in India, village by village. The commissions published extensive volumes of testimony, both oral and written, which are in turn prosaic and lyrical, combative and dull. It is easy to forget how small these commissions were, how dependent they were on knowledge provided by others. Scott-Moncrieff was joined by Thomas Higham, India's chief inspector of irrigation; by Denzil Ibbetson, chief commissioner of the Central Provinces; by John Muir-Mackenzie, chief secretary to the government of Bombay;

and by a sole Indian, Rajaratna Mudaliar, a member of the Madras Legislative Council. The five men met in Lahore, and over two periods of six months each, in 1901 and 1902, they traveled together across India. They covered more than five thousand miles; they interviewed 425 witnesses in 91 sittings. "This is pretty hard work," Scott-Moncrieff wrote to a relative in England, "listening to witnesses and asking questions for six hours is, I can assure you, rather fatiguing"; but he conceded that "we travel very luxuriously, generally in special trains" and, at every stop, "we are most hospitably entertained in the generous old Indian fashion."[7]

Finally they settled in Lucknow, by the banks of the Ganges, and wrote their report surrounded by meteorological charts, statements, petitions, and two years' worth of impressions.

Once again, they began with the rains: "Not only a main factor in determining the value of irrigation" but also "the primary source of all means of supplying it." The irrigation commissioners went back to Henry Blanford's writings, their main source on India's struggle with the monsoon. They drew attention to water's "unequal distribution throughout the seasons, its still more irregular distribution over the surface of the country, and its liability to failure or serious deficiency."[8] They turned, then, to India's geology, which had been Blanford's original concern. Each of India's main geological regions, the expanse of alluvial and crystalline soils, offered possibilities, and challenges, for irrigation. The commission's data showed that 20 percent of the cultivable land of British India was under irrigation by the turn of the twentieth century, from the large canals of the Gangetic plain to the 626,000 wells watering Madras Presidency.

They concluded that India offered "a wide but not unlimited field" for ambitious engineers. They listed the many obstacles, from the topography of the landscape to the cost of irrigation works. The British government retained the fixation on parsimony that had driven famine policy in the 1870s. In hundreds of pages of testimony to the irrigation commission, witnesses were asked, again and again, about whether investment in irrigation would ever pay for itself. The commissioners dreamed of a time when farmers would pay for water "what it is really worth"; they searched for ways to justify irrigation expenditure to the state, ever anxious about the charge of extravagance. They also saw as a problem water's very expanse, in relation to the lines of jurisdiction that ran through India. "The manner in which the various states and territories are intermingled," they wrote, were an "obstacle" to their visions. The commissioners expressed fulsome thanks to the princely rulers that had hosted them on their tour, but the patchwork of sovereignty that interspersed regions of direct British control with princely territories caused anxiety. The problem was enduring: the "only suitable site for a storage work," they observed, "may lie in a territory whose people would not only derive no benefit, but might even be put to considerable loss." They had diagnosed a source of hydraulic inequality that would sharpen over the century ahead.[9]

Among the resources the irrigation commissioners looked to was India's ocean of water under the ground. By the late nineteenth century the quest for subterranean water had attained new fervor. A breakthrough in technology brought with it a sense of untapped possibility: engine-driven pumps promised to reach much deeper underground than the manual ways that were little changed from Babur's time. After the famines of the 1870s, agricultural officials in the most arid districts encouraged the construction of wells by landowners. "The main difficulty . . . in well construction is not the discovery of water," wrote W. C. Bennett, director of agriculture in the Northwest Provinces, "the water level is known locally all over the Provinces." The problem was to determine where the soil could support deep wells, given the "extreme capriciousness and uncertainty" of the clay stratum beneath the surface.[10]

From the start, and in contrast with government-built canals, wells were in private hands; "policy," such as it was, involved encouraging landowners to dig wells. The government's role, if it had any, was limited to providing credit and information. Already by the 1880s, Bennett estimated that almost 58 percent of the irrigated area in the province he administered was under well irrigation; canals, which received infinitely more money and attention, accounted for only 24 percent. To his mind the importance of wells could not be overstated. They usually held enough water "to carry the people through one season of failure of the rains"—though he cautioned that if the rains should fail two years running, as they had in 1876 and 1877, the wells would run dry.[11] Water prospectors began to imagine the map of India with an added dimension—the underground. As well as roads and railways, wrote an engineer from the Bombay Deccan, it would help if maps could show the "*water* levels as well as showing depth of spring levels below the surface."[12]

By the early twentieth century, Madras was at the forefront of a water-mining boom. A keen observer of this economic revolution was Alfred Chatterton, a British engineer who had made his career in the Madras government. He served as superintendent of industrial education and insisted on the development of a department of industry in the province to encourage local manufacturing. He was quietly sympathetic to *swadeshi* initiatives to support local producers and boycott imports. To that end he opened a government pencil factory in Kurukkupet, which promised to boost local industry.[13]

Irrigation was among Chatterton's main interests as an engineer. He was the author of a book on lift irrigation (the extraction of water from wells), and the possibilities he saw before him, as he surveyed the Madras countryside, seemed limitless. On a tour of Trichinopoly he reported that there had been a "very large number of applications" for government loans for the purchase of engines and pumps. He appended a list of all of the firms supplying oil engines in Madras, a total of 125 in the province; the largest dealer was Massey & Co., which in 1904 supplied ninety-one engines of a total of 869 horsepower. Chatterton was impressed: "It is obvious from this that the oil-engine has come to stay in the Madras Presidency and that it is suited to the conditions of the country," he wrote. "I am sanguine enough to think that this is merely the beginning of things and that in the next year or two the use of oil-engines will increase very rapidly for irrigation work."

That year, the northeast monsoon had brought scanty rainfall; the new technology was put immediately to the test. "The engine was started 568 times," reported an agricultural station at Melrosepatnam, "and ran for 2074 hours." Most of the engines ran on liquid fuel—that is to say, petroleum—which was half or even a quarter the price of kerosene oil. The main source of supply was from the Borneo oil fields, imported by Best and Company and held in a four-thousand-ton storage tank in Madras. In Borneo, as in the fields of the Burmah Oil Company, a new era had begun.[14]

India tumbled into the embrace of fossil fuels. Very early in the twentieth century, oil unleashed a vision of plenitude based on the extraction of water—a vision that continues to shape contemporary India. "The oil engine and pump do work," Chatterton observed, "[utilizing] to an extent absolutely unknown previously the quantity of water available for irrigation." The abundance of water beneath the soil lay in "a succession of spring channels, one below the other." Their exploitation had once relied on a vast corps of voluntary or *corvée* labor, but now "much larger quantities of water could be abstracted from the river-beds by means of engines and pumps." Already there were three-quarters of a million wells scattered across Madras. Most of them were in the hands of small landowners. The potential for further development seemed vast. Chatterton evoked a terrain with "many deep old furrows filled with sand which mark the ancient course of river-beds" that were "full of water." Water engineers imagined a whole country underground—a network of vanished rivers still running with water, waiting to be forced to the surface. On this view, the "problem of subterranean water" was part of a broader problem of energy: its extraction depended on "distributing power cheaply . . . by wire ropes, compressed air or electricity."[15]

II

Nowhere in India was the quest for water more transformative than in Punjab. In a remarkably short time, Punjab rivaled the Gangetic plains and the river valleys of the southern peninsula as India's agrarian heartland. The impetus for the change came from British attempts to bring water to what they claimed as "wastelands" in western Punjab—a land at the limits of the monsoon, described by James Douie, an experienced district official in the province, as "in reality part of the great desert extending from the western Sahara to Manchuria." Here, "irrigation was not designed to assist agriculture and diminish the losses from seasonal vagaries, but to create [agriculture] where it did not exist."[16]

Between 1885 and 1940, the colonial government built nine Canal Colonies—townships of settlement each built around an irrigation canal—through a stretch of western Punjab between the Beas and Sutlej rivers and the Jhelum, covering an area of 13 million acres. The largest of them was the Chenab settlement, started in 1892; the most challenging for the engineers was the massive Triple Canal project, completed in 1915 through difficult terrain. More than a million people moved voluntarily to these new colonies from the more densely populated lands of eastern Punjab; they inaugurated a revolution in the production of wheat, cotton, and sugar. The Canal Colonies attempted to create a new landscape—and a new society. They were among the first of their kind, establishing a pattern that would become common throughout the world in the twentieth century: a state-financed and state-directed resettlement of people in the service of agricultural "development." The British officials who envisaged the project in the 1880s were ambitious. They sought to create a new sort of Indian village, populated by cultivators carefully selected for their vigor and ability, providing a new dynamo for agrarian capitalism.

Speaking to an audience at London's Royal Society of the Arts in 1914, James Douie—a career civil servant who had spent thirty-five years in Punjab, his roles including chief secretary to the government, and settlement commissioner—lauded the achievements of Punjab's engineers. In his mind, the "colonization" of "waste" land in the

Punjab was part of a global process. "When one speaks of colonization the mind turns at once to new countries," he conceded: "the prairies of the United States, or the pampas of Argentina." Many of those large movements of people, too, had involved the search for new sources of water to make arid lands productive. And in Punjab a similar movement was underway. Douie's account made clear how fully the Canal Colonies were the product of an engineer's vision backed by the power of the state—right down to "allotting land in complete squares." People "who love the irregular fields of the homeland, with their hedges of black and white thorn or wilding rose, may think these unenclosed rectangular fields dull," he admitted—but they were designed for utility. In a society where the rules of inheritance allotted an equal share of land to every son, "twenty-five equal rectangular fields can be divided, accurately, easily, and cheaply."[17]

The state reserved the power to decree not only the shape of each parcel of land but also the composition of its population. The language of science lent authority to ideas about physical capacity and heredity—many British officials were obsessed with categorizing India's people according to the size of their skulls. The impetus to conflate physical appearance with innate character was especially strong in Punjab, which had become the prime recruiting ground for the British army in the aftermath of the 1857 rebellion. To make this strategic shift appear natural, army recruiters, high officials, and a number of their local allies proffered an enduring myth that Sikhs—alongside the Gurkhas of Nepal and in contrast with "effeminate" Bengalis—were a "martial race."[18] Relating how he had made his choice of future colonists from among the large number of applicants in any given village, Douie was straightforward: "I looked at their chests," he said, as he proceeded to show his audience in London a series of slides of unclothed Punjabi bodies. One of his fellow officers had focused on the applicants' hands.[19]

The British gamble paid off. By 1915, the financial commissioner of Punjab wrote of Chenab district that "the land revenue of this tract . . .

exceeds any other district in India."[20] Punjab became the engine of agrarian growth in India. By 1931, 46 percent of all canal-irrigated land in British India was in Punjab; Madras came a distant second. Prosperity from cultivation spurred local industrialization with the rise of cotton ginning factories. With land grants to families who had served in the army, the British shored up loyal support in the region. Many local men saw themselves as protagonists in what one of them called "man's conquest over nature."[21] But social transformation was never free from tension. In 1907 protest erupted in the Canal Colonies against legislation giving the government sweeping new powers to regulate settlement and land use. Colonists complained about the inadequate supply of irrigation water to their fields; they chafed against a lower-level bureaucracy that accumulated arbitrary power; they objected to the infringement of customary rights in the name of scientific management.[22]

—

BRITISH OFFICIALS LIKE DOUIE SAW PUNJAB'S CANAL COLONIES IN light of a worldwide process of frontier colonization. He referred repeatedly to Saskatchewan and Manitoba, to Australia and the American prairie. He could just as well have looked across to China. In China's far northeast, a comparable process of peasant colonization was underway. It dwarfed the movement in Punjab. Between 28 and 33 million Chinese migrants moved to Manchuria after 1850; the movement accelerated in 1890, simultaneous with the development of Punjab's Canal Colonies. Many went to work in mines and on the railways, but most Chinese migrants to Manchuria went as cultivators. By the 1920s, the soybean made up 80 percent of the region's exports. A relatively small proportion of the migrants owned land on a freehold basis; many more leased their land or worked as sharecroppers. Expanses of Manchuria were owned by Chinese official organizations, or by private and semiprivate companies. Large landowners accumulated holdings. And if the Chinese state did not play as active a role in encouraging migration as did the British colonial government in northwestern India, it was far from absent.

As in Punjab, family was the "engine of migration" to Manchuria. Families in Shandong and Hebei sent young men to Manchuria as part of a diversified strategy for family survival. But there the expectation of return was almost universal, unlike in Punjab, where families moved permanently to new settlements and over shorter distances than their Chinese counterparts destined for Manchuria. As in Punjab, most Chinese migrants to Manchuria moved in small groups of kinsmen or fellow villagers. They went where uncles, cousins, or others from their villages had blazed a trail. The railway and the steamship made their journeys more affordable, and took them to places previously inaccessible. When this happened on a large enough scale, whole "villages across the sea" emerged, each resembling a northern Chinese village transplanted to Manchuria.[23] In historical perspective, these movements represent the final closing of a global frontier—the crescendo of a process of migration, settlement, and colonization that had begun a few centuries earlier.[24] From Punjab to the American West, colonization depended on water; in Manchuria, too, irrigation was vital to the expansion of settlement and production.[25] Everywhere, colonization displaced communities of people—often pastoralists—who were already there. Their livelihoods and their cultural habits were disregarded as these "empty" lands were settled in the name of agriculture and civilization.

The very designation of western Punjab as "waste" stripped local pastoralists of their claim to the land: this was no neutral description of a landscape, but rather a justification for what administrators and engineers wished to create. Pastoralists in western Punjab soon found their livelihoods under threat. Where once they knew the landscape intimately, now British engineers and new settlers had captured the water. In the words of a local official, many local people were "driven to migrate by the gradual impoverishment of their villages." Many other "calamities for future generations," as Alexander von Humboldt had called them, lay in store. One was malaria—changes in the hydrology of the land created conditions ripe for its spread. A local petitioner complained that the Canal Colonies had "injuriously modified the climate" so that "malaria always prevails there." Another enduring problem that emerged from Punjab's great experiment was waterlogging so severe that it made some lands impossible to cultivate.[26]

III

The transformation of India's waters went far beyond Punjab. Bombay, Madras, and the United Provinces were all sites of intervention and experiment spurred by the recent memory of famine. Hydraulic engineers built on the nineteenth-century schemes like the dams designed by Arthur Cotton along the Krishna, Godavari, and Kaveri rivers of southern India. In Punjab, the land allocated to the Canal Colonies was already in government hands. Elsewhere, a prolonged process of land acquisition was underway, facilitated by 1894 legislation making it easier for the government to take over private land for "public" purposes. While the 1894 Land Acquisition Act aimed to ease railway construction, it came into force just as water engineering projects proliferated across India. By the early twentieth century a long and painful process was underway, displacing families and whole communities to make way for water infrastructure projects—it continues to unfold to this day.

Sitting in the Maharashtra Archives in Mumbai, housed in the decaying Gothic splendor of Elphinstone College, I saw thick files full of petitions and disputes over land, water, and compensation. "Attempts were made to acquire the land amicably but the owners refuse reasonable terms," an official notice of 1889 read, and so it would be seized: "Permission may be granted to take possession of the land on the expiration of 15 days from the publication of this notice." That was just the beginning; just how much land was taken over this way, piece by piece? Some years later, in 1903, Sakharam Balaji, a farmer who stood to lose his lands, wrote in anguish to the local government office. He had lent support to the construction of a local dam by supplying it with materials, "straining my every nerve" in the process. He was then stunned to find out that the dam would drown his two hundred acres of land. "Money given to me by the Government in light of compensation will not suffice," he declared, for where else could he go? "Land if properly cultivated and repaired will last long," he said, and he had invested handsomely in his own. The archival file is complete with crossings-out and handwritten insertions as Sakharam (or perhaps his scribe) sought the right words. "My occupation is nothing but husbandry," he argued, "and my maintenance is solely dependent on this husbandry." His concern was well founded. Displacement meant more than the loss of land—it was a loss of livelihood, an uprooting of life. Sakharam was fortunate. The local government ruled in his favor and resituated the dam. But many others would suffer irreparable loss.[27]

Groups of landowners came together to protest their dispossession. The residents of Belgaum district in Bombay Presidency—"the most loyal and dutiful, but at the same time the poorest, subjects of His Majesty"—feared that a dam, linked to a series of canals, would "inundate almost all the lands of our valleys causing the residents to shift themselves for their lives to some other distant tract not yet known." They made a powerful claim, couched in the language of justice; they challenged the government on its own terms: "The inhabitants of the above-named villages have equal rights to secure the advantages of benign rule of British Government, as the villages which are to be profited by this canal." The petition ends with more than a hundred signatures in Marathi.[28]

THE PEOPLE OF BELGAUM WERE DISPLACED, IN PART, BECAUSE OF A cascade of new demands for water from India's growing cities. Whereas the decades from the 1840s to the 1880s had seen India reduced to a "colonial" economy—focused on the export of raw materials, and the import of manufactured goods—the tide had started to shift by the 1890s. While factories in India were concentrated in a handful of cities—Bombay, Ahmedabad, Kanpur, Coimbatore—Indian industrial capitalism hit its stride in the early decades of the twentieth century, boosted by the suspension of imports and the voracious military consumption of manufactured goods during the First World War. The French geographer Jules Sion observed after the war: "In Bombay, almost all the factories have Indian owners, directors, engineers. The country's capitalists compete with foreigners for mining concessions. Many vast plantations of tea and coffee are managed entirely by locals. This economic nationalism faces an obstacle in the entrenched habits and ignorance of the masses," he observed, "But it has already proved its vitality."[29]

With urbanization came new demands on water. Bombay was the first city in British India to have a municipal water supply, which started operation in 1860. But the city's growth quickly outstripped the scheme's capacity. In 1885, the Tansa project aimed to augment urban water supply; when it opened in 1892, it supplied the city with 77 million liters of water a day, though this still reached only a small proportion of its residents. The project came with unwanted consequences. Water supply in the absence of adequate drainage led to waterlogging in many neighborhoods of the city, contributing to the conditions that allowed the bubonic plague epidemic of 1896 to spread. As the city of Bombay grew—its population reached 1.2 million by 1920—its tentacles reached ever further into the countryside of Maharashtra in search of water for its residents.[30]

As industrialists grew more prominent, both within the colonial government and within the nationalist movement, their voices grew louder. Among their most pressing needs: water and power. The use of water for irrigation was "well understood," the Indian Industrial Commission reported in 1918, but now water had a "double object"—irrigation and electricity generation.[31] The era of "white gold," as hydroelectric power had come to be known in Europe and America, came to India.

India's first hydroelectric plant was built outside the Raj, in the princely state of Mysore, in 1903. The Sivasamudram Dam along the Kaveri River was built to supply electricity to the Kolar gold mines nearby, which were India's largest. India's biggest industrial firm, Tata and Sons, followed closely behind. Having made a fortune exporting opium to China in the first half of the nineteenth century, the Tatas were among the earliest to move into the cotton industry. Perennially in need of electricity to power their factories in Bombay, the Tatas looked to harness the heavy monsoon rains that fell on the Western Ghats mountain range. Theirs was already a self-consciously global enterprise. The Tatas' close ties with the United States meant that American models and expertise were foremost on their minds when they decided to build a hydroelectric plant of their own at Khopoli, in the Western Ghats, completed in 1915. Within three years, the plant was supplying Bombay's cotton mills with forty-two thousand horse-power of electricity, for twelve hours a day.[32]

One of the enduring heroes of India's attempt to transform its waters was an engineer named Mokshagundam Visvesvaraya. Visves-varaya was born to a poor Brahmin family in the princely state of Mysore in 1860. He studied engineering in Pune's College of Science, and served the government of Bombay for twenty-five years as an engineer in the Department of Public Works. Visvesvaraya was ascetic in his habits, a firm believer in hard work and self-help. His first big achievement was the construction of a new water pipeline in the town of Nasik. In the course of redesigning the water supply of the much larger city of Pune, Visvesvaraya patented a new system of automatic sluice gates, which remained in use for decades afterward; in keeping with his dedication to public service, he gave up any claim to royal-ties. From the outset, Visvesvaraya was obsessed with the "waste" of water. He grappled with "how to bring under control the irregular distribution of water to crops . . . and its wasteful use by cultivators." He found that "the cultivators were unaccustomed to control" by en-gineers and wise administrators. To assert that "control" over water would be a lifelong battle for Visvesvaraya.[33]

Visvesvaraya's horizons expanded in the first decade of the twentieth century, when he began to travel the world. His first trip abroad was to the country that impressed him more than any other: Japan. Visvesvaraya spent three months in Japan in 1898. He wrote a book based on his impressions, but "did not . . . think the time was opportune to publish it"; he was a civil servant, and not one for overt political confrontation. Yet seeing Japan had led him to conclude just how little the British were doing, in comparison, for the development of India. "Since all industrial progress in Japan has been achieved in comparatively recent years," he wrote, "she offers to India the most direct and valuable lessons obtainable in material advancement and reconstruction." The greatest lesson, he thought, was that the Japanese state had taken a direct and interventionist role in fostering economic development.[34]

Visvesvaraya's career took him to West as well as to East Asia. In 1906, the government of Bombay deputed him to the British protectorate of Aden, to investigate the city's water supply. Poring over the city's sanitary records, he found an alarmingly high death rate; he was quick to conclude that "a system of pipe sewers is the only satisfactory method." The authorities accepted his recommendation that a series of wells be dug to supply the city from the River Lahex. Upon retirement from government service—and perhaps frustrated by the racial glass ceiling of the Raj, which meant that he could never become chief engineer of Bombay—Visvesvaraya toured the world. He visited western Italy, Russia, and North America to study dams and irrigation techniques. Arriving in New York, he found "an association of Indian traders and businessmen there, men of energy, vitality and ambition."[35]

After a year of roving consultancy, Visvesvaraya was persuaded in 1909 to take up the position of chief engineer of Mysore. Among the princely states, Mysore was self-consciously progressive; a succession of dewans (prime ministers) had, with the Maharaja's encouragement, instituted educational and infrastructural reforms. A few years after his appointment as chief engineer, Visvesvaraya assumed the role of

Dewan: he made primary education compulsory; he invested in infrastructure and sanitation. Given free rein and a generous budget, Visvesvaraya scaled up his ambitions. He dreamt his masterpiece: the Krishnarajasagar Dam along the Kaveri River. It was Visvesvaraya's plan for a "multi-purpose" project—the dam would irrigate one hundred acres of land, it would provide power to the Kolar gold fields, and it would electrify the city of Bangalore. The problem, as we will see, is that the British had their own plans for the same river.

———

CHANNELING RIVERS AND RAINS, PUMPING WATER FROM UNDERground—there was a third frontier in view at the turn of the twentieth century: the ocean. Frederick Nicholson joined the Madras civil service in 1869, aged twenty-three and fresh from Oxford. Over the next three decades he would serve as collector of Tinevelly (Tirunelveli), Madras, and Coimbatore; he witnessed the great famines of the 1870s, which shaped his view of the world; he fell in love with the Nilgiri hills, where eventually he retired and lived until his death in 1936.[36] At the turn of the twentieth century, Nicholson turned his attention to the fisheries of Madras. He wrote in 1899 that "when we despair of food independent of climate for a rapidly-increasing population, of industries for non-agriculturists, of manure for deteriorating soils, we may thank God that we have yet got the fisheries to develop." "Food independent of climate" was precisely what India's quest for water sought to bring about—Nicholson saw this more starkly than most. "The sea yields its harvests in enormous quantities wholly irrespective of droughts and seasonal catastrophes," Nicholson wrote, "and the food, being highly nitrogenous and concentrated, is of extreme value." As insurance against a fickle monsoon, the fisheries were hard to beat. He told the Lahore Industrial Conference in 1909 that, when he had first arrived in India, "the time was not ripe for devising for the distant and vague harvest of the sea what was barely coming into contemplation for the harvest of the soil under foot"; but now, the time was ripe for "adding the harvest of the sea to the harvest of the soil." Nicholson made a crucial distinction: the government's interest in fisheries was purely as a source of food, and not as a source of revenue—which is how the British had treated India's land from the earliest days of the East India Company's rule.[37]

Nicholson traveled to Japan and to Denmark, seeking inspiration for the development of the fisheries of Madras. Like Visvesvaraya, Nicholson thought that Japan's example was most apposite for Madras. There, an "ancient" fishing industry had been transformed by "scientific foresight," backed by an energetic government and generous investment; Nicholson suggested that India's fishing industry in 1907 was where Japan's had been before the Meiji Restoration of 1867.[38] Nicholson created the Madras Department of Fisheries in 1907 and served as its honorary director. His genuine sympathy for poor fishing communities melded with a widely shared view that held "poor," "ignorant," and caste-bound hereditary fishers culturally inferior to landed cultivators. The result was a policy of gradualism. Nicholson doubted whether the rapid advance of fishing technology was possible in India; he preferred to build incrementally on existing practices using existing social structures; he was an enthusiast for cooperative societies as a way to foster a stronger sense of collective action among fishers.

Nicholson's successor, James Hornell (1865–1949), devoted years of his life to understanding the fisheries of India's eastern coast. At the turn of the twentieth century, after a decade working on the isle of Jersey, Hornell traveled to Ceylon to survey the marine fisheries there. From 1908 to 1924, he played a leading role in running the Madras Fisheries Department; he undertook detailed studies of coastal fisheries, on the economy of fishing and the changing composition of the catch; he developed a particular fascination with indigenous fishing vessels along the coasts of the Indian and Pacific Oceans, on which he published over a hundred articles in his lifetime. More than Nicholson, Hornell's caution about the wholesale technological transformation of India's fisheries was based on respect for fishers' traditions and their ways of life. In 1917, Hornell described the daily scene on the shore at Tuticorin, long a center of India's pearl fishing industry. "There is no wholesale fish market except the beach, there are no companies or large owners controlling each a number of boats, and while there are certainly some fish salesmen and traders, these men seldom or never keep any accounts," he reported. "The catch is usually thrown in a heap on the beach and the 'lot' as it lies is sold by auction—the buyers must appraise its value by eye, and make their bids accordingly."[39] But change was on the horizon; at sea, as on land, new technology and new commercial ambitions strained against the limits that British policies, and British interests, had placed upon them.

IV

The quest for water altered India's economic balance. For centuries, it had been a matter of common sense that India's wealth lay in its river valleys—they were the most densely settled, the most intensively cultivated, the most prosperous regions. They had always been the core regions of political power. Wealth and poverty, it so often seemed, were a function of geography. The meteorological divisions of India mapped onto the standard of living. "The most densely populated, and therefore the most fertile regions are those of silt deposit, the population becoming most dense towards the river deltas," wrote agricultural hydrologist Edward Buck in 1907.[40] It appeared an immutable fact of nature—but even by the time Buck was writing, the link between population density and agricultural productivity was no longer so clear. By the 1910s, the largest of Punjab's Canal Colonies brought more revenue to the government than any other district in India.

Artificial irrigation sparked an agricultural boom in the drier regions of India's west and south: a boom focused on the production of high value crops for export. Public investment in irrigation poured into regions and crops most likely to bring the state revenue and to bring farmers profit. Punjab and Sind in the northwest, parts of Gujarat and

Bombay Presidency in the west, and some parts of Madras—the western region, around Coimbatore, and coastal Andhra in particular—flourished. Those regions still remain India's most prosperous. The gap in productivity between these favored regions and the rest of the country has grown more marked in the second half of the twentieth century, but the roots of a fundamental reversal in India's economic geography lie in the water boom of the late nineteenth and early twentieth centuries.

The "ancient" zones benefited far less than the irrigated areas from new markets and new technologies. Riverine Bengal, the fabled wealth of which had drawn in the East India Company in the seventeenth century, now saw a protracted decline in its relative economic position within India. Agricultural productivity in the valleys of Tamil Nadu declined precipitously. A "tide of indebtedness" consumed smallholders in both regions, and also along the Gangetic valley. In both the boom regions and those left behind, the control of water as well as the control of credit concentrated land in fewer hands. Dry regions that had not benefited from irrigation did worst of all. By the 1920s, most agricultural land in India was still not irrigated, it was rain-fed. Already by the early twentieth century, irrigated lands produced four times as much as those that depended on rainfall. A recent survey by economic historians emphasizes the point: for most of India, most of the time, rainfall was the most important factor shaping agricultural output, and yields were probably among the lowest in the world.[41]

Two large commissions of inquiry in the 1920s—the Royal Commission on Agriculture in India, and the Banking Enquiry Commission, both on the model of the 1901 irrigation commission but on an even larger scale—revealed Indian cultivators' continuing vulnerability to climatic fluctuations. Irrigation had advanced rapidly, but most Indian farmers did not benefit from it. "Except in the north-west," the report on agriculture maintained, "the whole country is dependent on the monsoon and all major agricultural operations are fixed and timed by this phenomenon."[42]

The main response of Indian cultivators to this vulnerability was to borrow money. The scale of rural indebtedness in India emerged from the inquiry into India's banking system, a colossal exercise in information-gathering that went province by province, reporting its results in 1930 in dozens of thick volumes. "From the sowing of the seed to the sale of the products," one of hundreds of witnesses in the United Provinces of North India testified, "it is the indigenous moneylenders and bankers who enable the produce of the villages to be brought to the market." The monsoon shaped the rhythms of the money market. Cultivators' need for credit converged on certain crucial times in the season: credit for seed and manure in October and November, after the rains; credit to pay agricultural labor at harvest time. In many parts of rural India there was "an exaggerated alternation of over-work and unemployment." Account after account from rural India made the same point. "The annual rainfall is scanty and uncertain and irrigation is nominal," wrote a local official from Meerut district in the United Provinces; and "a peasant proprietor once entrapped by the *mahajan* [moneylender] can never extricate himself."[43] Many borrowers faced 24 percent compound interest. "Ninety-five per cent of the agricultural classes are in debt," wrote an administrator in Mathura district, "due to successive failures of crops in this district."[44] Only a few respondents disagreed. The collector of Kaira district in Bombay Presidency told the agricultural inquiry that monsoon failure was not one of the main causes of indebtedness: "The frequency of very bad seasons in which the cultivator would be left completely insolvent is not very great," he noted.[45]

The most creative response to the fact that most Indians still depended on the rains for their livelihoods came from J. S. Chakravarti, who worked for years in Mysore. He was a colleague of the great engineer Visvesvaraya; Chakravarti, too, flourished in the southern Indian princely state that was bolder than any part of British-ruled India in its approach to the problems of water. Chakravarti worked for Mysore's State Insurance Committee for many years, and in the 1910s rose to the position of controller and financial secretary to the government of Mysore. He advocated a system of drought insurance in preference over generalized crop insurance: while the failure of a particular crop could be down to bad practice or neglect on the part of individual farmers, a failure in the rains affected everybody. Where rainfall dipped below a certain proportion, say 35 percent, of average, cultivators would receive a payout. Chakravarti saw rainfall insurance as "intimately connected with three sciences, *viz.* economics, meteorology, and agriculture." His starting point was that "Indian agriculture is dependent almost entirely on rainfall" and that the "quantity of rain during the year and its distribution as regards time are almost the only essential factors" in determining the incomes of farmers; absence (or excess) of rain at certain critical periods in the growing season could be devastating. And what Chakravarti called the "rainfall factor" was "uncontrollable by human exertions." Only the state could carry out an insurance scheme on the scale he envisaged, though he hoped that private providers might eventually enter the market. The case for insurance was clear. "Agricultural insurance will also be famine insurance," Chakravarti declared, for "under the present circumstances, a famine in India does not generally mean grain-famine, but money-famine, due to enforced unemployment of the agriculturist." Chakravarti's scheme gathered dust; one Indian commentator observed toward the end of the twentieth century that Chakravarti's scheme was far in advance of what the World Bank had come up with by the early 1990s. The primary reason for this neglect is that a very different approach to mitigating climatic risk emerged in mid-twentieth century India—an approach that emphasized technological solutions to the problem of water.[46]

Reflecting on the state of India's development, the members of the Indian Industrial Commission wrote confidently in 1918 that "the terrible calamities which from time to time depopulated wide stretches of country need no longer be feared." In a monsoon climate, "failure of the rains must always mean privation and hardship," but it no longer need lead to "wholesale starvation and loss of life."[47] This conveyed a strong sense that something fundamental had changed in India over the first two decades of the twentieth century. The risk posed by climate had been mitigated by both policy (the early-warning system of the Famine Codes) and by technology (railways and irrigation). As long as India remained predominantly agrarian, some level of risk would remain, but the commissioners envisaged a future in which industrialization would provide new employment and greater security, as India's population moved from the countryside to the cities. In the 1870s, the idea that famine was inevitable in India prevailed among British administrators. By the 1920s, most observers believed that India had conquered famine. But anxieties about water did not go away.

V

India's engineers fought to assert their sovereignty over the monsoon; climate science demonstrated, in those same years, that the monsoon moved to planetary rhythms—rhythms far beyond human control. The pioneer of Indian monsoon meteorology in the early twentieth century was a brilliant mathematician and a modest man: "kindly, liberal minded, wide of interest, and a very perfect gentleman."[48]

Gilbert Thomas Walker was born in Rochdale, Lancashire, in 1868, the fourth child in a family of eight. He grew up in Croydon, just outside London, where his father was the borough's chief engineer. Gilbert received a scholarship to the elite St. Paul's School in 1881, and went on to a distinguished undergraduate career in mathematics at Trinity College, Cambridge. His talents were idiosyncratic. At Trinity he left behind him "the legend of his prowess in throwing boomerangs on the Cambridge Backs"—his study of their aerodynamics marked the beginning of his fascination with the physics of the atmosphere; he acquired the nickname "Boomerang Walker." In 1890, Walker suffered a breakdown in his health; he spent three summers recovering in Switzerland, where he developed a passion for skating. In time, both hobbies would nourish his insights into the world's weather. Returning to Trinity as a fellow in mathematics, Walker's work focused on electromagnetism. Formalizing his fascination with boomerangs, he also wrote an essay on the aerodynamics of sports and games. There was little in his background or experience to suggest that within a few years, he would be director of the Indian Meteorological Department.[49]

John Eliot was approaching retirement. His models for monsoon forecasts had grown more complex during his years in charge of the Indian weather service, but they were erratic in accuracy and had failed to predict the droughts of 1899 and 1900. Convinced, still, that he was on to something, Eliot searched for an able statistician as his successor, someone who could make sense of the profusion of pressure, temperature, and wind readings dispatched from observatories across the Indian and Pacific oceans.[50] Walker's reputation reached Eliot's notice, perhaps through shared Cambridge networks; the chance to develop a new field of inquiry, in an unfamiliar land, was attractive to the younger man. Before sailing for India, Walker toured meteorological observatories in Europe and the United States: this was his crash course in the science of weather. Visiting field stations in the Midwest, Walker was especially impressed by the sophisticated techniques deployed by the US Weather Bureau.

Walker arrived in India just three years after the last major famine. Eliot's failures had dented public and official confidence in meteorology. In 1904, Walker took over as chief. He kept a low profile for the next four years: marshaling resources, recruiting staff, bolstering a skeletal meteorological department. In those early years in Simla—the Himalayan summer capital of the Raj, where British officials rushed to escape the pre-monsoon heat—Walker found time for his two beloved hobbies. He was often spotted flinging boomerangs on Annandale, the only stretch of flat ground in Simla. He designed a low canvas screen to keep the ice cold on Simla's skating rink—it was so effective that, by January, "the ice was too hard to be skated on with pleasure," and the rink's owner asked him to remove it. Walker's mind leapt to make new connections. Many years later he revealed that his experience with the ice rink had led him to understand the "extreme transparency of the air to heat radiated from the ground during the very dry winter periods" over North India.[51]

From these beginnings came a vital breakthrough in climate science.

WALKER'S APPROACH TO UNDERSTANDING AND PREDICTING THE monsoon was resolutely empirical. He relied on a vast "human computer"—that is, on the labor of his Indian staff led by Hem Raj—to process the numbers. "The relations between weather over earth are so complex," Walker felt, that "it seems useless to try to derive them from theoretical considerations"—the monsoon was too complex.[52] Rather, Walker sought to amass and analyze as much weather data as he could, from all over the world. This had been Blanford's challenge: to determine how far the monsoon's causes as well as its effects were confined to "India and its seas." Blanford's first forecasts relied on Himalayan snowfall; while he was increasingly aware of remote influences on India's climate, he assumed a closed system. Eliot moved to incorporate influences from across the Indian Ocean into his model, but he misunderstood the relationships at work. Walker, with more statistical tools (and more staff) at his disposal, broadened his parameters. His team processed a quantity of data that would have been inconceivable a generation earlier. The numbers told a clear and a new story: they suggested that "the monsoon system extended to a pan-oceanic," even planetary, scale.[53]

Walker took aim, first, at the desiccationists. He showed that there was little evidence to support their claim that human activity, and particularly deforestation, had modified India's climate in the nineteenth century. However much the denudation of forests may affect the climate and the soil moisture of a particular locality, the scale of the monsoon system far outstripped such local influences.[54] As his thoughts turned to monsoon prediction, Walker looked west of India.

The Nile had long been on the minds of India's hydraulic engineers, as comparison, inspiration, or competition; Walker turned his attention to the annual Nile flood for different reasons. "Inasmuch as the Nile flood is determined by the monsoon rainfall of Abyssinia," he wrote, "and as the moist winds which provide this rainfall travel in the earlier portion of their movement side by side with those which ultimately reach the north of the Arabian Sea" so there was a "tolerably close correspondence" between the extent of the Nile flood and the strength of the Indian monsoon. This was "seasonal foreshadowing" at work—a term Walker preferred to the more confident "forecasting."[55]

Walker's statistical prowess paved the way for his most startling discovery.[56] Mining data from across the world, Walker noticed that "there is a swaying of pressure on a big scale backwards and forwards between the Pacific Ocean and the Indian Ocean, there are swayings, on a much smaller scale, between the Azores and Iceland, and between the areas of high and low pressure in the North Pacific." The "Southern Oscillation," as Walker named it, had a "much greater" influence on "world weather" than the other two.[57] He called these pivotal areas of high and low pressure "centers of action." Walker had identified an inverse relationship of atmospheric pressure at sea level across the Pacific Ocean, measured by readings from stations at Darwin and Tahiti. The usual pattern was for high pressure in Tahiti and low pressure in Darwin, driving the winds from east to west. The pressure contrast across the Pacific drove the storied westerly "trade winds"; but they were prone to periodic reversals that could last for one or two seasons. This was the mystery at the heart of Walker's findings. Changes in the location and intensity of the "centers of action" shaped the world's climate—but what prompted these changes?

Walker's immediate challenge, as head of the Indian weather service, was to determine how these "centers of action" affected India. The broad contours of the picture were clear by the early 1920s. "Abundant Indian rains," he wrote,

. . . tend to be associated with low pressure in India, Java, Australia and S. Africa; with high pressure in the central Pacific Ocean (Samoa and Honolulu) and South America (Chile and the Argentine); with previously scanty rainfall in Java, Zanzibar, Seychelles and South Rhodesia; and with low temperature in the Aleutian islands.

He sought correlations in time as well as in space; there were "foreshadowings"—in Zanzibar or in the Aleutian Islands—of dearth or plenty in India; Walker probed the "lags" of a season, or two, in the relationships he discovered.[58] But the forces at work proved elusive: "I cannot help believing that we shall gradually find out the physical mechanism by which these [oscillations] are maintained," Walker said in 1918. A few years later he told the audience at his presidential address to the Royal Meteorological Society that "variations in activity of the general oceanic circulation" would likely be "far reaching and important."[59] It would take another forty years for it to become clear just how "far reaching and important" the oceanic dimension really is.

The role of official meteorologist had left Walker little time for basic research; he took what opportunities he could. The First World War brought new challenges. The meteorological office had always scrabbled for resources; during the war, Walker's deputies were transferred out. G. C. Simpson and Charles Normand were sent to Mesopotamia in 1916, where Normand took charge of military meteorology. Their expertise formed part of a wider infusion of British Indian personnel into Iraq to accompany an even larger number of troops: hydraulic engineers and entomologists, together with a great many vessels built in Indian shipyards.[60] In their absence, it fell to Hem Raj to keep Indian meteorology running. A veteran Indian officer in the department with a "photographic memory" for weather charts, Hem Raj oversaw the department's day-to-day operations. Walker later paid warm tribute to "R.B. Hem Raj, who sacrificed his life in the cause of the allies by concealing a serious illness in order that he might continue his important assistance in an under-staffed office."[61] Even as research ground to a halt in Walker's office, the war brought significant advances in global meteorological understanding, thanks primarily to the work of Vilhelm Bjerknes in Bergen—together with his son Jacob and Halvor Solberg—on the development of midlatitude cyclones. Bjerknes and his team brought a metaphor from the battlefields to their understanding of weather "fronts," as they illuminated the dynamic interaction of warm and humid air currents with polar currents. It would be Jacob, five decades later, who finally identified the El Niño phenomenon, expanding on Gilbert Walker's insights.

However wide he cast his statistical net, Walker remained a keen observer of his surroundings. He developed an interest in the flight of vultures and kites, which he watched through a telescope in Simla. He noticed that the birds knew where to look for updrafts, allowing them to ascend to up to two thousand feet without flapping their wings. This insight inspired Walker to take an interest in the physics of cloud formation; he even wanted to take up gliding upon his return to England, and was sad to find that "at 65 his reactions were too slow to allow him to be a successful glider pilot."[62] Walker left India in 1924, twenty years after taking over the Meteorological Department. He became professor of meteorology at Imperial College, London. Freed from the practical responsibilities of the Indian weather service, Walker turned his attention in the 1920s to developing his understanding of what he called "world-weather"—to which he saw the Southern Oscillation as crucial. Walker remained eclectic in his interests and his methods. In 1927 he warned of the danger of over-specialization: "There is, today, always a risk that specialists in two subjects, using languages full of words that are unintelligible without study," he wrote, "will grow up not only, without knowledge of each other's work, but also will ignore the problems which require mutual assistance."[63] There is a hint, here, of the quest to defend meteorology as a science, against those who would see it as mere observation. However global his perspective became, Walker never left the monsoon behind.[64] One of his successors as director of Indian meteorology, Charles Normand, explained many years later why this was. "The Indian Monsoon," Normand wrote, describing Walker's work, "stands out as an active and not a passive feature in world weather." India, on this view, was a driving force in the world's climate. The irony was that "Walker's worldwide survey ended by offering a promise for the prediction of events in regions other than India," since India's experience seemed "more efficient as a broadcasting tool than an event to be forecast."[65]

Soon after the war, the advent of long-distance flights between Europe and Asia in the 1920s shed new light on atmospheric dynamics while demanding more comprehensive forecasts for aviators. From an observatory in Agra, the Indian weather service could now send balloons up to twenty thousand feet, which returned measurements of upper air conditions. Flight through clouds was giving scientists "insight into cloud formation"; aerial photography provided a new perspective on the vertical dimension of the weather—"we see how different a cloudscape seen from above is from one viewed from the ground."[66]

———

EVEN BEFORE THE WAR'S END, THE PROSPECT OF POLITICAL CHANGE was in the air in India. Under pressure from the eruption of mass political protest in India, to which we will return in the next chapter, the reforms enacted under the Government of India Act of 1919 devolved many responsibilities to the provincial level—a system known as "dyarchy." It expanded representative government, though electorates remained small. In parallel to this, a change in personnel began to be felt across the colonial administration—a process of "Indianization," as it was called, in the bureaucracy. Lower-level judges and immigration officials, health inspectors and government scientists were more likely, after 1920, to be Indian rather than British men—but men they mostly remained. And there were limits to how high Indians could rise, which in turn fueled middle-class support for the nationalist movement.

Meteorology moved in the same direction. In the early 1920s, new Indian officers were hired to senior positions in the Meteorological

Department: G. Chatterjee, of Presidency College, Calcutta, took over the Upper Air Observatory in Agra; S. K. Banerjee, a noted mathematician, joined the department in Simla—he would go on to become the first director of the Indian Meteorological Department after independence. In Bengal, a young statistician called Prasanta Chandra Mahalanobis, fresh from Cambridge, was hired to work at the observatory in Calcutta; his impact on the path of India's economic development would be immense. In 1921, Gilbert Walker wrote to Delhi, insisting that "it is natural for the government to insist that a serious attempt be made to find and train Indians who shall be capable of carrying out the work of the Department," noting that there was little to distinguish the abilities of British from Indian staff. "On political grounds," he noted, "it is obvious that this policy of Indianization must be loyally accepted." J. H. Field, Walker's successor, went further. In 1925, Field pleaded for more resources—he wanted six new posts, and a more energetic program of research. He justified his demands by insisting on the talent of Indian meteorologists, and by pointing to the new demands that would arise from air links between Europe and Asia. "India has now for the first time an opportunity to show what an Indianised department can do," he declared. "The opportunity is magnificent and unique: if the controlling department of the government will only rise to the height of this occasion and give me what I ask, it is to be expected that my Indian staff will justify demands as an outstanding example of efficiency in running their own concern." In an age of austerity Field did not secure the resources he had asked for, but the "Indianization" of the meteorology department was underway. The Indian officials who joined the department in the 1920s were "nationalists to the core," writes D. R. Sikka, director of the Indian Meteorological Department in the late twentieth century. This did not prevent them from being "loyal to the department," believing it to stand above politics. But in the 1920s and 1930s India's rainfall became ever more deeply political.[67]

IN HIS ACCOUNT OF ISAURA, ITALO CALVINO WROTE THAT THE city's water gods inhabited the whole vertical expanse of water. They lived, he wrote,

> in the buckets that rise, suspended from a cable, as they appear over the edge of the wells, in revolving pulleys, in the windlasses of the norias, in the pump handles, in the blades of the windmills that draw the water up from the drillings in the trestles that support the twisting probes, in the reservoirs perched on stilts over the roofs, in the slender arches of the aqueducts, all the columns of water, the vertical pipes, the plungers, the drains, all the way up to the weathercocks that surmount the airy scaffoldings of Isaura, a city that moves entirely upward.[68]

The fictional Isaura is timeless, its waters unchanging. In early twentieth-century India, technology transformed water in every dimension. Electric pumps extracted water from the depths and balloons measured the moisture of the upper atmosphere; big dams harnessed the descent of rivers, for irrigation and flood control and to generate electricity. A vision of India that "moves entirely upward" sat alongside the older (and flatter) maritime conception of India at the heart of an imperial web of sea-lanes. The next chapter will turn to the story of how struggles for water were intensified by Asian nationalisms in the 1930s.

WATER AND FREEDOM

A N ANCIENT DICTUM—THAT THE CONTROL OF WATER CONFERRED political power—acquired new meaning, and new urgency, in an age of nationalism. From India to China, water was at the heart of programs for political renewal and national development in the 1920s and 1930s. The rising generation of leaders in Asia included engineers, architects, and physicists, alongside lawyers and schoolteachers. Many of them felt that the conquest of nature in the early twentieth century had not gone far or fast enough. For inspiration they looked to the world's rising powers. They studied the New Deal in the United States. American technological modernity was epitomized by the Tennessee Valley Authority, which gathered together previously disparate approaches to flood control, river navigation, electricity generation, soil conservation, irrigation, and public health. Asian nationalists drew lessons from the breakneck industrialization and the colossal engineering schemes of the Soviet Union, not least because the Soviet Union was also a major Asian country that had attempted to reengineer landscapes that resembled those of China's and India's northwestern reaches, and has done so at a pace unprecedented in global history.

In India, in China, across Southeast Asia, nationalist movements were unstable social coalitions. Their leaders struggled to create a sense of unity and purpose, while acknowledging the fractures of social and economic inequality and addressing regional disparities. Many divisions emerged over the control and sharing of resources— among which water was often the most vital. The 1920s and 1930s were characterized by deepening contacts and solidarities among anticolonial and revolutionary movements across and beyond Asia. But when it came to tangible material questions, like sharing water, they began to draw firm lines around their respective domains. In these decades between the world wars, the seeds were sown for water conflicts that would intensify in the second half of the twentieth century after Asian nations won their freedom from colonial rule.

I

As we saw in the last chapter, India's landscape was reshaped by the quest for water in the first two decades of the twentieth century. In the accounts of engineers the construction of canals and dams, and the pumping of underground water were purely a technical process, outside politics. Few of them, British or Indian, departed from the assumption that the British colonial government would lead the charge. But the rise of nationalism raised new questions about who would benefit from these changes in India's land and water.

Indian nationalism emerged as a powerful mobilizing force in the first decade of the twentieth century. The Swadeshi movement arose in protest against a 1905 British plan to partition the province of Bengal, ostensibly for economic reasons, but also to divide what the colonial government perceived to be a threatening locus of political organization. "Swadeshi," meaning "home-made," began with the boycott of

British goods in favor of locally produced products, but it burgeoned into a diverse movement that included those—branded "terrorists" by the British—who advocated the violent overthrow of the colonial government. The protesters achieved their immediate goal: the British revoked their decision to break up the province. But the Swadeshi mobilization was transient and fragile; it splintered into mutually hostile factions. It was largely confined to the province of Bengal, and even there, it was dominated by elite Hindus to the exclusion of Muslims. The Swadeshi movement mirrored similar uprisings elsewhere. Across Asia, the early twentieth century saw a wave of boycotts and demonstrations and strikes. The same year the Swadeshi movement began, Shanghai witnessed a widespread boycott of American goods in response to the wave of violence and discriminatory legislation directed against Chinese immigrants in the United States in the early twentieth century. By the 1910s, these stirrings of unrest had turned into mass movements.

In India, the most effective and visible political leader was a lawyer named Mohandas Karamchand Gandhi. Born to a merchant family in Porbandar, a port town in Gujarat on India's western coast, Gandhi spent decades outside India. He studied law in London between 1888 and 1891. There he came under the influence of the spiritualist Theosophy movement; he discovered vegetarianism; he experienced a political and spiritual awakening that led him to the study of Indian philosophy and religion. In 1893, Gandhi took up an offer of a job as a lawyer in South Africa. He quickly came to lead protests against the race-based exclusions and restrictions faced by the Indian community in South Africa—a diverse group that included Gujarati merchants and traders, concentrated in Durban and Johannesburg, as well as indentured workers from Tamil Nadu and Bihar, who worked on the sugar plantations of Natal. Gandhi, like most of South Africa's Indian community, supported the British in the South African War, a brutal contest between English and Afrikaner settlers. Hopes that the British would reward Indian support after the war proved short-lived. The rapprochement between the English and the Afrikaners in the postwar settlement led to a tightening of restrictions on the Indian community, including a requirement for them to carry identity cards ("passes")— though, always, the colony's African majority faced discrimination that was far worse. Immersed in reading Tolstoy and Thoreau, Gandhi experimented with communal living at a settlement named Phoenix. He started a printing press. He honed his political tactics—a form of nonviolent civil disobedience that he dubbed *satyagraha* ("struggle for truth.")

During his South African years, Gandhi developed a critical account of British rule in India. He published *Hind Swaraj* in 1909, a treatise that took the form of a dialogue with an imagined reader. Gandhi took aim not only at the violence and tyranny that underpinned British rule in India but also, more radically, at its material effects. "India's salvation consists of unlearning what she has learnt during the past fifty years or so," he wrote. "The railways, telegraphs, hospitals, lawyers, doctors and such like have all to go, and the so-called upper class have to learn to live consciously and religiously and deliberately the simple life of a peasant."[1] Gandhi concluded that "machinery is the main symbol of modern civilization; it represents a great sin." In rejecting "telegraphs, hospitals, lawyers [and] doctors," Gandhi was being provocative; he aimed to shock his readers into asking questions about the ultimate ends of India's embrace of industrial modernity.

Gandhi's analysis stood at odds with the rush to secure India against vulnerability to nature—a process which, we have seen, involved many Indians alongside British water engineers and administrators. Over the years, Gandhi developed further the idea that India's freedom lay in living with the rhythms of nature. He was repulsed by India's cities, though his vision of the country as an agglomeration of "village republics" was largely a myth drawn from the writings of British orientalists like Henry Maine. As a symbolic figure, as a tactician, Gandhi was unrivaled within the Indian nationalist movement. His economic ideas remained marginal. They stood as a quiet counterpoint to the powerful tune of more technology, more control,

more progress. Most of Gandhi's associates and many of his follow-
ers had a different view of what India needed—the "simple life of the
peasant" was precisely what they aimed to relegate to the past.

When Gandhi returned to India in 1915, he hurled himself into
political activity. He had already acquired the honorific "Mahatma"
("great soul"); his reputation as an effective organizer and powerful
speaker had traveled with him from South Africa. He began on a
small scale, interceding on behalf of indigo workers in Champaran,
Bihar, who were protesting their exploitative working conditions. By
1917, Gandhi was India's preeminent politician. He jolted the Indian
National Congress to life, expanded its membership, forged a coalition of rural and urban supporters. Gandhi rejected the class-based
mobilization of the left in favor of an emphasis on conciliation;
among his supporters were India's largest industrialists, including
the Birla family. Gandhi launched his mass Non-Cooperation Movement in 1919, in protest against the slow pace of political reform in
British India and directed in particular against the prolongation of
the state's wartime emergency powers. A campaign of protests and
boycotts, fasts and vigils, lasted until 1922, when Gandhi called it
off following an act of violence in the small town of Chauri Chaura,
where Congress supporters had attacked a police station. A cycle of
repression and concession would unfold over the subsequent two
decades. The British government of India locked Gandhi and his
lieutenants up in prison on many occasions, interspersed with periods of negotiation.

The nationalist upsurge spanned Asia. The Non-Cooperation
Movement in India raged at the same time that China saw an out-
pouring of social and political protest against the territorial concessions that Japan, victorious on the Allied side, had gained in China
after the First World War. A wide coalition of youth and students and
activists came together in a loose grouping known as the May Fourth
Movement. In Vietnam and in Indonesia, too, the 1920s saw the rise
of new political and social movements directed, respectively, against
French and Dutch colonial rule. In all three countries, unlike in India,

communism emerged among the most powerful and most compelling of political movements.

Asia's nationalist movements spoke the language of freedom and sovereignty, and it is on the richness and multiplicity of these concepts that historians have focused their attentions. But there was always a strong material underpinning to the ambitions of Asia's nationalists. It is here that the history of nationalism intersects with the battle to bring unruly waters under control. Nationalist leaders needed water, mineral resources, and fossil fuels to realize their plans for industrialization, to make good on their promises of an end to hunger and poverty. A new sense of confidence crept into visions of Asia's future. Consider this contrast: In 1909, the imperial finance minister of India had characterized each of his budgets as a "gamble on the rains," conveying a sense of fatalism about the power of nature over economy and society. Twenty years later, Jawaharlal Nehru—Cambridge-educated lawyer and scion of an elite Allahabad family, son of pioneering nationalist Motilal Nehru, and by the 1920s Gandhi's most trusted younger colleague, and one of India's most influential and charismatic politicians—declared that "modern science claims to have curbed to a large extent the tyranny and the vagaries of nature."[2] Nehru was clear about the material urgency behind every vision of freedom. "Our desire for freedom is a thing more of the mind than the body," Nehru said, but most Indians suffered "hunger and deepest poverty, and empty stomach and a bare back." For the masses, "freedom is a vital bodily necessity."[3]

In China, as in India, water was a vital ingredient of freedom. The control of water was essential to China's emergence from a century of humiliation at the hands of imperial powers, which had culminated in 1911 with the collapse of the Qing dynasty. Sun Yat-sen, architect of China's republican revolution, applied himself to the problem of China's development, even as rival regional polities tore China apart. In *The International Development of China,* Sun set forth an expansive vision of China as what he called an "economic ocean" for the world. His book was replete with maps of rivers diverted, maps of rail lines laid, maps of ports dredged and electricity generated.[4] Water was at the heart of his vision. Sun told a meeting in Guangdong in 1924: "If we could utilize the water power in the Yangtze and Yellow rivers to generate one hundred million horsepower of electrical energy, we would be putting twenty-four hundred million men to work!" Sun predicted that "when that time comes, we shall have enough power to supply railways, motor cars, fertilizer factories and all kinds of manufacturing establishments."[5]

In contrast with India, where the focus had long been on irrigation, Chinese river engineering in the early twentieth century focused on flood control. Though China, too, had suffered from the great droughts of the 1870s and 1890s, it had also experienced disastrous river flooding on a scale unknown in India. By the 1920s, the Yellow River, famously silt-laden, posed a particular challenge—a challenge embraced by an international corps of engineers. Two Americans, John Freeman and O. J. Todd, played a central role; their Chinese protégés included Li Yizhi (1882–1938), whose stature in China was akin to Visvesvaraya's in India. Li studied in Berlin and then visited hydraulic projects across Europe; he was aided in his work by a new cadre of Chinese graduates from MIT and other leading American institutions. "To manufacture cotton into yarn, to grind grain into flour, to light cities and otherwise modernize this part of Shansi," Todd wrote, "will be part of the benefit that these Yellow River Falls may confer on the nearby country."[6] His ambition found many echoes across China, in India, and in other parts of the colonized world. But beneath that ambition was an enduring sense of fragility.

———

TO HARVEST WATER WAS TO REDRESS THE INEQUALITIES OF nature—to even out the uneven reach of the monsoon, to ensure against the particular unpredictability of the rains in the places that needed it most. But water was also an engine of inequality between people, between classes and castes, between city and country, between regions. The command of water underpinned the accumulation of land. The control of water was a source of power; its absence, a source of enduring exclusion. In the first three decades of the twentieth century, water was at the material heart of many struggles for freedom—but freedom for whom?

The question arose forcefully in the western Indian town of Mahad, near Poona, in March 1927. The local Dalit community—those excluded from the Hindu caste system, once known as Untouchables, whose daily lives were marred by residential and occupational segregation as well as by violence and material deprivation—were denied access by upper-caste Hindus to a local tank containing drinking water. Although a court had ruled that this exclusion was illegal, it continued—as it did in countless towns and villages across India, as it still does today. Dalit leader Bhimrao Ambedkar—a brilliant lawyer from a poor family in western India, who had received scholarships to study at the London School of Economics and Columbia University—led a march to the tank. He drank a symbolic cup of water from the reservoir. The retaliation from local caste Hindus, who felt their social dominance under threat, was brutal and immediate. Dalits were attacked; many lost their jobs. "We now want to go to the Tank only to prove that, like others, we are also human beings," Ambedkar declared, as he launched a *satyagraha* with four thousand volunteers. At the last minute, he called off the movement, trusting in the courts to deliver justice for his community. It took a decade for Ambedkar's trust to be vindicated, when a further ruling insisted the tank be opened to all—contradicting the caste Hindus' claim that it was private property, and therefore that they were free to exclude whomever they chose from the tank's waters.[7]

In the broadest terms there remained a tension at the heart of the Indian nationalist movement. As one political theorist has described it, it was a tension between, on the one hand, "social freedom from caste domination," and, on the other hand, an overriding emphasis on the immediacy of "political freedom from colonial rule," deferring or subsuming those other struggles.[8] Ambedkar and Gandhi would find themselves on opposite sides of that debate, coming into conflict in the 1930s over whether Dalits should have separate representation in the legislative councils of British India, which India's Muslims already had. It is no coincidence that both leaders invoked the symbolic as well as material power of water. For his part, one of Gandhi's most effective and iconic campaigns was the "salt march" to the sea at Dandi, in 1930. Choosing the British salt tax as the symbolic focus of his *satyagraha,* Gandhi observed that "next to air and water, salt is perhaps the greatest necessity of life."[9] The vital properties of salt linked the coastal ecosystem with the lives of millions inland. Gandhi's was an argument about climate and society—the poorest, who labored outdoors in the heat, were most in need of salt. Where Ambedkar's march on the tank had drawn attention to water as an indicator of profound social inequality, Gandhi used it as a symbol of unity. Within and beyond India, competing claims on water and resources escalated in the 1930s.

II

How far could Asia's environmental inheritance be molded? What was the potential of technology to transform Asia, to make use of water and to make water available to all? Contending answers to these questions played out in the decades between the two world wars. Iron confidence in the conquest of nature, expressed by engineers and scientists and nationalists, alternated with a sense of vulnerability before nature's power and its unpredictability. As new knowledge of the monsoon became more widely known, climate itself provided a new way to think about Asia, its boundaries—and its future.

"I wish to treat the monsoon as a way of life," wrote Japanese philosopher Watsuji Tetsuro (1889–1960) in the late 1920s; this was "something that a hygrometer cannot do." And so the fullest expression of the idea that the monsoon constituted the essence of India came not from a European but a Japanese observer. Watsuji was a scholar of Japanese ethics and aesthetics. He translated the works of Søren Kierkegaard into Japanese. He traveled to Germany in 1927 to study with Martin Heidegger.[10] His journey took him by way of Southeast Asia, India, and the Middle East. During and after his journey he wrote *Fūdo*, loosely translated as "climate"; it was Watsuji's response to Heidegger's *Sein und Zeit* (Being and Time). *Fūdo* was not translated into English until 1961. It is unlikely that the book was widely known in India. But India was central to the book's argument that climate shaped culture, society, and history. *Fūdo* is unusual in contrasting India's climate not primarily with Western Europe's, but with Japan's and China's. Watsuji's work was part of a larger intellectual and political movement in Japan to think about Asian societies— their similarities and contrasts—in light of European domination of the world, and in light of Japanese ambitions for regional supremacy.

The humidity of a monsoon climate, Watsuji believed, "does not arouse within man any sense of a struggle against nature," unlike in desert lands. The "distinctive character . . . of human nature in the monsoon zone," he insisted, "can be understood as submissive and resignatory." This was in part because of the monsoon climate's doubleness: it "typifies the violence of nature" with its huge storms, "the power is so vast that man is obliged to abandon all hope of resistance"; but this is "a threat filled with power—a power capable of giving life."[11] In Watsuji's eyes India represented the most extreme manifestation of a monsoon climate. "It is the rainy season, brought by the monsoon, that has done most to create the resignation of the Indian," he noted. He observed that "over two-thirds of India's 320 millions (a fifth of the world's population) are farmers and grow their crops thanks to the monsoon" and so "whether it is late, whether it lasts its due time" are "matters of great moment." India's masses, Watsuji argued, had "no means of resistance against nature." There was "no escape for India's people from such insecurity of life." This insecurity brought about "a lack of historical awareness, a fullness of feeling and a relaxation of will power."[12]

This was a familiar pattern of argument: a familiar set of stereo-types about Indians as lazy and emotional. Nineteenth-century British liberals claimed that Indians lacked the rationality for self-government. They were too close to nature. Watsuji drew on this intellectual tradition; but in his writing we also see a distinct sense of Japan's historic mission to "save" Asia from European domination and from its own backwardness. "The people of the South Seas have never made any appreciable cultural progress," he declared, but "there would be startling advances if some way were found to break this mold and set this teeming power in motion." The resignation of Indians, he wrote, "prompts in us and draws out from us all our own aggressive and masterful characteristics." It was "on such grounds that the visitor to India is made to wish impulsively that the Indian would take up his struggle for independence." A struggle, by implication of this circular reasoning, that could only be guided by peoples whose climates had endowed them with different traits.[13] Watsuji implied that the Japanese were better placed to lead this charge than Europeans. Westerners could never truly understand the monsoon, whereas Japan had its own experience of tropical climates on the southern fringes of the archipelago, and on its model colony of Taiwan. Watsuji was not alone. Between the wars, many Asian students, scientists, and political leaders contemplated the relationship between nature and power, between nature and empire, between nature and nation. Watsuji Tetsuro concluded that for India's future, "change depends upon the conquest of climate."[14] Stripped of its moral, even spiritual, connotations, that conquest was, ultimately, a question of technology.

A less abstract perspective, but one that shared Watsuji's concern with how climate and ecology shaped culture, came from the Bengali sociologist and economist Radhakamal Mukerjee, a professor at Lucknow University who devoted much time and many pages in the 1920s and 1930s to the problems of rural India. Mukerjee was deeply concerned with water. In recent years the eccentric and eclectic Mukerjee has been recovered by historians as a prophet of

an ecologically sensitive and localist approach to development—but
he cuts an ambiguous figure. He was a committed eugenicist; he absorbed the racial and environmental determinism of his time and then
inverted it, calling, for instance, for *lebensraum* for the "teeming millions" of India and China.[15] Nevertheless, his was a rare voice of
concern about India's environmental balance at a time of rapid development—and his concerns were more tangible and specific than,
say, Gandhi's. "Man, tree, and water cannot be regarded as separate
and independent," Mukerjee wrote; he decried "crimes" against nature that would in turn "[let] loose destructive forces." Wise development, Mukerjee argued, would pay heed to the "natural balance
of man with the organic and inorganic world around him." Only
in that balance could human society find "security, well-being, and
progress."[16]

Mukerjee's prescription for India's future came from his close study
of the riverine landscape of his native Bengal. Drawing on the work
of the Russian anarchist geographer Léon Metchnikoff, who in 1889
published a wide-ranging history of riverine civilizations (including
the Ganges valley), Mukerjee thought of river basins as living entities.
Each river was "a synthesis or epitome of all the possible environmental variations and influences"; each river's "properties, colorations
and varied taste," as well as its "plastic or destructive power," was
a product of climate and geology. Mukerjee's diagnosis was that the
vital force of Bengal's rivers had been eviscerated by more than a century of British rule. He observed the deterioration of soil quality from
overintensive cultivation. Others had observed this worrying trend,
and ascribed it to the pressure of population. But the root problem, as
Mukerjee saw it, was that "agriculture comes to be influenced rather
[more] by the state of the market than by an arranged succession of
crops which may replenish the soil." Pressing on the ecology of land
and water, the demands of the colonial state and capitalists for the
products of the soil had left the Bengal delta "moribund." But where
did the roots of its revival lie? For Mukerjee, as also for the British
hydraulic engineer William Willcocks—famous as the architect of the

first Aswan Dam on the Nile River—part of the answer lay in recovering and reviving local traditions of irrigation and water management. For others, as we will see, only the wholesale transformation of nature by technology would match the scale of the challenge.[17]

Another part of Mukerjee's concern echoed the debate of the early twentieth century about India's place in the world—a debate that ranged across many fields of science and politics—over whether India was better seen as a bounded territory or as part of an oceanic realm. Of all of the ways that human beings had "gained a gradual mastery of the waters," he argued, "by far the most significant development is trade by sea." India's maritime connections, Mukerjee observed, "usher[ed] in an oceanic civilization superseding the fluvial." The resources of the river valleys were "narrow and limited" in comparison with oceanic commerce, which "extends as wide as the world." The more that traffic on the sea-lanes sucked up the produce of the river valleys, the sharper their decline became. Demand from distant markets upset what Mukerjee called "ecological balance." But he was optimistic. He felt that the excesses of "oceanic civilization" were now apparent; he looked forward to the moment when "man becomes more agriculturally inclined than ever before and atones for his past neglect."[18] He was to prove prescient—though the motive force of a return to agriculture, and a revival of the river valleys, was not atonement so much as necessity.

IF THE INTEGRATION OF INDIA WITH REGIONAL AND GLOBAL MARkets had placed new demands upon soil and water, the collapse of those markets in the 1930s created new dilemmas. For the first two decades of the twentieth century, many rural communities in India had relied on resources from overseas to survive—exports from the rice fields of Burma, and the remittances that came back to India from the wages of migrant workers in Burma, Malaya, and Ceylon. This was the key to a puzzle that historian Christopher Baker confronted in a brilliant and neglected 1981 essay on the economic integration and subsequent disintegration of Asia. In India, as in China and Java, the 1920s marked the "critical point" when "land ran out," Baker wrote. What demographers have struggled to explain is that, despite dire warnings in the 1920s, no Malthusian crisis ensued. To the contrary, population growth gathered pace even as agricultural yields declined year after year. The answer, Baker saw, lay in the interconnected regional economy that provided a lifeline for the densely settled agrarian heartlands of southeastern India or southern China, providing new opportunities for long-distance migration for their young men and a smaller but still significant number of young women. The expansion of rice cultivation along the Irrawaddy, Mekong, and Chao Phraya river basins after 1870 added around 14 million acres of new rice-growing lands in fifty years.[19] The opening of this final frontier of cultivation was accompanied by vast migration from India and China to Southeast Asia. More than 20 million passenger journeys traversed the Bay of Bengal, and a similar number the South China Sea, in the half century after 1870. Migrants went to work on the rubber plantations and tin mines of Malaya, on the tobacco fields of Sumatra, on the docks and in the mills and factories and on the streets of the growing port towns of Singapore and Rangoon, Penang and Surabaya. Many of these journeys were temporary, their pattern circular. Violence was never far from the experience of migrant workers; they traveled under a variety of arrangements and agreements, founded on debt. But Southeast Asia provided a horizon of opportunity, however fragile; year after year, the number of new arrivals in Southeast Asia outstripped the number of people heading back home.[20]

The global economic depression of the 1930s changed everything as it disconnected the regional economies of South and Southeast Asia. The depression made the inequalities of colonial capitalism starkly visible. Frustration about rising unemployment and intolerable debt found an outlet in anti-immigrant sentiment; mass political movements began to speak of redistribution. The collapse of global commodity markets led to a reversal of the flows of migration that had become entrenched over sixty years. The number of Indians departing Burma and Malaya exceeded the number of arrivals between 1930 and 1933; the same was true of the Chinese throughout Southeast Asia, despite the fact that China in those years suffered both from civil strife and from escalating Japanese military intervention. More than six hundred thousand people left Malaya between 1930 and 1933. They had to fend for themselves when they returned home. The Indian government's agent in Malaya noted that repatriation to India in times of distress "is proving less and less effective as a remedy against unemployment." Tamil workers in Malaya received no relief, and "their suffering is merely transferred from Malaya to South India."[21] John Furnivall, Burma-based British scholar, administrator, and Fabian socialist, wrote with prescience in 1939 that "we can already see that 1930 marks the . . . close of a period of sixty years, beginning with the opening of the Suez Canal, and, although less definitely, the close of a period of four hundred years from the first landing of Vasco da Gama in Calicut."[22]

III

When the colonial government of Madras opened the Mettur Dam along the Kaveri River in 1934, it was for a brief moment the largest dam in the world. It had been in the works for almost two decades. The dam could "boast of controlling works that leave those at Assouan [Aswan] well behind," one newspaper report declared, revealing how far water engineering had become a global endeavor. Mettur was three times the Aswan dam's length, standing 5,300 feet long, 176 feet high, 171 wide, and boasting a sixteen-foot roadway on top. The idea invoked, again and again, was control. "Rivers in India are not all tidy instruments," the columnist observed. India's rivers had a will of their own: "not many of them are content just with carrying water from mountain to sea," he wrote, "they love to spill it on the way . . . to damage while they enrich the lands they flow through." He drew a clear lesson: "They do not restrain themselves, and must be restrained."[23] Not all observers were so sanguine. In a handwritten note on a file in the Tamil Nadu archives in Chennai, a civil servant who signed off as "SA" took a dim view of what seemed to be the hubris behind the Mettur Dam:

> The Superintending Engineer's report is too self satisfied, or takes too much for granted, the infallibility of the officers of the department and the rank ignorance and prejudice of the ryots [farmers]. I do not think the position in Tanjore district is quite so very simple as the . . . report makes it out to be. While I am behind no one else in my admiration of the skill of our engineers who have built the great dam at Mettur and have succeeded in opening out the possibility of providing efficient irrigation facilities for large tracts of Country, the problem in Tanjore district has yet to be studied with sympathy and local knowledge.[24]

But SA's was the voice of a minority. The "restraint" of water, more than any other single solution, promised to address so many of the problems that came together to create a sense of agrarian crisis. Declining soil fertility, falling crop yields, the closing of overseas frontiers for migration and the depression's shock to trade—all of these combined with a broader sense of the enduring unpredictability of a monsoon climate. But every scheme to control water had the potential to create conflict between users upstream and downstream, between beneficiaries and losers. Since those unequal benefits often fell on either side of a political boundary, attempts to control water sharpened awareness of borders. Everywhere, as claims multiplied on river water for irrigation and power, so too did efforts to claim water as territory. The Mettur Dam had run into just this problem—that is why it was so long in the making. The Kaveri River flowed through both Madras Presidency and the princely state of Mysore. Mysore was quicker than Madras to attempt to harness the river, thanks in large part to the work of the engineer M. Visvesvaraya. But if Visvesvaraya's plans for his Krishnarajasagar Dam were realized, the British claimed, Mettur would not have enough water. The tangle that ensued became the first, and certainly not the last, territorial dispute over water in modern India. The first treaty between the governments of India and Mysore over water dated back to 1892; at that time of agricultural intensification, it was already clear that conflict might lie ahead. Unable to reconcile their dispute over Krishnarajasagar, both sides went to arbitration by the imperial government of India and agreed upon a technical solution: the tribunal decided on the exact quantities of water that Mysore and Madras were entitled to. Neither side was entirely satisfied, but they signed an agreement in 1924.[25] The distribution of Kaveri water has continued to haunt the Indian states of Tamil Nadu and Karnataka since 1947—a point of recurrent conflict, both within and outside of the courts.

Many of Asia's leaders believed that centralized planning would balance the needs of different constituencies. Planning would address the conflicts that arose between regions and communities; it would distribute resources in the most equitable and efficient way. Water resource planning took its place alongside economic planning in China as well as in India. In 1933, and in the aftermath of disastrous floods two years earlier, Chiang Kai-shek, who was by that time in command of a large part of China, assembled the Yellow River Conservancy Commission. Just as Indian engineers started to imagine the uses of water beyond irrigation, so China's planners too moved toward multipurpose water projects. Foreign engineers were drawn to the challenge that the Yellow River's control posed. Eminent German hydrologist Hubert Engels set up a Yellow River research center in Dresden; the League of Nations, too, lent its support and expertise to the Yellow River commission.[26] China was an independent republic in the 1930s but faced a growing threat from the territorial expansion of the Japanese empire. In India, the Congress party—which won large majorities in the elections of 1937, held under an expanded franchise—began to think about India after British rule, even if the arrival of freedom seemed to lie in the distant future. The Congress party's National Planning Committee, convened in 1938 by Jawaharlal Nehru, brought together a coalition of left-leaning nationalists, Gandhian thinkers, industrialists, and scientists including Radhakamal Mukerjee. It saw itself as a state-in-waiting. The group formed several subcommittees, of which one dealt with "River Training and Irrigation," chaired by Nawab Ali Nawaz Jung, chief engineer of the princely state of Hyderabad. The committee reported that "it is important that our rivers should be developed to the greatest possible extent and effectively utilised." It was a task that could not wait: "Conservation of water by storage," they concluded, "has become a matter vital to the future" of India.[27]

———

THROUGHOUT THE 1930S, INDIAN AND CHINESE PLANS TO CONtrol water proceeded with each oblivious to the other—and Chinese plans were soon consumed by the crisis of war with Japan. It would be a long time before their river engineering projects put them on a collision course. But there were portents of trouble to come. In the early 1930s, there was a flare-up of tension on the fringes of British and Chinese control—on the border between Burma, still ruled as part of British India, and Yunnan. A secret British intelligence file went into great detail on "Chinese Claims to the Irrawaddy Triangle"; it was filled with correspondence and translated pamphlets and newspaper articles, all deployed as evidence that the Chinese state was making a "fantastic claim . . . to the whole of Burma north of latitude 25°35N, right up to the Assam border." On the Chinese side, William Credner, a geographer sympathetic to Chinese nationalism and based at Sun Yat-sen University, undertook an expedition to the Irrawaddy triangle along with three Chinese officials in 1930. They sought to address the "long-outstanding question of the undemarcated northern and southern sections of the Yunnan-Burma boundary," left undefined in a treaty of 1894. They protested successive British military expeditions in the area, which the British justified on the grounds of suppressing a local slave trade. The Chinese party "advanced far into the uncivilized and remote districts," he wrote, and undertook the "plentiful collection of information"—not only on the "boundary question," but also on the "topography of the region." For now, the border was important as a symbolic marker of Chinese sovereignty: "It is hoped that Yunnanese of all classes will unite in striving to prevent the territory from being treated as a British colony again," an intercepted Chinese memorandum declared. But there was also a hint, in the close attention to landscape and the flow of rivers, that frontier regions would become vital for other reasons, too: for their water and their mineral riches.[28]

The question of borders arose, in a different sense, in India's fisheries. By the 1930s, V. Sundara Raj had succeeded James Hornell as the first Indian director of the Madras Fisheries Department. Unlike his predecessors, he looked forward to the wholesale transformation of India's fisheries by technology. Writing at the height of the Depression, as the regional economy had contracted and patterns of inter-regional migration had reversed, Sundara Raj worried that the Ceylon government had begun "deep-sea fishing experiments" with a trawler, in what he saw as water belonging to Madras. He pointed to Malaya, too, and the "great awakening in these sister states"; his concern was that "other Governments will exploit the Madras fishing grounds." He repeated his request, denied the first time around, for a trawler and a cutter, to commence his own deep-sea exploration. Sundara Raj saw "intensive ocean research and exploration of ocean grounds" as a global trend—he cited examples from Japan, Canada, and the United States.[29]

IV

From the late 1930s Asia was embroiled in war. China's experience of war was most prolonged, and most traumatic. Beginning with the annexation of the northeastern Chinese region of Manchuria in 1931, the Japanese empire advanced, propelled by the actions of local military commanders. Japan's rulers eyed China's mineral resources, its strategic position, and its territorial expanse. General

Chiang Kai-shek's very success in gaining control over China presented a threat to Japanese ambitions, which had been well served by China's internecine strife in the 1920s. In 1937, simmering conflict erupted into full-scale war. Under pressure of conflict, the best laid plans for the development of water resources went awry, with catastrophic effects. In June 1938, retreating Chinese troops breached the Yellow River dikes in Huangyuankou, in Henan Province, to stop the Japanese advance on the Nationalist stronghold of Wuhan. It was, in the words of one historian, "the single most environmentally damaging act of warfare in world history." Its dikes breached, the Yellow River rushed southeast, spilling into the Huai River system and drowning tracts of flat land on its way. More than eight hundred thousand people were killed, and 4 million displaced, by this desperate act of hydraulic sabotage.[30]

The war in Asia spread in December 1941, when Japanese forces simultaneously attacked Pearl Harbor and swept through Southeast Asia. Within a year, the Japanese empire had absorbed a region that had, since the nineteenth century, been divided among imperial powers. They conquered British-ruled Malaya and Burma, the Dutch East Indies, French Indochina, and the American-ruled Philippines. The fall of Burma brought the threat of a Japanese invasion to India's borders.

Though Indian territory saw little fighting, it became a vast supply base and center of operations for the Allied war effort in Asia—and Indian troops constituted a sizeable contingent of Allied forces in every theater of war. The war also transformed Indian politics. Incensed that the British government had declared war on India's behalf without consulting Indian politicians, the Congress party resigned from the provincial governments it had controlled since the elections of 1937. After the failure of negotiations with a British delegation led by Labour politician Sir Stafford Cripps in August 1942, Gandhi launched another mass campaign of civil disobedience—the Quit India Movement. Parts of North India became ungovernable. The British responded by dropping bombs on civilians to quell the revolt.[31] In their search for support as Congress party leaders languished in prison, the British turned elsewhere. The war boosted the power and the standing of the Muslim League and its leader, Muhammad Ali Jinnah, who had in 1940 passed the "Pakistan resolution," calling for the establishment of a homeland for India's Muslims—though how, where, and when were questions left deliberately unclear. The British were forced to concede that India would gain freedom, in some shape or form, after the war.

———

AS HISTORIAN SRINATH RAGHAVAN HAS SHOWN, THE WAR LED TO A vastly expanded role for the state in the economy, laying the groundwork for the apparatus of planning in independent India.[32]

Among other fields, the war gave a boost to meteorology, as India became a hub of military aviation. The Indian Meteorological Department grew fast: its budget trebled between 1939 and 1944, and it established a new base of operations on a thirty-acre campus along Delhi's Lodhi Road. It proved difficult to find and train enough staff to keep pace with the expansion of facilities. Some of India's leading meteorologists suffered loss and hardship during the war. Most of the staff of the Burma Meteorological Department were Indian; and when Japanese bombing raids began on Rangoon, they joined the exodus of up to a half-million Indian refugees—most of whom walked back to India, through jungle and mountains, into Assam. The director of Burma's weather service, S. C. Roy, walked from Rangoon to Imphal. One of his deputies, S. N. Ghosh, survived the long trek only to be killed in a Japanese bombing raid on the Indian border. The war saw the recruitment of a new cadre of meteorologists in India—the generation that would staff India's weather service after independence. The meteorology department had three times as many staff by the end of the war as it did at the start. In 1944, Charles Normand retired as its director after thirty-one years working for the department; his successor, S. K. Banerji, was the first Indian to head the meteorological service. The war saw the beginning of aircraft weather reconnaissance over the Bay of Bengal, through a series of flights between Madras and the Andaman Islands. It also witnessed a breakthrough in communications technology. The India Meteorological Broadcast Center was established at the Royal Air Force base in Nagpur, in central India. The Royal Air Force and the United States Air Force installed the first teleprinters in India for the transmission of weather data.[33]

The development of meteorology was oriented by military needs. In weather forecasting as in medicine, civilian applications for the new technologies were a low priority. However much the new technologies promised, India's experience of the war shattered the complacent assumption, pervasive by the 1930s, that nature had been conquered.

WHEN THE JAPANESE INVADED BURMA IN 1942, BRITISH INDIA lost 15 percent of its total rice supply. In some areas that took large imports of Burmese rice, like Madras, the shortfall was overcome by local production. But in Bengal a long-term decline in the rural economy came together with natural disaster, compounded by wartime political bungling, to cause a catastrophic famine—the first in India since the early twentieth century. The return of starvation to Bengal came as a traumatic shock. From the time of the Indian Industrial Commission in 1918, most observers took for granted that famine had been confined to India's past. In the 1920s and 1930s, nutritional scientists and health officials began to think about food as a way to enhance life rather than simply to sustain it—their concern moved from absolute starvation to malnutrition. "The days when we could cast the blame on the gods for all our ills are past," Nehru had written in 1929.[34]

During the winter monsoon of 1942, a fearsome cyclone struck eastern Bengal, flooding fields and destroying crops. "In violence and devastation it surpassed any other natural calamity that befell this country," one contemporary account observed, "it forced from the Bay a high tidal wave" that reached 140 miles per hour. The cyclone "swept the standing crops, blew off the roofs, uprooted most of the trees, demolished the huts"; the floods that followed "washed away nearly [three quarters] of the livestock, and some 40,000 human beings."[35] Unnerved by the prospect of a Japanese invasion from Arakan, local officials imposed a scorched-earth policy, denying local cultivators the boats they used to transport rice. Internal divisions paralyzed the Bengal government. Driven by Winston Churchill's animus toward India, the British cabinet ignored every warning. They continued to export Indian rice to feed troops in other theaters of war. They refused to deploy Allied ships to send relief to Bengal.[36] As shortages intensified, the most vulnerable people—landless laborers, fishers, women, and children—starved. Calcutta's relative wealth sucked in from rural Bengal rice that could have fed those in dire need.[37]

The vulnerability of Bengal's poor, like their debts, had compounded over decades. During the Depression, smallholders unable to repay loans had lost much of their land. The productivity of Bengal's lands had declined in the twentieth century as railway embankments stemmed the flow of rivers and invasive water hyacinth choked streams. By 1942, the crisis was acute. As scarcity closed in, patrons deserted their sharecroppers, choosing to pay them in cash rather than in kind just when inflation made rice unaffordable. Families abandoned their weaker members. Hit by the successive blows of a loss of imports, "boat denial" by the state, a devastating cyclone, and a lack of relief, the economy and society of Bengal collapsed.[38]

Even the conservative *Statesman* newspaper of Calcutta published photographs of starving children and abandoned corpses—scenes reminiscent of the 1870s and the 1890s when the great El Niño droughts had combined with the churning effects of capitalism to deliver disaster. These images met with stony-faced inaction by the British government. This time, Indian observers held the British government directly responsible for starvation. "It was a man-made famine which could have been foreseen and avoided," Jawaharlal Nehru wrote from Ahmednagar jail. He was sure that "in any democratic or semi-democratic country, such a calamity would have swept away all the governments concerned with it." But just as disturbing was the callousness of wealthy Indians. Nehru expressed disgust at the "dancing and feasting and a flaunting of luxury" in Calcutta while millions starved. S. G. Sardesai, a Communist activist, decried the "unbridled profiteering" of hoarders and speculators, and argued that "total mobilization means vigorous and just procurement of the genuine surplus from rural areas, vigorous price controls, and total rationing in cities."[39] When they finally secured London's commitment to relief in the autumn of 1943, Indian officials had to raise the alarm that Bengal's continued starvation could endanger the war effort.

When the Stanford University economists V. D. Wickizer and M. K. Bennett examined Asia's rice economies in 1941, they surveyed the wreck of what had once been an integrated system. In their analysis, they used the term "Monsoon Asia" as "a convenient designation for a specific group of countries in which monsoonal climatic conditions profoundly influence both agriculture and economic life." "Monsoon Asia" was bound together by climate, by the direction of the winds, and by the trade in rice, but divided by empires. Wickizer and Bennett witnessed "Monsoon Asia" splintering further as a result of depression and now war. They wrote of their hope for "a reversal of the recent trend towards economic nationalism." Their recipe for regional sustainability was for a return to the free commerce in rice, augmented by capital investment. But their projection of "unfavorable" conditions proved much closer to the eventual outcome. "If peace should come with important territorial changes in Monsoon Asia," they argued, "changes in the political composition of Monsoon Asia following the termination of present wars might readily result in a rather sudden shift and re-orientation, completely reversing the tendencies of the past decade or more."[40]

The most enduring political consequence of the Bengal famine was the decisive rejection of any postwar return to the old ways of unregulated markets and inter-regional trade in rice. Indian planners and politicians, technocrats and populists, all emphasized the need for self-sufficiency in the future. Water was vital to their plans. Starvation's return to India scarred Nehru's generation of leaders. Having asserted that national sovereignty would alleviate the problem of starvation, Nehru and his contemporaries were haunted by the prospect of failure. "We live continually on the verge of disaster in India, and indeed disaster sometimes overwhelms us," Nehru wrote. The same year, Patna University economist and demographer Gyan Chand declared that "ours is a death-ridden country. We might very well adopt the human skull as our national emblem."[41]

———

AS THE WAR APPROACHED ITS END, THE EXPERIENCE OF FAMINE IN India—and also in China and Vietnam—came together with the force of rising expectations. In the eyes of many Asian observers, only the wholehearted embrace of state planning, wedded to powerful technology and under the control of nationalist rather than colonial forces, would address the colossal vulnerability of Asia's people to privation and starvation, both of which the war had laid bare.

Even British planners began to contemplate large schemes to transform water. The Bhakra Dam, in Punjab, was first proposed in 1944. It was a monument to British plans for India's postwar reconstruction at a time when few believed the Raj would collapse so quickly after the war's end.[42] The project had its skeptics. "For advertisement reasons some authorities in India have published optimistic forecasts of the time in which they propose to construct high dams," one official scrawled in an archival note. "There is sound opinion, unbiased by connection with these projects, which considers these forecasts fantastic."[43] But speed and scale were what Indian nationalists wanted.

One of the voices in favor of a planned conquest of India's rivers was the scientist Meghnad Saha. Saha was born in 1893 in an East Bengal village, to a lower-caste family without education or resources, and with several children to feed. His scientific aptitude was evident from childhood; he won a series of scholarships that led him to Calcutta University in the 1910s. He studied in England and then in Germany, returning to a position at Allahabad University, one of India's most distinguished institutions. Saha's pioneering contributions to astrophysics gained him widespread recognition, notably for his paper on "Ionisation in the Solar Chromosphere." By the 1930s he was no longer content to confine his work to the laboratory. He founded the journal *Science and Culture* to reach a wider public; he became a missionary for scientific development, and a strident critic of Mahatma Gandhi's suspicion of modern technology. "We do not for a moment believe that better and happier conditions of life," he wrote, can be secured by "reverting back to the spinning wheel, the loin cloth, and the bullock cart."[44] One of Saha's central concerns was with water, and his dreams of water were grand.

Saha's essay on "Flood," published in *Science and Culture* in 1943, at the height of the war, envisaged wholesale environmental transformation. He described the decline of the Damodar River: diverted by railway embankments, its course had moved toward Calcutta, and now it threatened to inundate the city. Drawing on a global range of examples and references—the 1913 floods in the Miami valley, and, above all, the Tennessee Valley Authority—Saha argued that the key lay in a "radical solution" to make the Damodar "a perennial" rather than a seasonal river: that is, to "liberate" it from the monsoon. He argued for the adaptation of the American approach: "to regard the whole river basin as a unitary area, and coordinate plans of flood control with those of irrigation, development of backward agricultural areas, development of hydroelectric power, and improvement of navigation."[45] Elaborating on his plans in another article the following year, written together with his colleague Kamalesh Ray, Saha expressed optimism: "Nature, vested interests and thoughtless managements made a once prosperous valley a wilderness, but Nature, Man and Science can again make it a smiling garden," they wrote. Saha was scathing in response to those, like Radhakamal Mukerjee, who had argued that restoring forests and local efforts at soil conservation would strip the Damodar of its destructive power. Saha called the claim that deforestation affected rainfall "absurd"—a claim for which there was "not a single iota of positive proof." If changes in forest cover and land use had any effect at all on local climate, Saha argued, they "must be extremely small compared to the huge monsoon currents which are responsible for the precipitation on the Damodar Valley." Working at the cutting edge of planetary science, Saha was well aware of new work on the monsoons, and their integration with other parts of Earth's climate. The scale of India's climate was so vast as to render any local modifications in the water cycle trivial in importance. Saha pointed out that rainfall in the Damodar valley was determined by "atmospheric conditions in the Bay of Bengal . . . which are generally thousands of feet in depth", "local conditions" could have little effect upon them.[46]

Rainfall was beyond human intervention. But human intervention to transform the landscape could neutralize the threat posed by uncertain rainfall, securing rivers from the alternating lack and excess of water. And here, Saha was confident about the future. "We are fortunate to live at a time when the large scale experience of thousands of dams constructed in the USA since 1915 are at our disposal," he said; he believed that the global circulation of ideas and technology, a process of learning, would come to India's aid. In valorizing the American and Soviet models, Saha indicated that his dreams for India extended beyond anything the sluggish British colonial state could carry out. He envisaged the construction of dams in eastern India that would last for "hundreds of years."[47]

—

IN THE 1920S AND 1930S, WATER BOTH CONNECTED AND DIVIDED Asia. A new awareness of the dynamics of climate made clear the extent to which Asia's coastal arc shared vulnerability to powerful cyclones that crisscrossed its seas. In the fields of geography and climatology, the idea of "monsoon Asia" arose to highlight the common rhythms of rural life governed by extreme seasonality. To think of

Asia as bound together by water in every dimension—the rains, rivers, and seas—was to suggest that material conditions transcended the borders between empires. But those borders hardened in the decades between the wars. The depression of the 1930s broke many links in the chain that held monsoon Asia together: barriers to movement proliferated and migration patterns were overturned; commodity markets collapsed and the trade in rice declined. These reversals made the question of who controlled water all the more important.

The Second World War provided new tools and revived old fears. The trauma of famine and social breakdown met a newfound confidence in the power of state planning and big technology to reshape economy, society, and the environment. The next chapter turns to the struggles waged by India and by other independent Asian nation-states to understand and conquer water.

SEVEN

RIVERS DIVIDED, RIVERS DAMMED

ETWEEN 1945 AND 1950 THE MAP OF ASIA WAS REDRAWN. THE war overturned imperial rule in Asia. The prestige of European powers in South and Southeast Asia never recovered from their rapid collapse before the Japanese advance in 1942. Economically ruined by the war, European empires could not afford to hold on to their colonial territories by force without American backing—which was forthcoming only when it furthered US interests in the deepening Cold War. Most importantly, emboldened and militarized Asian leaders refused to contemplate any return to the old order.

In the war's aftermath, new states were forged from the ruins of empires. In 1947, British India was divided into the independent states of India and Pakistan: a bloody partition along religious lines that cost millions of lives and ruined millions more. In 1948, Burma and Ceylon gained their independence from the British in mostly peaceful transfers of power—but Burma immediately faced multiple internal

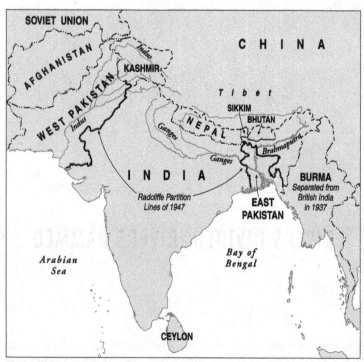

The partition of India in 1947, showing the division of the Indus and Ganges basins.

revolts, by Communist guerrillas and by the Karen and Kachin ethnic minorities. The Dutch tried to hold Indonesia by force. They were driven out by Sukarno's nationalist forces in 1949 after a protracted anticolonial war that also saw failed uprisings by Communist, Islamic, and regional secessionist forces. In Vietnam, the Communist leader Ho Chi Minh declared independence in 1945, moving into the vacuum left by the sudden surrender of the Japanese at the end of the war. But the French were determined to return to Vietnam, and with growing American support they waged war against the Viet Minh until Ho's victory at the battle of Dien Bien Phu in 1954. In East Asia, too, the war's end brought revolutionary change. The civil war between Chiang Kai-shek's Nationalist forces and Mao Zedong's

The borderland between India and China, showing the Brahmaputra/Yarlung Tsangpo river.

Communist army culminated in Mao's victory and the foundation of the People's Republic of China in 1949, together with a de facto Nationalist state on the island of Taiwan.[1]

Asia's political transformation was so rapid, so dramatic, so violent, that few at the time gave any thought to its environmental consequences. With the benefit of hindsight, we can see that those consequences were profound. The impact of these midcentury partitions and borders on Asia's waters are a vital, and neglected, part of any history of the second half of the twentieth century. We have hardly begun to reckon with their effects, both positive and negative, on the lives of a significant proportion of humanity.

Dam building, more than any other project, epitomized Asia's new leaders' confidence in their ability to tame nature. India had fewer than three hundred large dams at independence; by 1980, it had more than four thousand. Dams were the single largest form of public

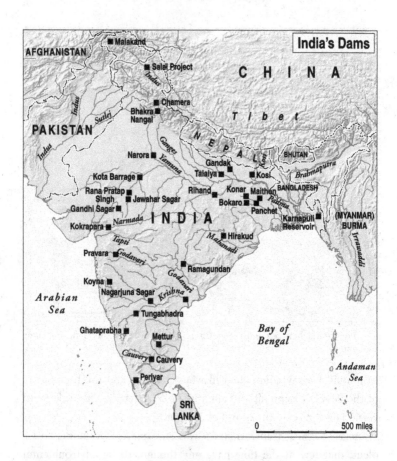

investment in modern India, swallowing considerably more government expenditure than health care or education. Dams promised to
liberate India from the capricious monsoon; they promised finally to
free it from the specter of famine that had struck so often, and so
harshly, in the colonial era. India was not alone: the enthusiasm for
dam building was global. Under Mao, China built large dams on a
scale that eclipsed India's efforts: an estimated twenty-two thousand
after 1949, almost half of all the large dams in the world. Along the
Mekong River, dams formed part of the American strategy to shore

up the anti-Communist state of South Vietnam following the French defeat at Dien Bien Phu.

These projects proceeded in parallel; by the 1960s they came into contention. Dams tried to make rivers conform to political borders by impounding their waters and diverting them to serve the needs of national development. But as multiple projects and competing ambitions arose, upstream and downstream, dams made tangible the material interdependence that transcended borders. At the time of independence, few in India had thought much about the fact that many of India's rivers originated in Chinese territory. Only when both sides' ambitions for river development swelled did the cross-border flow of rivers appear as a threat.

In the postcolonial age, large dams carried enormous symbolic weight. They epitomized dreams of development. More than any other technology, they promised the mastery of nature. In the global history of dam building, India played a pivotal role. India's experience exemplified the scale of the challenge facing the Third World, but also the scale of ambition that new states upheld. Because of the unevenness of the monsoon, India's rulers were obsessed with water. So, too, were the legions of foreign experts who arrived to help India's quest. Unlike China, India benefited from aid from both sides in the Cold War: India's developmental plans were a terrain for competition between the United States and the Soviet Union. India's water engineers received advice from around the world, and in turn they shared their expertise through the United Nations and other international bodies. The broader influence of India's addiction to large dams was cultural as much as it was political, as Indian cinema captured the imagination of viewers across Asia and Africa. Some of the most iconic Hindi films of the age were set against a backdrop of India's struggles for water; their stories resonated far beyond India.

The conquest of water by concrete behemoths came at enormous cost. Some of those costs were evident from the outset—the displacement of people from their homes, the flooding of villages and forests by new reservoirs. Others became clearer with time. Few in the 1950s or 1960s could see just how fundamentally dams would transform Asia's ecology of water. It was in that era that Asia's states and peoples started on a collision course toward the water-related crises they face today.

|

Partition was a particular kind of British decolonization, which came about as an attempt to engineer, in an extremely compressed period, nation-states with clear and decisive ethnic majorities out of previously heterogeneous colonial territories. It was implemented first in Ireland, and then twice in the 1940s: on the Indian subcontinent, and in Palestine. While political tensions between the Indian National Congress and the Muslim League ran high in the 1930s, it was only after the end of the Second World War that it became likely that India's future would be a divided one. Until then, the League's claim to represent all of India's Muslims—divided by language and region, by class and politics—rang hollow. In the aftermath of war, a spiral of violence supercharged political negotiations, and accelerated the timetable for independence. The British, fearing entanglement in an Indian civil war, and reeling from economic crisis, sought to leave as quickly as possible, no matter what the cost. Last-ditch negotiations failed when the Congress leadership were unwilling to concede to Muslim League leader Muhammad Ali Jinnah's demand for a weak federal government with power resting in provincial hands. On June 3, 1947, British prime minister Clement Attlee announced the plan to partition the subcontinent into India and Pakistan. Lord Louis Mountbatten was appointed the last viceroy of India, charged with overseeing the division. The job of drawing the border fell to Cyril Radcliffe, a lawyer with no prior experience of India. Closeted with a small boundary commission, supplied with maps and census returns, his task was to draw a line to carve off the Muslim-majority areas of Punjab and Bengal from the rest of British India, thereby creating the western and eastern wings of Pakistan—divided by more than one thousand miles of Indian territory. The location of the border was not announced until the day after independence, on August 15, 1947.[2]

Nobody had predicted the colossal scale of upheaval that followed. In just over a month, between September and October 1947, more than 849,000 refugees entered India on foot. A further 2.3 million crossed the Punjab border by train. Trains were attacked by armed mobs on both sides of the border—their packed carriages became chambers of death. The Indian and Pakistani armies, by mutual consent, crossed the frontier into each other's territory to lead convoys of refugees back to safety. Arriving refugees were settled in what had been deemed "evacuee property"—many who had sought temporary refuge from the violence returned to find that their homes had been seized, their departure construed as an intention to emigrate. South Asia's cities swelled with new arrivals, Delhi and Karachi and Calcutta above all. Around 20 million refugees crossed Radcliffe's border, more than half of them in Punjab.[3]

———

THE SIFTING OF INDIA'S RELIGIOUS MAJORITIES AND MINORITIES IN 1947 was also, as one historian describes it, a "division of nature."[4] Radcliffe himself was aware of the problem in Punjab: his border, he recorded, was "complicated by the existence of canal systems, so vital to the life of Punjab but developed only under the conception of a single administration."[5] His solution pleased nobody. Canals were severed from their headworks. In Punjab, Partition broke the carefully planned canal network laid down over a half century. Bengal had no need of the intricate irrigation systems of arid Punjab, watered as it was by the monsoon and by the Himalayan rivers. But there, the border tried to contain a naturally volatile waterscape. As the geologists and bridge builders of the nineteenth century had found, Bengal's rivers changed course suddenly; *chars,* or sandbanks, emerged with the deposit of silt and vanished with the coming of floods. The *chars* were so fertile as to be desirable land for cultivation—if they arose along the riverine borders, were they now part of India or Pakistan? For those who inhabited this braided landscape of land and water, the answer had vital consequences.[6] The Bengal border ran through the sacred Ganges and the turbulent Brahmaputra. In 1947 there was little infrastructure to stem the flow of this water, but there were many plans in place. What would happen in the future, when engineers on both sides eyed new ways of harnessing the waters?

For some, the unity of nature was set against the human divisiveness of Partition. The socialist Rammanohar Lohia wrote of his astonishment that Nehru was willing to divide India's great river basins out of political expediency.[7] Saadat Hasan Manto, the most incisive and enduring chronicler of Partition in fiction, turned to the problem of water in his 1951 short story "Yazid."[8] The story's opening image is almost shocking: "The riots of 1947 came and went. In much the same way as spells of bad weather come and go every season." In his first two sentences, Manto evokes the indifference of nature to human suffering; he signals the insignificance of human folly faced with the cycle of the seasons; he also draws the suggestion that Partition's violence may have been as "natural" as the rains—a message colored with irony, since it runs counter to so much of Manto's fiction, which depicts Partition as the monumental consequence of petty and all-too-human decisions. The most memorable exchange in the story takes place between the sage village midwife, Bakhto, and Jeena, wife of the protagonist, Karimdad. One day Bakhto arrives with news that "the Indians were going to 'close' the river." Jeena is nonplussed: "What do you mean by closing the river?" When Bakhto replies, plainly, "they will close the river that waters our crops," Jeena laughs in disbelief: "You talk like a mad woman . . . who can close a river; it's a river, not a drain."[9]

PARTITION AFFECTED EVERY PART OF GOVERNMENT, EVERY INSTITUTION. The Indian Meteorological Department, too, was divided in

1947. One of the pressing tasks for the partition of Indian meteo-rology was the exchange of observational data—all original records relating to the weather of Pakistan, wherever in (undivided) India they were held, were transferred to the new Pakistani meteorologi-cal service. Both sides held that climatological data were "records of common interest," as if to acknowledge that the monsoons respected no human frontiers. They supplied each other with duplicate cop-ies. And then there was the question of the instruments upon which weather science rested. These, too, were divided: the Indian Meteo-rological Department reported that "out of the stock of instruments and stores held at [headquarters in] Poona and Delhi, stocks between 20 to 25% of each item were to be given to Pakistan." A simple list conveys a deep rupture:

> As a result of the partition, 2 type A Forecast Centres, 1 type C Fore-cast Centre, 5 Auxiliary Centres, 8 Aerodrome reporting stations, 3 Radio-sonde stations, 14 Pilot Balloon Observatories, 82 surface observatories, and 1 seismological station were transferred to the Pakistan Meteorological Service.

Incidentally, the list also conveys how dense the infrastructure of me-teorology in British India had become by the end of the war. There is a poignant sense of how hard meteorologists fought to keep doing their work, regardless of the chaos and violence around them. "In-terim arrangements were made," they noted, "for the issue of storm warnings, etc. for certain regions falling in Pakistan."[10]

Like meteorologists, engineers and economists looked at the ma-terial knots tying the two new states together and many of them believed that a future of cross-border cooperation was inevitable. Just a year after the event, C. N. Vakil, a professor of economics at Bombay University, wrote a pamphlet on *The Economic Conse-quences of Partition*. It was prosaic in the face of colossal upheaval. The facts, he thought, made it "easy to appreciate the need for an agreed economic policy between the two Dominions now and in the future"—that both states formally remained Dominions within the British Empire until 1950 provided a measure of political cover for negotiations to take place. Against "the atmosphere of communal bitterness as well as increasing mistrust," Vakil believed that "fundamental economic forces in the two Dominions are likely to work in the direction of mutual inter-dependence." But he acknowledged the real possibility that "political forces" would win out; he saw that India and Pakistan could end up in a state of "economic warfare." The darker edge to his pamphlet came in his wish to inform "the layman" of the economic "weapons" at India's disposal if "warfare" it was to be.[11]

IMMEDIATELY AFTER PARTITION ENGINEERS ON BOTH SIDES MUD-dled through. In the midst of crisis they kept the water running. In December 1947, the chief engineers of East and West Punjab signed a Standstill Agreement to maintain supplies to the Bari Doab, one of the Indus River canals ruptured by the border: the headworks fell on the Indian side of the border and most of the canal in Pakistan. When Punjab's Canal Colonies were built they had been conceived as a unitary system, its hydraulic parts each useless in isolation; now the engineers had to improvise. And then the water stopped. On April 1, 1948, the day the makeshift agreement expired, India shut off the water supply to the canal. The fears that Manto depicted in his fiction mirrored historical events—"it's a river, not a drain." But the rivers, too, were national now.

The sudden stoppage raised alarm on the Pakistani side. In the midst of the spring sowing season, the waters to the Upper Bari Doab and Dibalpur canals stopped, disrupting cultivation and threatening the harvest. Residents of Lahore saw the canal that bisected their city empty of water: before their eyes was a visceral sign of Pakistan's vulnerability. East Punjab's engineers, on the Indian side, shut off the canal water to Pakistan without the approval of the central government in Delhi; Nehru himself worried that "this act will injure us greatly in the world's eyes."[12]

The conflict over Indus waters joined the territorial conflict between India and Pakistan over the princely state of Jammu and Kashmir, whose Hindu ruler had chosen to join India rather than Pakistan under considerable pressure from the Indian side, and against the wishes of the territory's predominantly Muslim population. The tussle over Kashmir erupted into military conflict within months of Partition, following the invasion of the territory by Pathan militia from the northwest with covert support from the Pakistani state. Upon Kashmir both India and Pakistan projected their anxious sense of truncation, in the sense that both sides ended up with less territory than they thought they ought to have, as a result of a hasty partition that satisfied nobody. Both sides came to see control over Kashmir as a vindication of their founding ideologies: for India, the extension of a secular and democratic polity; for Pakistan, the achievement of a Muslim homeland in South Asia. The views of Kashmiris, then as now, were ignored. But the Kashmir dispute also had a hydraulic dimension: of the five tributaries of the Indus River, one, the Jhelum, originates in the Kashmir valley; another, the Chenab, flows through Jammu. Gnawing at India and Pakistan, through their inability to find a solution in Kashmir, was a quest to control the state's water.[13]

Pakistan and India each made their case before a global audience. The dispute over the Indus attracted international attention because it seemed like just the sort of water conflict that many others could face as the map of the world was redrawn. Pakistan's delegation to the United Nations declared in 1950 that "the withholding of water essential to an arid region to the survival of millions of its inhabitants" was "an international wrong and a peculiarly compelling use of force contrary to the obligations of membership in the United Nations."[14] The Indian argument, by contrast, was that India "has the right under the Partition, as also in equity, over the waters of rivers flowing through her territory." India's lawyers also advised that the provisions of international law were ill-suited to "the case of two countries, which have come into existence from the partition of a previously existing national unit"—which rested on the idea that British India was a "national unit," a strange claim for Indian nationalists to make just a year after independence.[15]

Both sides used Partition to bolster their arguments. India insisted that Radcliffe's line had given Punjab's richest agricultural lands to Pakistan, including the Canal Colonies—lands farmed mostly by Sikh and Hindu cultivators who now found themselves uprooted as refugees in India. Partition had "disrupted [a] unitary system of canal irrigation and therefore the entire economy of the area," executed with "complete disregard of physical or economic factors." In this light, the Indians argued, it was their prerogative to make best use of the water resources that remained. India's advocates portrayed eastern Punjab as the victim of a partition that had been imposed upon it "to satisfy the ideology of Mr. Jinnah and his Muslim League." With little warning, East Punjab "found itself an economic unit, and a very much underdeveloped area." Its survival depended on wresting control over "the life-giving waters from the Himalayas" that had, through British canals, "been unfairly diverted to increase the prosperity of distant tracts" that now lay across the border. The Pakistanis retorted that India "wishes to make a desperate attempt to escape the economic consequences of partition"—which Pakistan, as a new state, had no choice but to face. The Pakistani submission to the tribunal of arbitration gave Partition a material as well as an ideological dimension: "Apart from religious and cultural considerations, one of the main objects of partition is to enable the residents of the two Dominions to use and develop their economic resources for their own benefit." They closed with a goad: "East Punjab should have the courage to face the economic consequences of a political standing by itself."[16]

Each Punjab "found itself an economic unit"—the phrase suggests that this happened as one might "find oneself" in an unfamiliar destination after getting on the wrong train. Divided provinces, like the divided nations of which they were part, had to stand alone where once they were part of a larger whole. The vogue for planning demanded a simplified model of the economy upon which plans could be made. This cemented the vision of an Indian economy set apart from the whole web of connections that tied India to Southeast Asia and beyond.[17] Partition stymied many plans: it struck at the mutual dependence of the jute growers of eastern Bengal and the export houses of Calcutta, at the ties between coal producers of eastern India and Pakistan's factories, at the carefully calibrated use of water by farmers along the length of the canals of Punjab. These new "economic units" unleashed a desperate competition for water: the precondition for every vision of prosperity.

In both India and Pakistan, Partition generated a sense of loss and a feeling of vulnerability. Following India's water stoppage, water engineering became an urgent priority in Pakistan. Pakistan's engineers designed a new canal project known as the BRBD (Bambanwala-Ravi-Bedian-Dibalpur); it would run parallel to the partition border, a "canal designed to sever Pakistan's [water] supply from India." A volunteer corps of laborers rallied to the cause of the new canal as an act of national defense—it came to be known as the Martyrs' Canal.[18] For India, the loss of the productive agrarian lands of western Punjab hastened the push to develop its eastern reaches. Plans for a large dam at Bhakra, on the drawing board since the early twentieth century, now became a priority. Old fears of famine had never gone away; they were reactivated by Partition. Eastern Punjab needed new sources of water to keep its most vulnerable districts secure from a failure in the rains.

The Indian water stoppage lasted a matter of weeks. Negotiations between the two sides resumed at the end of April 1948. In exchange for payment, India agreed to continue supplying canal water to Pakistan for an unspecified period, during which "alternative sources" would be developed. Both sides clashed repeatedly, their claims often directed at international observers. Pakistan proposed international arbitration; India insisted it was a domestic matter. In 1951 the American David Lilienthal—a senior official in the Tennessee Valley Authority, and now a globetrotting development consultant—toured India and Pakistan; he took a particular interest in the Indus water dispute. Lilienthal contrasted "politics and emotion" with "engineering or professional principles." He described how Partition, driven by emotion and not by reason, "fell like an ax" upon the Indus basin. But, he added, "the river pays no attention to Partition—the Indus, she 'just keeps running along' through Kashmir and India and Pakistan."[19] He wrote a long piece for *Collier's* magazine warning the American foreign policy establishment that Kashmir was "another Korea in the making." Lilienthal had proposed that water management be removed from the political realm. He had faith in the shared training and professional camaraderie of India's and Pakistan's water engineers; he believed in their ability to work together for a "cooperative," technical, solution. In Partition's aftermath, such apolitical solutions were as attractive as they were unrealistic.

———

THE PARTITION OF INDIA MARKED THE BEGINNING, NOT THE END, of the division of Asia's waters. A year after Mao Zedong's army overpowered Chiang Kai-shek's Nationalists in the Chinese Civil War, inaugurating the People's Republic of China in October 1949, Chinese forces invaded Tibet. The annexation of Tibet was a thorn in India's relations with China. Many Indian politicians, including members of Nehru's cabinet, urged him to take a hard line, but Nehru opted for a path of conciliation, recognizing that India was not in a position to take any action. What went almost unremarked at the time was that the annexation of Tibet in 1950 also gave China control over much of Asia's freshwater. The Indus was divided between India and Pakistan—but its source is on the Tibetan Plateau, which was now ruled as part of China. From the Tibetan Plateau flow the Brahmaputra (known in Tibet as the Yarlung Tsangpo), the Salween, the Mekong, and also the Yangzi. The source of the great rivers still seemed, in 1950, remote, wild, untouched by the modern world. It is no surprise that water was mostly invisible through the process of dividing Asia into modern nation-states. In the second half of the twentieth century, water resources would become increasingly important to the process of marking and laying claim to the earth, increasingly pivotal to conflicts between Asian states. In 1950 water was not, or not yet, a cause of conflict except between India and Pakistan. But their effects on shared water resources would be among the most far-reaching consequences of Asia's midcentury territorial disputes.

Still, the power of nature, paying no heed to new borders, was on full display in August 1950, when—on Indian Independence Day, the fifteenth—a powerful earthquake tore through the borderlands straddling India, East Pakistan, Tibet, and Burma. The earthquake was one of the ten most powerful ever recorded, caused by the collision of two continental plates. Its epicenter was in Rima, Tibet—and came just three months before the Chinese invasion—but the bulk of the damage fell on the northeastern Indian state of Assam. Even as politicians were busy redrawing the map of Asia, the earthquake altered the landscape and devastated human lives. As if to underscore the remoteness of the earthquake's epicenter from centers of political power, relief was slow to arrive. The earthquake blocked the course of many tributaries of the Brahmaputra, changing the river's course. The British botanist and explorer Francis Kingdon Ward was traveling in Tibet at the time, and penned one of the few eyewitness accounts from the earthquake's epicenter. He wrote that "the immediate result of the earthquake was to pour millions of tons of rock and sand into all the main rivers . . . displacing millions of cubic feet of water."

Every scheme to engineer water had to contend with the instability of Asia's mountain rivers; with the growing confidence of postcolonial engineers, caution began to be set aside.[20]

II

In 1951, India carried out its first census after independence. It was at that time the largest census ever undertaken in the world. The average life expectancy in India stood at just 31.6 years for men, and 30.25 years for women.[21] In the United States at the same time, that figure

was 65.6 years for men, and 71.4 for women. For every 1,000 live
births in India at the time, more than 140 infants died. This was an
indictment of two centuries of British rule, since the "abstract number which is the average human life span," as philosopher Georges
Canguilhem noted, revealed much about "the value attached to life
in a given society."[22] In China, after more than a decade of war, life
expectancy was no higher. Nothing illustrates so plainly the magnitude of the challenge before the governments of Asia's new states. For
India and China, as for Pakistan and Burma and countries all along
the great crescent, harnessing water was a priority in their quest to
transform the conditions and expectations of life.

In his introduction to the census, commissioner R. A. Gopalaswami pinpointed what he saw as a turning point in India's population
history, around 1921. Until then India's population had grown slowly.
The terrible famines of the late nineteenth century, the prevalence of
infectious diseases like plague and malaria, the devastating and rapid
toll of the influenza epidemic of 1918, which killed between 12 and
13 million people in India—taken together, they produced a grim toll
of premature death and debilitating illness. After his account of the
influenza, Gopalaswami's narrative reaches its pivot: "We now reach
the turning point," he wrote, where after 1921, "we hear no longer
about abnormal deaths." From that point on, he argued, famine and
mass epidemics ceased to be the killers that they were in India. Some
of the credit he gave to the mobilizing power of Indian nationalism,
some to administrative improvements that came from lessons the
British had learned from earlier disasters. Gopalaswami presented a
picture of India's climate that was no longer the threat that it was:
"Though the usual cycles of vicissitudes of the seasons continued and
the brown and yellow belts of the country continued to suffer from
droughts . . . there was no extraordinary calamity" in the years between the wars. Good policy was matched with good fortune, since
"nature also seems to have been kindlier" in the two decades after
1921, with fewer major droughts. The Bengal famine of 1943 was a
devastating reminder that famine could return to India. But it did not

invalidate the longer-term pattern, and prevailing wartime conditions made it exceptional; rather, Bengal "gave a sharp jolt" to India's leaders, and reminded them of the need for vigilance.[23]

Gopalaswami went on to consider the implications of India's rapid population growth for the country's future. The most alarming statistic, in his mind, was that the area of cultivated land per capita in India had fallen by 25 percent in thirty years: land had run out, and yields were declining. He ended his discussion on a note of foreboding, placing India's experience in global context. It could well be the case, he wrote, that "we are passing through the last stage of that exceptional phase in the growth of mankind in numbers which was introduced mainly by the opening up of the New World and partly by the creation of a world market."[24] Gopalaswami steered clear of Malthusian alarm. Like many of his generation he believed in the power of the state, in the marriage of wise planning with technology, to solve social and economic problems.

There were three possible responses to the challenge of feeding India. The independent Indian state tried them all. To the extent that the productivity of the land suffered from the very small plots of land held by the majority of cultivators, land redistribution seemed a promising solution. Soon after independence, Nehru's government proposed to abolish the *zamindari* system, the practice by which large landowners acted as intermediaries between the state and cultivators, wielding the power to collect taxes, a key feature of both Mughal and British administration. Although it had to confront some vested interests, *zamindari* abolition, which had to be approved state by state, was relatively straightforward—in the political culture of free India, *zamindars* epitomized the old feudal order that independence was meant to sweep away. But at most 6 percent of land in India changed hands under these reforms. The chief beneficiaries were most often farmers who were already relatively well off. Any energy behind land redistribution in India fizzled by the mid-1950s: by that time rural landowners had cemented themselves as an important constituency in Indian electoral politics; they closed ranks to defend their interests.[25]

The second approach, implemented with considerable success, was for the state to intervene more actively in the food economy. A commitment not to intervene in markets had been a shibboleth of British administration in India after the end of the East India Company. As we have seen, that iron confidence in markets shaped the British approach to the famines of the 1870s and 1890s, in which so many millions of Indians died. But the Second World War reversed that faith abruptly. India saw the rise of an elaborate apparatus of food control that lasted well beyond independence. The American T. W. Schultz, one of the pioneers of development economics, remarked in 1946 that "no country in the world, with perhaps the exception of Russia, has gone so far [as India] in controlling basic food distribution." By 1946, close to eight hundred cities and towns were covered by the rationing scheme. In 1947 the wartime "Grow More Food" campaign was resurrected, and in 1949 the government of India set the campaign's goal as the attainment of national self-sufficiency in food grains by 1952.[26] In the late 1940s and early 1950s, the Indian state purchased 4.3 million tons of food from its own farmers, and just short of 3.5 million tons abroad. An elaborate network of transportation and storage was established, providing the skeleton of India's public distribution system that, to this day, remains vital to the food security of hundreds of millions. As we will see, when serious drought threatened northern India in the mid-1960s, and Maharashtra in the early 1970s, the state's food distribution proved its worth and averted famine.

But the approach that received by far the most attention, and the most funding, was the quest to intensify agricultural production—to grow more food on the same amount of land. Gopalaswami observed that this could be done, in Indian conditions, by increasing the spread of double cropping—in which two crops were grown each year, one in the winter and one in the summer—through the expanded use of fertilizer; and, above all, by expanding year-round irrigation to free agriculture from dependence on the monsoon. In 1951, India launched the first of its five-year plans for economic development—

influenced by Soviet central planning, but maintaining a mixed economy. Fifteen percent of total expenditure under the first plan went to irrigation, and a majority of that to what were called "major and medium irrigation projects." At their apex stood the large multipurpose dams that would transform India's rivers in the years to come. The mastermind behind the Indian planning commission was the brilliant Bengali statistician Prasanta Chandra Mahalanobis, who had spent some years in the 1920s working for the meteorological department at Alipore observatory in Calcutta, and who had published a statistical essay on floods in Orissa.

THROUGH THE LATE-NINETEENTH-CENTURY CLIMAX OF IMPERIAL globalization, Asia had been reconfigured laterally, as steamships and railways connected distant places. Now, however, vertical space came to matter more. Engineers began to think about the gradient of each river's fall so they could be harnessed for hydropower; geologists aimed to measure the depth of water resources in underground reservoirs. The reorientation of Asia along a vertical axis—as if the map were now drawn in three dimensions—had everything to do with the conquest of water. And the conquest of water, ultimately, promised the enhancement of life and the diminution of premature death.

III

Inaugurating the Hirakud Dam on the Mahanadi River in eastern India, just a year after he became prime minister, Nehru described the scene as a "fascinating vision of the future which fills one with enthusiasm." He wrote that "a sense of adventure seized me and I forgot for a while the many troubles that beset us"—the "troubles," that is, of mass refugee movements after Partition, of multiple insurgencies, of hostilities with Pakistan, of governing a new and heterogeneous nation. The sight of Hirakud convinced Nehru that "these troubles will pass" but "the great dam and all that follow from it will endure for ages to come."[27]

The Indian state wasted no time pursuing its ambitious agenda. At independence, India had only thirty dams higher than thirty meters. Most colonial works of hydraulic engineering had been on a smaller scale: fifteen- to twenty-meter-high tanks and bunds, linked to a network of canals. During the Second World War, the colonial state began to think big—and the plans of the 1940s paved the way for India's largest hydraulic schemes in the 1950s. The largest of them were large water storage works at Bhakra, in Punjab, Hirakud in Orissa, Tungabhadra and Nagarjunasagar in the Deccan, and the Damodar valley project in Bengal. They epitomized the imperative of multipurpose development. Each scheme promised a cluster of benefits: year-round irrigation; water storage to even out the concentration of the monsoon rains; embankments to prevent flooding during torrential rain; river navigation; and hydroelectric power. Each scheme had its own priority among these uses—the Damodar project was designed with flood control primarily in mind; Bhakra's particular symbolic importance came from its role in compensating for India's loss of Punjab's Canal Colonies at Partition, by creating a new hydraulic infrastructure around the dam complex. The projects were technically demanding as well as expensive. The Damodar valley project owed most to outside funding and expertise: it received a US$18.5 million loan from the World Bank in 1950, and its technical advisors included David Lilienthal. For the most part, India's large dams were funded by the state's tax revenues, their costs recouped later through an "improvement cess," a levy on landowners who benefited most from the dams. But large dams very often ran over budget and behind schedule.

The large schemes were the most visible manifestation of India's attempt to remake nature. Many smaller projects sprang to life after independence: irrigation dams in the Vindhya hills, the Pykara hydroelectric scheme in Tamil Nadu, the Sarda Canal in Uttar Pradesh, the Sengulam project in Kerala. Beyond these lay innumerable canals restored, power lines laid, tanks and irrigation channels built or resurrected. Many of these schemes built on colonial precedents. What was new was their scale, but also the language in which they were justified. As a perceptive visitor to India observed at the time, the colossal projects "stand out for their quality of newness," even if their impact on water availability and power generation was less than the cumulative impact of many smaller initiatives. The mega-projects marked a true departure: "dependable in the worst monsoon, dynamic in the most backward region." Above all, "they stand for something India could not build, and did not will, before she became a nation."[28] The moral fervor behind the quest to harness India's waters came from this sense of historic opportunity. So, too, did the planners' willingness to force through their schemes, whatever the cost.

The Bhakra Nangal project, in Punjab, was India's most prominent engineering scheme. It stood 680 feet high, the second-tallest dam in the world at the time; it consumed 500 million cubic feet of concrete. Bhakra's location in the partitioned province of Punjab added resolve and poignancy to its promise of a more secure future. The dam had first been proposed in 1944, and construction began soon after independence. Facing the future, India's water engineers began with a familiar problem: "One of the chief characteristics of rainfall is its unequal distribution over the country," they wrote, and "another important characteristic is the unequal distribution of precipitation over the year."[29] They set out to free India from the seasons.

A. N. Khosla (1892–1984) stood behind many of India's plans to harness water after independence. He was the first chair of the Central Water and Power Commission of India, a graduate of the Roorkee College of Engineering and a stalwart of the Punjab irrigation department.[30] He imagined the future in, by his own admission, "fantastic" terms. In an address delivered on All-India Radio in 1951, Khosla declared that "it will be no idle dream to contemplate the linking up of the Narmada with the Ganga through the Sone, or with the Mahanadi over the Amarkantak plateau, and thus connect the Arabian Sea with the Ganga and the Bay of Bengal right through the heart of India."[31] Old dreams of reshaping India's geography—Arthur Cotton's dreams—gained new life after independence.

Dreams of hydraulic engineering were inseparable from dreams of freedom. Kanwar Sain, Khosla's successor as head of India's water authority, wrote that "the river valley projects constitute the biggest single effort since independence to meet the material wants of the people, for from irrigation springs ultimately the sinews of man, from power the sinews of industry." He voiced the hopes of many of India's planners and architects when he declared that the dams "are indeed the symbols of the aspirations of new India, and the blessings that stream forth from them are the enduring gifts of this generation to posterity." His words were followed, in a public information pamphlet, by page after page of statistics: kilowatts generated and projected, hectares irrigated, gallons of water stored, tons of concrete expended. Large numbers were a form of rapture.[32] For Khosla and for Sain, as for so many of their peers, Bhakra was the showpiece.

Jawaharlal Nehru addressing a large crowd at the dedication of the Bhakra Dam, October 1963. CREDIT: Bettmann/Getty Images

Jawaharlal Nehru and Zhou Enlai celebrating the New Year on a special train taking them back to Delhi after a visit to the Bhakra Dam, December 31, 1956. CREDIT: Bettmann/Getty Images

Opening the Nangal Canal in 1954, Nehru's reverence was palpable. "What place can be greater than Bhakra Nangal," he wondered, "where thousands of men have worked or shed their blood and sweat and laid down their lives as well? Where can be holier than this?" Nehru spoke at length in Hindi and more briefly in English, before a crowd of thousands. "We talk about Mother India," he said, and now "Mother India is in labour, producing and creating things." At the time, India was ablaze with demands for the creation of linguistic states out of the composite provinces inherited from British India. "We talk so much about changing the provinces, expanding them, shortening them, disintegrating them," Nehru said at Nangal, his irritation unconcealed. "I do not mind our people getting terribly excited about it," he said, "and forgetting the major things." But Nehru made a clear contrast between the "major things"—"Making a new India . . . putting an end to the poverty of India"—and what he called "petty disputes." Nehru turned to the theme of revolution. "A revolution does not mean the breaking of heads," he insisted; of India's own gradual, nonviolent revolution, he declared that, with independence, "we finished it in a way in the political sphere . . . we have to continue it in the social and the economic sphere."[33] Bhakra became a symbol of India's ambitions. It was an obligatory stop on the itinerary of every official visitor. On the last day of 1956, Nehru took Chinese premier Zhou Enlai to Bhakra—the two had built a rapport over the previous two years, though it would turn out to be short-lived. "These are the new temples of India where I worship," Nehru told Zhou. "I am deeply impressed," Zhou replied.[34]

The excitement of the Bhakra project was captured in a 1957 documentary made by the government of India's Films Division. It was produced by Ezra Mir, born Edwyn Meyers: an Indian Jewish filmmaker who began by making propaganda films for the British during the Second World War, and who went on to make seven hundred documentaries in the 1950s and 1960s. The Films Division was charged with bringing "new India" to life on screen. Its productions included films about India's freedom struggle and profiles of political leaders and musicians. Above all, the films dramatized the quest for "development"—for health and water, food and education. Public information films played in cinemas across the country before the feature films that people had bought tickets for—they reached an audience of millions.[35]

The Bhakra film was an epic. Its visual language came from a tradition that had circulated globally during the war: Soviet and Pathé newsreels suggested the form, the genre, the structure in which India's propagandists worked—and which they made their own. The narrator's voice is clipped and serious. The soundtrack begins with nineteenth-century European music: brassy and bright, like a march. "For centuries," the narrator begins, "gazing upon the parched lands of Punjab and Rajasthan, we have dreamed of reclaiming the desert." The struggle at the heart of the film is clear from the outset, as suddenly the background music turns to an Indian folk theme and the film cuts to a scene of women lining up at a well, whose "search for water, " we are told, "was never-ending." The solution, the film's narrator declared, lay in the "unused, wasted" waters of the Sutlej River, flowing down from the Himalayas. And then "at last," as if inevitable, "the decision was taken—the Sutlej must be tamed." The film cuts to the image of a young, studious-looking engineer, slide rule in hand: he is the protagonist of this drama. The narrator spins out superlatives, the music turns to a fanfare of trumpets, the camera shows us Bhakra's sheer size: it is "massive," "stupendous," "mighty," "a miracle." It held out for India "the promise of an exciting, dramatic future."[36]

The film was finished before the dam. It shows us a worksite of ceaseless, noisy activity: the rumble of drills; the explosion of blasted rock; the clatter of a conveyor belt ferrying bucket after bucket of material up to the dam; the clink of hammers and the heft of spades as imported machines—some of them reassembled, piece by piece, on location—dovetail with the oldest kinds of human work. India's dams were a lucrative source of contracts for foreign engineering firms like Hazra & Co. of Chicago, who provided material, equipment, and many consultants. But dam building was also a spur to local industry; the largest cement factory in Asia came up to feed the "colossus" that was Bhakra. Toward the end of the film, we see the changing of shifts at the end of a working day. Darkness descends on the site and "the lights of the great dam are switched on, glowing like stars." At moments in the film, the dam seems to transcend technology, evoking a deeper and more ancient sense of wonder—it was a "miracle."

Until the last few minutes of the film, the only voice we hear, apart from the narrator's, is a brief clip of Harvey Slocum—the straight-talking, autodidact American dam builder who supervised Bhakra's "army" of Indian engineers, and who died suddenly on site in 1961. But then, finally, our focus shifts to the makers of Bhakra. At the end of a shift, "men emerge from every corner." We see a worker arriving

Harvey Slocum, the American dam builder who supervised construction at Bhakra. CREDIT: James Burke/Getty Images

home as his three children run to greet him. As he enters his modest quarters, his wife rises to offer him water, a smile on her face. It is the first intimate or domestic scene in the film—the first sight of a woman, or of children; in the background the orchestra is replaced by a simple folk melody played by a flute. The parting message is one of national unity: Bhakra represents the nation. "Never in the long history of India," the narrator declares, "have there been so many men from different parts of the country working together for a common purpose." The effort of creating Bhakra united Indians where the politics of region and linguistic identity divided them. The film closes with the voices of the workers, each speaking his own language. B. Srinivasan faces the camera; he speaks in Tamil without captions or voiceover;

The Bhakra Dam under construction. CREDIT: James Burke/Getty Images

his speech is stiffly awkward before the camera. "My name is Srini-vasan. I have worked here for a year and a half. My home is in Madu-rai district." We hear from workers in Punjabi, Marathi, and Bengali, some of their words drowned out by the noise of construction.

India's new dams attracted thousands of visitors from near and far. They became landmarks on the landscapes they had altered beyond recognition. Their clean lines and monumental size reminded some observers of Buddhist stupas. Aesthetically, as well as symbolically, they were the "temples of new India." But most Indians in the 1950s had never visited a big dam. Most Indians did not read the *Indian Journal of Power and River Valley Development,* the pages of which told a heroic story of India's hydraulic adventures. The place where

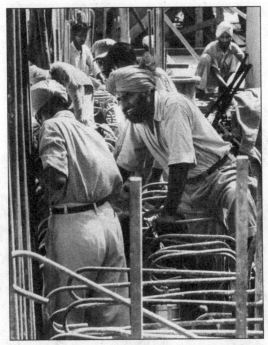

Construction workers on the Bhakra Dam site came from all over India and had many specialized skills. CREDIT: James Burke/Getty Images

most Indians encountered the grandeur of India's water projects was on screen. Public information films made an impression on many minds, but the feature films of Hindi cinema really captured people's hearts—they, more than any pamphlet filled with statistics, gave India's hydraulic revolution emotional content. And in the process, they reached beyond India's shores to make India's dams a symbol for hope and progress across postcolonial Asia and Africa.

Where many a foreign consultant ended up "a White Man's Burden character," as David Lilienthal described himself in his diary, one American observer of India's water projects was admiring, even at times uncritical: Henry Hart, a University of Wisconsin political scientist.[37] Having worked as a "minor administrative officer" in the

Tennessee Valley Authority as a young man in the 1930s, having seen a "newly-harnessed river brought to life" and the rural southern land-scape of his childhood transformed, Hart lived in independent India between 1952 and 1954. He sought the answer to a question that weighed on many American minds as the Cold War intensified: "Can a revolution . . . be *built*?" Hart traveled the length of the country, funded by a Fulbright fellowship and a grant from the Ford Foun-dation. His book, *New India's Rivers,* remains the most detailed and sympathetic account we have of India's great hydraulic experiment.

Hart's eye was drawn to the workers who were reduced to "an army" in so many depictions of India's dam fever; he dedicated his book to "all who died without seeing the new India they built." And where many observers saw only the romance of technology, Hart's romanticism sought artisanal skill. At the Tungabhadra Dam site, for which Hart saves his richest description, he encountered Vellu Pillai, a fifty-three-year-old stonecutter who chiseled granite from the quarry to exacting specifications. Hart discovered that Vellu Pillai came from a family of stonecutters in Thanjavur. His had been "a life of dressing stones for well-linings and walls"—for "ten prosperous years," he had carved temple deities. Evoking the ruined capital of the Vijayanagar empire, just a few miles away, Hart saw the dam resurrecting ancient skills. "The design itself was novel," he wrote," but the teamwork of brain and hand was a renaissance."[38] Hart speaks to no women workers, but their presence and their labor are visible in his account and especially in the photographs that illustrate his book; their stories remain untold.

The centripetal force of India's water projects drew workers, skills, and materials from across India—and beyond. Hart observed that some of the earliest workers on the Tungabhadra worksite were Telugu porters who had arrived from Burma during the war; they were among the half-million Indian refugees who had walked through the hilly jungles back to India when the Japanese bombs fell. Many died on the arduous journey on foot across the mountains into Assam. In the 1920s and 1930s, Burma was a frontier for Telugu-speaking migrant workers from coastal Andhra: they provided much of the labor on Rangoon's docks, they pulled rickshaws, laid roads, worked in the rice mills.[39] As long-established patterns of migration shut down during and after the war, new ones opened up along India's river valleys. From the moment the war ended, the large projects' capacity to generate employment was a strong argument in their favor. Water projects were themselves a way to prevent the resumption of Indian labor migration overseas, which so many Indian nationalists had come to oppose as exploitative. "At present a great deal of Indian labour is being sent to Burma and other places," noted a water engineer surveying the Ramaprasadasagara scheme along the Godavari. "An irrigation scheme of this magnitude would prevent the exodus of labor."[40]

Some workers traveled to the dam sites in groups. Teams of porters, masons, and stonecutters arrived, some of them following an old practice of migration from place to place in search of work. Others ended up building the dams of India after a chance encounter. In a wartime hangar along the Tungabhadra River, India's first factory for the manufacture of sluice gates was overseen by Mr. Eswariah; his "sheer mechanical intelligence" had been discovered by the dam's chief engineer, Srirangachari, in a "Madras highway repair shop."[41]

On many accounts of India's dam projects, as in the public information film on Bhakra, a sense of mission drove the workers. There can be no doubting that idealism animated the efforts of many who toiled on the dam sites. But so, too, did the need to earn a living. For only so long could appeals to sacrifice on behalf of the nation mask poor pay and harsh conditions. The workers on the Hirakud Dam learned how quickly the postcolonial state was willing to deploy force to keep construction going. In 1954, Hirakud's workers established a union of their own to challenge the officially recognized association of workers. They grappled over pay rates and the rhythms of work. After a breakdown in negotiations, the district magistrate ordered armed police to disperse a group of workers who were headed toward the chief engineer's residence, reportedly with the intention of harming him. Following a *lathi* (baton) charge, fifty workers were hospitalized; two died the following day.[42] In Hart's account, which reflects the official view, the strike was the work of "agitators" affiliated with the Communist Party. As he tells the story, India's climate itself played a role. "On any great outdoor work built in a monsoon climate," he declared, "the hot, pre-monsoon months are the tense season." And so, in the end, the strike appears in Hart's account as a mere interruption: "On the Monday after the bloody Friday, men began going back to work. By Wednesday, the dam began to rise again, full speed."[43] It is as though India's dams were a juggernaut, with a life and force of their own. That is certainly how they must have seemed to those who stood in the way of the engineers' plans.

Not every commentator was unequivocally in favor of the large dams. The Bengali journalist Kapil Prasad Bhattacharjee was among the earliest to call into question the approach of the Damodar Valley Corporation. As a student in Paris in the 1930s, he was influenced by the work of French hydrologists. Schooled in the economic nationalism of Dadabhai Naoroji and Romesh Dutt, Bhattacharjee argued that the Damodar valley project would perpetuate a colonial effort to keep India poor by keeping it an agrarian economy. He worried for the future of Calcutta as a port; for Bhattacharjee, the worst effect of the Damodar projects would be to silt up the Hooghly River. He felt that more could be achieved by repair and restoration—the "proper maintenance of old canals, tanks, lakes"—than through expensive projects dependent on foreign engineering expertise. Voices like Bhattacharjee's were in a very small minority in the 1950s, not only in India, but all over the world. He worried about the economic and the ecological effects of dam building; even Bhattacharjee had little to say about their human consequences.[44]

———

As India sought to tame its rivers, the dispute with Pakistan over who controlled the Indus intensified. It did not take long for David Lilienthal's idealistic vision of water beyond politics—his proposal for the shared development of the Indus basin—to fail. In early 1954, the World Bank proposed an alternative solution, which gave shape to the treaty that eventually came about. In place of Lilienthal's idea, the bank proposed to finish the job of Partition by dividing the water of the Indus and its tributaries completely, if not neatly, between India and Pakistan. The eastern rivers—the Beas, the Ravi, and the Sutlej—went to India; the western rivers, the Chenab and the Jhelum, went to Pakistan. Since the latter, too, originated in Indian territory, India would have the right to use their water for irrigation, transportation, and power generation, up to a limited volume. The negotiations dragged on for six years: India's position of strength as the upper riparian continued to trouble Pakistan. But much remains obscure about the process. Despite a commitment to greater transparency, the bank declassifies documents on a case-by-case basis; many of my requests were denied, as they contained "classified material provided by member states." Dam building remains a sensitive subject—its history raises uncomfortable questions about the present and the future.[45]

As ambitions for river development intensified, older water disputes reasserted themselves within India. British India had always existed amid a patchwork of other forms of sovereignty—the absorption of the former princely states into independent India raised its own problems where water was concerned. The Kaveri River dispute between the princely state of Mysore and the British-ruled Madras Presidency dates back to 1891, when the Maharaja of Mysore first proposed to make use of the river's water in his domains. The failure of the two sides to agree on how to share the Kaveri's water stalled M. Visvesvaraya's great Krishnarajasagar Dam, which began to rise only after a 1924 agreement that apportioned fixed amounts of the river's capacity to each state. In 1956, the internal map of India was redrawn: some of the old provinces of British India were broken up into new states, their boundaries corresponding roughly to the boundaries of linguistic regions. The old Madras Presidency was divided into Tamil

Nadu, Karnataka, Kerala, and Andra Pradesh, each with its own state legislature—and each with a new set of claims upon the central government for resources. In this context, the Kaveri dispute resurfaced. The government of Karnataka reopened the argument, arguing that changing political and economic realities demanded a revision of the earlier agreement. In particular, the ambitious plans each state had to expand irrigation—under the auspices of India's overarching five-year plans—spawned a new set of claims upon the Kaveri's waters. The Kaveri dispute remains India's longest-running water conflict; it is far from resolved. Competing claims to ownership over water, in the postcolonial era, were not only national: they were just as likely to be regional.[46]

IV

Mehboob Khan's 1957 melodrama, *Mother India,* remains one of the world's best-known films. It is, among other things, Indian cinema's great water epic. The film opens with the shot of an aged Radha, the film's protagonist and the eponymous "Mother India," touching a clump of earth to her mouth; she raises it above her head, hands trembling. Behind her: tractors, power lines, roads—the churn of progress. A procession of construction equipment roars in the foreground, drowning out the music. The camera pans to a shot of a large dam rising. A jeep arrives in the village, full of khaki-clad, white-capped men—functionaries of the ruling Congress party. They tell Radha that the new dam will bring water to her village; they want her, as the community's most respected elder, to inaugurate it. She refuses in all humility, until, head lowered, she allows the politicians to lead her to the dam. They place a ceremonial garland around her neck. At the very moment she is about to pull the lever to open the dam's gates, the film slips into flashback. Radha's life comes to symbolize the struggle of the Indian nation for freedom.

Early in the film Radha's husband is maimed in an accident; scenting her vulnerability, the predatory moneylender makes advances to Radha, which she rebuffs. The capricious power of nature over Radha's life and livelihood is a recurrent theme. In one of the film's central song sequences, we see Radha working in the fields with her children, pausing to feed them a meager meal of porridge, eating none herself. But she is proud, unbowed. As the song comes to an end, storm clouds build in the sky. The sky darkens, the screen crackles with lightning, the wind rises and the rain pours. The family's makeshift shelter collapses. Floods destroy the village. The crops lay ruined. Radha's youngest child dies. Even in extremis, Radha makes her surviving children reject the moneylender's offer of food. Redemption comes when the villagers fight back the flood; they come together to harvest a good crop the following season. In their dance, the camera zooms out to show us the massed villagers form the shape of a map of undivided India.[47] Radha's song implores the villagers not to abandon their land.

In opening the film with the large dam, and a vision of prosperity, *Mother India* suggests that India's victory was, in part, a triumph over the monsoon. These were the freedoms that India fought for, and won: freedom from want, freedom from exploitation—and freedom from the vagaries of nature.[48] The publicity pamphlet for the film used a quotation from the German Orientalist Max Mueller to encapsulate its central message: "If I were to look over the whole world to find out the country most richly endowed with all the wealth, power, and beauty that nature can bestow—I should point to India."[49] In *Mother India*, vulnerability to the weather is confined to the unhappy past; it represents an old, unchanging India, juxtaposed against an India where technology and political freedom would triumph over nature.

Mehboob Khan (1907–1964) was born in Baroda state and moved to Bombay as a young man to work for a noted horse supplier to the film industry; his first job was repairing horseshoes. He rose as a producer in the era of silent film. Khan was an active participant in the Progressive Writers Movement, a group of left-leaning writers, dramatists, and film producers who forged an Indian art that reflected the social conditions of the country. In common with many others of his generation, Khan's strong commitment to Indian nationalism melded with an outward-looking internationalism: he was voracious in his absorption of artistic inspiration from diverse places; he believed in a sense of shared struggle, across what would come to be known as the Third World, against imperial exploitation. The logo of his production company incorporated a hammer and sickle, a symbol that was discreetly removed before the film was submitted for the Oscars.

From the start, Khan envisaged an international audience for *Mother India*. His initial working title was *This Land Is Mine*, and he wanted to work with Sabu Dastogir, an Indian actor successful in Hollywood; the plan fell through, but the ambition to reach the wider world remained.[50] *Mother India* was not the first Hindi film to strike a chord with audiences across Asia, Africa, and Eastern Europe. Raj Kapoor's *Awaara* (1951) was a global blockbuster, the story of an unfeeling and autocratic judge and his estranged son, who becomes a vagrant. The film's strong message of social justice, its memorable theme song ("Awaara Hoon"), and its visual splendor all compelled audiences across Asia and Africa; reportedly, Mao Zedong was a fan. Mehboob Khan followed Raj Kapoor's success with *Mother India,* which was popular across Francophone and Anglophone West Africa, in Ethiopia, across Asia and the Middle East, and in the Soviet Union and Greece. Egyptian audiences were enthusiastic about *Mother India,* which had immediate resonance in another country where dams symbolized modernity.

The film's influence endures. Anthropologist Brian Larkin describes a scene in northern Nigeria in the 1990s, where Lebanese distributors had been importing Hindi films for four decades. "It is Friday night in Kano, and *Mother India* is playing at the Marhaba Cinema," Larkin writes. "Outside, scalpers are hurriedly selling the last of their tickets to the two thousand people lucky enough to buy seats in the open-air cinema of this city on the edge of Africa's Sahel desert." Throughout the screening, "people sing along to the songs in Hindi, they translate the dialogue into Hausa and speak the actors' lines for them." Forty years after its release, *Mother India*'s appeal transcends generations.

"I have been showing this film for decades," a local distributor told Larkin, "it can still sell out any cinema in the north."[51]

How did a Hindi film about one woman's lifelong struggle against nature and exploitation prove so resonant across so much of the world? We are used to thinking of "development" as something imposed from on high, by all-powerful states upon unsuspecting populations. Since the 1980s, a lot of writing about "development" in the postcolonial world has been heavy with irony: we know, now, how so many of those schemes turned out. Their costs are too evident; their consequences, intended and unintended, have mounted.[52] But beneath the grand plans were also simple dreams of a better life. *Mother India* touched millions of people because it told a humane and humanizing story about dreams of water and plenty, dreams of security—dreams of a future that was better than the past.

V

A darker reality lurked behind the glossy dreams of dams and plenty. On a rock of colonial-era legislation, the Land Acquisition Act of 1894, millions of lives in independent India foundered. Passed to foster railway development, the law gave the state the right to compulsory purchase of land for "public benefit"—the law of "eminent domain," a version of which most modern states have retained. The postcolonial state pressed the act immediately into service. Bhakra, Hirakud, the Damodar valley: each of these projects began in 1948; each of them needed more than one hundred thousand acres of land. So, too, did the new steel plants of Bhilai, Rourkela, and Durgapur.[53]

Even before the end of the war, intensive discussions were underway about the numbers of people who would be displaced by the Bhakra Nangal Dam. One Punjab official wrote in February 1945: "We cannot obviously allow the whole scheme to be wrecked by a few obstinate people who may refuse to move." If the "majority" were "not strongly averse to the proposal," then those against it would "if necessary [be] ejected by force."[54] Pieces of paper came to bear immense value; a Punjab government official dismissed cultivators' fears of dispossession by pointing out that "they appear to have no right in the lands they cultivate at present." He could "sympathize whole-heartedly" with those "ousted from their ancestral lands," but the "cold truth" was that "their interests cannot be allowed to impede" a project "which, with its irrigation and hydroelectric potentialities, is likely to carry so many benefits to wide areas of the country."[55] The architects of Bhakra were under no illusions about the scale of displacement the project would cause. However attractive the alternative land offered villagers, one official wrote, "it is extremely doubtful whether it will be a satisfactory form of gilt-edged security."[56] Where could people go? How would they recover their livelihoods? And what of their bonds of community—their bonds with the piece of the earth they knew best? Too often, irreversible projects ploughed ahead with those questions unasked.

Displacement never went unchallenged. In December 1950, ten thousand villagers from an area soon to be submerged by the Tungabhadra Dam came together in a public meeting. They issued a list of demands, topped by their demand for adequate compensation for their loss of land and income. Among those who needed to be resettled were not only registered landowners, but also "houseless coolies, weavers, and ryots [cultivators]" who faced the "difficult present circumstances." Two years later, the government offered them a better deal: 30 percent over and above the amounts awarded by the Land Acquisition officers, but only to those "who withdraw their cases from the civil courts." In this rare willingness to compromise lies a clue to just how many people fought their dispossession in the courts of independent India. A few months after the meeting of Tungabhadra villagers, a touring official wrote to Hyderabad's minister of revenue: "I told the people gathered in a pretty large number at Manur that they should not stand in the way of the construction of the project but should consider it a great sacrifice on their part, since by the sufferings, if at all, of a small number the country is going to prosper."[57]

This was the crux of the utilitarian argument—the greatest good for the greatest number—that authorized India's great water projects at any cost: in families displaced, in villages drowned and futures ruined. Through most of the 1950s, India's courts agreed. When the Maharaja of Darbhanga, Kameshwar Singh, tried to resist the acquisition of part of his Bihar estate, the High Court ruled that "the Legislature is the best judge of what is good for the community . . . and it is not possible for this Court to say that there was no public purpose behind the acquisition contemplated by the impugned state."[58] Convinced of its mission, the Indian state felt little need even to document the displaced.

The true scale of displacement would only become clear later on. Through the painstaking work of activists like Walter Fernandes and scholars like Sanjoy Chakravorty, working through fragmentary archives and court cases, we now have a sense of the numbers of people that India's dams chased from their homes. From independence to the present day, that number is likely to exceed 40 million. It bears repeating: 40 million people in India have been displaced by dams alone. More than 50 million people have been displaced by the state's development projects writ large. India's *adivasis* have borne the brunt of this displacement, and they are also least likely to have received any form of compensation. One reason that almost every official figure vastly underestimates the numbers displaced is that they count only those who own land that has been acquired by the state through compulsory purchase. They do not include the large number of landless people in India who "depend on the acquired land for income"—tenants, wage laborers, service providers. The latter group have seldom if ever received compensation. The other way in which the true losses sustained by displaced communities exceed official calculations is that dams often submerged common property—forests, grazing grounds, and other grounds considered "wasteland" by the state. The commons were already under strain by the end of the nineteenth century, more likely than ever to be enclosed and possessed as private property—but the poorest groups in society, *adivasis* above all, still depended on them in the 1950s. Through a painstaking assembly of data, Chakravorty has estimated that water projects are by far the greatest cause of population displacement in independent India. The "core problem," he writes, is that "the population that benefitted from the development projects is fundamentally different from the population that was displaced or disrupted." The benefits of large dams have gone downstream; power from hydroelectric plants has gone to cities and factories and to farmers who "still have land [on which] to run their pumps."[59]

The environmental consequences of large dams have been considerable: forests have been drowned, soils salinated, rivers blocked in midflow, deltas starved of silt, natural drainage hindered—leading, ironically, to more severe flooding. But the sheer scale of the impact of large dams on Asia's ecology of water would only become clear toward the end of the twentieth century.[60]

Nehru himself began to have a change of heart at the end of the 1950s. "For some time past," he said, speaking about India's dam fever, "I have been beginning to think that we are suffering from what we may call 'disease of gigantism.'" He proceeded to tell his audience of engineers that "the small irrigation projects, the small industries and the small plants for electric power will change the face of the country, far more than a dozen big projects in half a dozen places."[61] But a massive failure in the monsoons precipitated a shift in political priorities.

VI

The United Nations' Economic Commission for Asia and the Far East (ECAFE) started life in Shanghai in 1947. The commission set about collecting information on economic conditions across Asia. Its first survey came out in 1948—an "incomparably more thorough and informative account of the economic life of the region" than any ever attempted—and painted a picture of a continent in ruins after the war.[62] ECAFE's original regional members were Pakistan, India, Burma, Thailand, the Philippines, and China. At a time when much of

Asia remained under imperial control, "nonregional" members—the imperial powers and the Soviet Union—had a major presence, as did "associate," nonvoting, members that were still under colonial rule or postwar occupation—Ceylon Malaya, Vietnam, Cambodia, Laos, Hong Kong, the Republic of Korea, and Japan. The intersection of Asia's postwar revolutions with the early Cold War created immediate political tensions. After the Communist victory in the Chinese civil war in 1949, ECAFE decamped from Shanghai to Bangkok; like the rest of the UN system, it recognized the Republican government in Taiwan as the rightful representative of China.

Asia in the 1940s was torn three ways: between colonial powers still clinging to power, newly independent states hungry for the technology and financing that would allow them to realize their grand plans, and the new superpowers in competition for their allegiance. ECAFE's deputy director, the American C. Hart Schaaf, made the bold claim that, amid these tensions, ECAFE was primarily the creature of new Asian states. He insisted that "in the most extensive and populous region on earth," the "most conspicuous political fact" was "a new dynamic nationalism"; he suggested that ECAFE had not only "witnessed" but also "facilitated" this movement. One indication of this stance was ECAFE's appointment of an Indian director, P. Lokanathan, a Madras University economist who had been editor of the influential periodical *Eastern Economist*. Invoking the judgment of an imagined "future historian of ECAFE," Hart Schaaf argued that ECAFE's role in the region would in time be seen as vital.[63]

Asia faced a "revolution of rising expectations" that cut across revolutionary and nonrevolutionary, capitalist and Communist states. ECAFE embraced water as a major regional priority, and one that could bring people together: it was an area where tangible results could be achieved. In 1950 the organization convened a meeting in Bangkok on flood control in Asia. Hydraulic engineers, city planners, and hydrologists from across the region came together to compare notes; through the mediation of international agencies, "often . . . a particular national project becomes regional." In his search for a metaphor that would explain the benefits of regional cooperation, Hart Schaaf turned to "the concept of an act or event which sets in motion a cumulative process of great momentum," adding that it was "around [this idea] that Lord Keynes and others have constructed their thinking about the 'multiplier.'"[64] ECAFE's achievements were piecemeal, perhaps relatively small, but they were not insignificant. The agency epitomized the ideal known as "functionalism," which Lilienthal had put forth in response to the Indus dispute: that technical matters, from public health to water management, could be removed from the political arena and solved cooperatively by experts whose professional camaraderie transcended geopolitical fault lines. But the politics of national sovereignty, like the geopolitics of superpower rivalry, could not be avoided. The very absence of the People's Republic of China from the ECAFE table made the organization's claims to a comprehensive regional perspective ring hollow. In 1954, for the first time, ECAFE's *Economic Survey* included "Mainland China," compiling statistics from official sources in the PRC: it was among ECAFE's most keenly read publications, as Asian planners, economists, and statesmen had the occasion to compare their countries' rates of "progress" with that of the revolutionary behemoth.[65]

The curiosity of Asia's leaders and engineers about China—fascination combined with suspicion—went beyond the capacity of an ECAFE report to satisfy. Nowhere more so than in China's largest neighbor, India. The government of India had been among the first to recognize the People's Republic of China in 1949. Nehru's first ambassador to China, the historian K. M. Panikkar, was keenly impressed by China's revolution, and by Mao. The Chinese government's invasion and annexation of Tibet in October 1950 alarmed and surprised India, which regarded a semi-independent and friendly Tibet as a buffer along its border with China. Nehru came under pressure from others within the Congress to take a harder line; but a realistic sense of India's inability to intervene prevailed. Relations between the two countries warmed in the 1950s in line with Nehru's foreign policy of nonalignment with either bloc in the Cold War. Among the Indians most keen to learn about developments in China were the country's leading hydraulic engineers. In May 1954, Kanwar Sain, chairman of India's Central Water Commission, and K. L. Rao embarked on an official mission to China to inspect and to report back on China's water projects—flood control in particular. Sain and Rao were among the first outsiders to see, firsthand, China's hydrological experiment; and because they were favorably disposed toward the Chinese government, they had extensive access to information.

Sain and Rao arrived in China on May 4, 1954, and stayed two months. They spent much of their time on the water, making many parts of their journey along the Yangzi by boat. Sain and Rao were chief protagonists of India's own colossal efforts to control water; the scale of work in China dazzled them. They undertook their tour of China half a century after the Indian Irrigation Commission had traveled through India in search of water. Theirs was part of the same quest: the famines of the late nineteenth century had unleashed in India a desperate and continuing search for sources of water to mitigate dependence on the monsoon. Sain and Rao were trained in the colonial tradition of Indian water engineering. But now they represented an independent nation, and for inspiration they looked not to Europe but to revolutionary China. Consider the contrast between the two tours: where the irrigation commission traveled with a retinue of servants in a specially chartered train, Sain and Rao were given strictly limited foreign exchange. They were accompanied by just two interpreters and two officials. They were impressed by the modesty, the lack of ostentation they saw—and by the absence of the tight social hierarchies of India, to say nothing of caste. Another thing puzzled Sain and Rao. For generations, Indian intellectuals had believed that the most fundamental bond between India and China was Buddhism. But "to our repeated questions about the Buddha" came a standard response: "Chairman Mao is our Buddha." In a new era, India and China needed a new language, a new basis for their interaction—it came naturally to two hydraulic engineers to see that basis in the shared problem of water.[66]

Distinguishing his perspective from the negativity about China that he had imbibed reading the accounts of "foreign diplomats," Sain commented on the high standards of sanitation he saw everywhere. An austere Indian engineer in the tradition of Visvesvaraya, Sain noted with satisfaction "an absence of headlines in the newspapers highlighting murder, scandals, or disgraceful lives." Most of all, he was impressed by the "clear-cut vision" state officials had of a "new China of their dreams," which in turn instilled in the people "unbound faith and confidence in the wisdom, goodness, and creative policy of the government." Sain contrasted the Shanghai he had seen in 1939, on his way home from the United States—a cosmopolitan and decadent city, in his mind—with the city he saw in 1954. "The bright lights had gone out," he wrote, but he meant this as a compliment. Shanghai now looked "more typically Chinese," shorn of Western concessions and colonial settlements. Like so many other port cities along the littorals of the Indian Ocean and the South China Sea, it was now "integrated [more] closely with the economy of the interior rather than dependent on foreign luxury trade."[67]

Sain's and Rao's report contains an extended list of every Chinese official they met, from ministers to field engineers to water scientists at the College of Hydraulic Engineering for Eastern China, in Nanjing. They were struck by the quality of China's hydraulic engineers. They delighted in the firm emphasis given to technical education in China. They praised the Chinese capacity for improvisation, building huge dams from local materials when imports were in short supply. Traveling through China, Sain's and Rao's thoughts turned naturally to comparisons with India. There were clear differences in the challenges each country faced. One sharp contrast between India and China was climatic—once again, what made India distinctive was the monsoon. "Unlike India, hemmed in by the Himalayas," they wrote, "China is open to Central Asia"; this meant that, in the summer, "China unlike India is not the single objective of the air circulation of a whole ocean." China received "less heavy and less concentrated rainfall" than India, and its rain was "much more equally spread across the interior." By contrast, China's rivers were more menacing than India's, more prone to burst their banks. India's great need was irrigation; China's was flood control. Both countries eyed an industrial future, and the promise of hydroelectric power attracted them both.[68]

India and China shared a sense of urgency. They shared a conviction that water held the key to security and prosperity—these translated into an addiction to mammoth projects. In both countries, bigger was better. Just as the pamphlets of India's Central Water Commission and the documentaries of its Films Division extolled the pace and the scale of dam building in India, so too did the Chinese state and its engineers take pride in their compression of time. In the five years since the liberation, they boasted, "250 major and thousands of minor irrigation projects" had begun in China, adding 9.2 million acres of irrigated land. Most striking to the Indian visitors were the "remarkable speeds of construction" China had achieved through the mobilization of labor on a scale "unknown in recent times."[69]

Soon after the establishment of the People's Republic in 1949, Mao's government made the Yellow River a priority. Known for centuries as "China's sorrow," the Yellow River was notoriously prone to flooding. Within a year after the liberation, both the Huai and Yellow rivers experienced catastrophic floods. In the interests of national reconstruction, they had to be tamed and conquered. Where India drew expertise and aid from both the Americans and the Soviets, and where Nationalist water engineers in China in the 1920s and 1930s had maintained close links with the United States and Germany, after 1949 Chinese hydraulic engineers combined Russian technical assistance with local ingenuity. In parallel with India's race to build Bhakra Nangal, the most ambitious Chinese dam was the Sanmenxia on the Yellow River.

Like Bhakra, the Sanmenxia had its origins in the 1930s; like Bhakra, a sense of urgency that followed the revolutionary upheaval of the 1940s brought it to the top of the agenda. The dam was located near the border between Shanxi and Henan, designed by Soviet engineers. The Soviets proposed a concrete gravity dam across the Yellow River, with a reservoir 360 meters above sea level. The initial plan would have displaced more than 800,000 people from their homes, and flooded 3,500 square kilometers of land. As plans for the dam went into circulation, between 1955 and 1957, Chinese experts debated it at length. During a fleeting moment of political openness under Mao's "Hundred Flowers" campaign, hydraulic engineer Huang Wanli—trained as a meteorologist at Cornell and Iowa in the 1930s, and then an aide at the Tennessee Valley Authority—raised the alarm. He argued in favor of a lower dam, with a smaller reservoir. He hinted that the Soviet plans had not undertaken a detailed analysis of costs and benefits. He felt that a smaller scheme, which displaced fewer people, would be less risky. He feared that the dam's design was no match for the heavy loads of silt that the Yellow River carried: the danger, as he saw it, was that the dam's reservoir would silt up, making the dam useless—or, worse, dangerous.

The Hundred Flowers campaign was short-lived; as people spoke more freely, Mao disliked what they said. His retribution was swift. Huang was condemned and humiliated. He was deemed a "rightist" and sent for "re-education."[70] His predictions proved uncomfortably accurate. Within a few years of the Sanmenxia Dam's completion in 1960, it was clear that its reservoir was clogged with sediment. As Sino-Soviet relations soured after 1960, it became easier to blame a faulty Soviet design for the dam's problems. Acknowledging the scale of the problem, Zhou Enlai ordered a reconstruction and renovation of the scheme, at huge cost. At the same time, the human cost of the dam was immense. Just as Bhakra and Hirakud began the decades-long displacement of Indians by large dams, Sanmenxia led to the forcible relocation of an estimated 280,000 people.

Other aspects of China's experience had no parallel in India, as Sain and Rao were quick to notice. The mobilization of labor in China was on a scale unknown in India, and this also set China on a path to water engineering quite different from the one established by their Soviet allies. Mao had prevailed in the Chinese civil war because of his stunning success in mobilizing popular support and enthusiasm—first in the vanguard of the anti-Japanese resistance, and then deployed against Chiang Kai-shek's Nationalist army. This commitment to mass action never left Mao. It shaped deeply his government's approach to water management. The first achievement of the people's energies, mobilized at the county level, village by village, was the two-hundred-kilometer People's Victory Canal, linking the Yellow River and the Wei. The *People's Daily* ran features on the exertions of model workers who had broken records and distinguished themselves in devotion to the cause. By the mid-1950s, as China's collectivization drive gathered pace, Mao's government laid ever-greater emphasis on irrigation from the ground up. This fervor reached a peak during the Great Leap Forward, when every county was set to work building its own dams and irrigation ditches. Zhou Enlai made a rare official acknowledgment of the disastrous consequences of this approach in 1966, when he said, "I fear that we have made a mistake in harnessing and accumulating water and cutting down so much forest cover . . . Some mistakes can be remedied in a day or a year, but mistakes in the field of water conservancy and forestry cannot be reversed for years."[71]

Of course, much of this took place after Sain's and Rao's visit. They saw no signs of danger in China's quest for water. In table after table they compared China with India—how much concrete their dams consumed, how much water their reservoirs could hold. The speed of canal excavation was where China's achievements were most dramatic in comparison with India's. The Indian engineers' conclusion was wistful: "In India, where similar human force is available, it should be possible to attain similar speeds . . . by proper organization and creation of enthusiasm among the people"; they chose not to mention that "proper organization" would have demanded a level of coercion that the Indian government was unwilling (or unable) to muster.[72]

The most revealing part of Sain's and Rao's report is a verbatim record of a speech by China's director of water conservancy, given at the end of their stay, in which he sought to address the questions that had arisen during their visit. Director Hao positioned the People's Republic firmly within an ancient tradition of water management in China. "The record of exploitation of water by the Chinese people," he wrote, "dates back to ancient times." But under the "corrupt" and "feudalistic" rule of a decaying empire, compounded by the failures of the Nationalist government, "the hydraulic constructions [of China] were seriously ruined owing to long years of negligence." The Communist state claimed the mantle of imperial power over water—it was a revival as well as a revolution.[73] China's archives of water control were on display everywhere the visitors went: at Tsinghua University, they were shown an eight-hundred-year-old text "containing excellent plans" of the Yellow River. Hao spent as much time telling his Indian visitors about small projects as he did extolling the gargantuan ones. His was a story of repair and renovation as much as creation. He spoke of the myriad ponds and dikes that conveyed water to the fields of southern China. He spoke of the spread of simple technology—a "Liberation-type waterwheel" that outdid the age-old technologies still in use.[74]

Sain and Rao returned to India filled with enthusiasm. For all that they grasped the complexity of China's approach to water, their message back home was a simple one: it was a message of scale, speed, and control. Their main "lesson" from the Yellow River for the management of the notoriously flood-prone Kosi River of Bihar was the lesson of centralized command; China's intensive emphasis on small projects fell by the wayside in their accounts. Soon after his return Sain was summoned by Nehru, and "closeted with him for about an hour." Nehru, soon to depart for China himself, asked for Sain's impressions. Listening closely, he pressed harder: how would Sain describe China in one sentence? Sain recalls that he gave Nehru an unscripted answer: "At present China is behind India in every field, but I feel that at the rate they are progressing, China may be ahead of India in 10-15 years." Nehru "made no comment," Sain remembered, "but I could see from his face that he did not relish this reply."[75]

India's most eminent water engineers returned from China with a sense that the two countries shared fundamental problems, and that there were lessons they could learn from China. But there were ominous portents, too. In his speech, Director Hao had described how China's water projects had been "extended to the border regions of our fraternal minorities and helped to promote national unity." There was never an attempt on the Chinese side to disguise the fact that water was intrinsic to political power. The conquest of water meant the conquest of space. With the control over water came the projection of state power over peoples with a different vision of water's uses: the people of Xinjiang, the people of Tibet. Unspoken was the thought that some day the "border regions" in question may include China's borders with India.

Sain and Rao faced a problem when they returned with the first-ever maps of China's water projects to be seen outside China. The Indian Ministry of External Affairs was rankled by what Sain and Rao had missed: the maps they had been given by Chinese officials claimed as Chinese territory a large swath of the borderland that the Indian state saw as integral to India. The maps were destroyed, redrawn to accord with India's understanding of its territorial boundaries. Sain later wrote in his memoirs that he was grateful this had been caught before the volume was published—if it had not, it would have been a source of "great embarrassment" a few years later, when India and China went to war over just those borders.

One of Sain's enduring impressions of his trip had been "how the Chinese people loved and admired the Russians." "The bookstalls are generally full of Russian books and journals," he wrote; Russian expertise was offered without condescension and without strings—or so it seemed. He had the opportunity to see for himself a year later, when he led an Indian delegation to the Soviet Union, sponsored by the United Nations' Technical Assistance Administration. Again, the visit of Indian technical experts was followed by an official visit by Nehru. Sain wrote a sweeping account of just how rapid the Soviet Union's economic progress had been since the revolution. Inspecting its hydraulic projects, he concluded that "the interests of the power engineers have been accorded pride of place," a dominance he traced back to Lenin's emphasis on electrification as the key to socialism— flood control and irrigation, India's and China's other great needs, were less valued. But Sain's conclusion was clear: China had much more directly to teach India than the Soviets.

———

AS CHINA'S EXPERIENCES INSPIRED INDIA'S ENGINEERS, SO INDIA'S experiences became a model for the rest of Asia. A year after his visit to the Soviet Union, ECAFE's director, Lokanathan, commissioned Sain to join a UN mission to survey the Mekong. Like the Brahmaputra, the Salween, and the Yangzi, the Mekong originates on the Tibetan Plateau. In the twentieth century, it has been Asia's quintessential "transboundary river," running through China, Burma, Thailand, Laos, Cambodia, and Vietnam before spilling into the South China Sea. The Mekong, the ECAFE commission noted with understatement, was "a perennial river of great importance." That importance was clear to the US government, which maintained an escalating financial and military presence in South Vietnam after the end of the French-Vietnamese war in 1954, which had resulted in the division of the country. The US Bureau of Reclamation, the domestic agency responsible for hydraulic engineering, had a global presence by the 1950s. Its engineers surveyed Thailand, the Philippines, and Indonesia, looking for hydroelectric potential. They had surveyed the Mekong, too, but their initial verdict was lukewarm.[76]

Convinced that a substantively international effort was needed, the UN went back. Sain was joined on the commission by Y. Kubota, the president of Japan's Nippon Koei corporation; G. Duval, a former colonial official; M. Sakaita, an engineering geologist from Japan; and the Dutchman W. J. van der Oord, a navigation specialist. They toured the Mekong in April and May 1956. The commission placed its faith in two large hydroelectric projects, one on the Tonle Sap in Cambodia, the other at Nam Lik in Laos. "The prick has gone too deep to be halted"—this is how Sain described his sense that large-scale hydraulic engineering was inevitable, now, in the Mekong as elsewhere in Asia, given the bold claims that had been made on behalf of big dams, and given the hunger for progress and development that he saw wherever he traveled. The following year, Sain joined another commission co-ordinated by ECAFE and led by Raymond Wheeler, former chief engineer of the US Army Corps of Engineers. Wheeler's account of the mission harked back to the language of colonial exploration. "There were no maps of the country," he wrote, "we had to make them . . . Nobody had any data on river flow, or even any idea how to keep data." Wheeler described the Mekong as "truly a virgin river." Historian David Biggs notes that the commission proposed "a cascade of hydroelectric dams and irrigation schemes in the valley from the Chinese border southward to the Mekong Delta." The Mekong commission signified an opening for private interests who stood to profit from the dam-building rush; Japanese, Korean, and Taiwanese companies, in particular, stepped in with materials and personnel.[77]

The Mekong commission was quickly overshadowed by the escalation of American involvement in Indochina. The US Bureau of Reclamation placed its faith in what it called "impact type projects," the grandest of them being the Pa Mong Dam, upstream of the Laotian capital, Vientiane. Vast, ambitious, planned to the last detail—the dam never materialized, as the United States became mired in military conflict in Vietnam that engulfed Vietnam's neighbors as well.[78] There was a close bond between American support for dam building in Asia and American strategic imperatives in the Cold War. But Kanwar Sain—a patriotic Indian engineer at the pinnacle of his profession, enamored of China but with close personal and professional links to the US Bureau of Reclamation—chose to spend a decade of his career with the Mekong commission, trying to coordinate the development of Asia's most international river. In his memoirs, he hints that the material reward of working for the United Nations was one clear incentive. But his motivations went deeper than the money.

Sain believed, like so many of his generation, that taming the waters was a goal beyond national sovereignty—and beyond ideology. Working for ECAFE alongside many former colonial civil servants and engineers now turned development consultants, Sain held to a vision of Asian nations working together to claim their rightful place in the community of nations. In a memoir that is detached, even clinical, in tone, a rare moment of emotion comes when Sain describes his "pilgrimage" to the site of Angkor Wat, in Siem Reap, Cambodia, while on his first Mekong mission: "I was very much moved by the ancient glory and culture of India reflected in Angkor Wat," he wrote.[79] Just as many of India's water engineers presented their "new temples" as standing within an ancient historical tradition of water engineering, so Sain appealed to a deep history of cultural exchange across borders to provide ballast for his vision of an Asia united by water—or by water engineers.

In a sense, Sain's faith was eventually vindicated. The Mekong commission outlasted the American war. It received a new lease of life in the 1990s and now stands as one of the most important, if not always the most effective, river-regulating bodies in the world.

VII

The "multiplier" that ECAFE invoked to justify its work on cross-border river valley development could have the opposite effect: as projects and ambitions escalated, so did the potential for conflict. After a high point of warmth in their relationship in the mid-1950s—the era of "Hindi-Chini *bhai bhai*" (India and China are brothers)—the territorial conflict between India and China intensified. The unmarked and mountainous frontier between India and China became contentious as both states intensified their presence in the borderlands. New infrastructure brought border regions within easier reach of Beijing and Delhi; military forces were stationed there; migration from the plains brought new settlers, often ethnically distinct from the people who inhabited the uplands. The spark for conflict was the construction of a Chinese road linking Xinjiang and Tibet—a road that passed within what India considered to be its territory. Indian intelligence did not find out about the road until 1957, by which time its construction was well advanced. India insisted on the sanctity of agreements made under the British; the Chinese charged that India now stood as the beneficiary of British imperial aggression. In a pained and lengthy letter to Zhou Enlai, which was later published by the Indian government, Nehru countered that "the boundaries of India were . . . settled for centuries by history, geography, custom and tradition." He turned, then, to water: "The water-parting formed by the crest of the Himalayas is the natural frontier" between India and China, "accepted for centuries as the boundary."[80]

Water was not, by 1960, perceived as a source of conflict, but recently declassified Indian sources show that there were fears about the future. Rumors were rife. In exile in India after the failed Tibetan uprising of 1959, the Dalai Lama raised the alarm in a public meeting. He charged that the Chinese state was "planning to build high dams across the Brahmaputra and Indus group of rivers in the Tibet region," and that the Chinese "had these schemes in view ever since they came to Tibet." He asked, pointedly, "how far such projects undertaken unilaterally would be in the interests of India." The Indian foreign ministry responded cautiously to a report on the Dalai Lama's speech. "We have . . . no information so far about any proposal of the Chinese government to construct dams across the Indus or Brahmaputra before the rivers leave Tibet," one official wrote, but he saw the "necessity of being alert in this manner." Indian officials were well aware that "there is a great fall in the Brahmaputra just before it enters India" with "potential for power and irrigation." But they were reassured by the thought that it would take "huge resources to make anything of it"; any plans the Chinese had "will certainly take a long time."[81] The Indian trade mission in Gyantse, Tibet—which clearly doubled as a source of intelligence—concurred. They reported to Apa Pant, the chief political officer in the Indian protectorate of Sikkim, that "construction of dams and reservoirs on the river is likely to involve huge resources including manpower, which the Chinese authorities will be able to utilize only after they have brought in large numbers of Chinese for settlement."[82]

The Indian Ministry of External Affairs was concerned enough to involve colleagues in the irrigation department. In a letter marked "top secret," K. K. Framji, chief engineer of India's irrigation department, reassured the foreign ministry that "substantial or imminent diversions by China for irrigation purposes in the Tibet region do not appear to be practicable." The construction of storage dams for power generation might even benefit India as they "would be helpful in mitigating floods in Assam or East Pakistan." But he then raised a darker prospect: "If the Chinese hydro-electric schemes are so projected as to divert substantial quantities of Brahmaputra flows away from the present course into adjoining valleys," this would be "a significant loss of valuable water resources to India, and even more so, to Pakistan." He concluded on a hopeful note: "No doubt we will be given timely information regarding any observed or reported activities towards any such diversion."[83]

Events soon overtook these concerns about the future of water. In 1959, India infuriated the Chinese government by granting asylum to the Dalai Lama; from that point, tensions on the border between India and China ran high. Both sides built up their military forces along the border; India pursued a "Forward Policy," stationing troops north of the McMahon line, the 1914 frontier that had marked the boundary between Tibet and British India. Taking the Indians by surprise, Chinese military forces launched attacks on both the eastern and western flanks of the border region on October 20, 1962. As the world was transfixed by the Cuban missile crisis, Indian and Chinese forces fought in the high Himalayas. But it was no contest: the Indian military was no match for Chinese forces, who won decisive victories. A month after the offensive, the Chinese declared a unilateral cease-fire and withdrew their forces to the "line of actual control"—or the de facto border.

The war with China marked a humiliating defeat for India. The Indian army was ill-equipped, ill-prepared; China's invasion seemed to mock the effort Nehru had put into fostering good relations between the two countries. Nehru's political legitimacy at home was battered. With hindsight, 1962 appears as the beginning of the end of the Nehru era in Indian politics. In the opening pages of his first novel, *Such a Long Journey*, set in Bombay of the 1960s and 1970s, Rohinton Mistry evokes a widespread sense that "the war with China froze Jawaharlal Nehru's heart, then broke it. He never recovered from what he perceived to be Chou En-lai's betrayal."[84] Nehru's frank, even desperate, plea for American military assistance in the war dented his commitment to nonalignment in the Cold War. India's defeat on the international stage coincided with a rising chorus of criticism at home, raising questions about the economic and political strategy Nehru had pursued in the fifteen years since independence. Was there a better way?

JAWAHARLAL NEHRU DIED IN 1964. HIS WILL AND TESTAMENT EXplained why he wanted his ashes to be scattered in the Ganges upon his death. "I have been attached to the Ganga and the Jumna rivers in Allahabad ever since my childhood," he wrote, "and, as I have grown older, this attachment has also grown." The Ganges, he wrote, "is the river of India, beloved of her people": bearer of "her hopes and fears, her songs of triumph, her victories and her defeats." Evoking the identity of the river with the very geography of India, Nehru wrote that each glimpse of the Ganges "reminds me of the snow-covered peaks and the deep valleys of the Himalayas, which I have loved so much, and of the rich and vast plains below, where my life and work have been cast." Nehru was adamant that "my desire to have a handful of my ashes thrown into the Ganga at Allahabad has no religious significance." Water still had imaginative power to evoke the sacred, to shape nations' perceptions of their limits. The Ganges remained the essence of India, the Himalayas India's natural boundary, even in an age when India's "new temples" were large dams.

EIGHT

THE OCEAN AND THE UNDERGROUND

IN *THE LIVES OF OTHERS*, SET IN BENGAL IN THE LATE 1960S AND early 1970s, Neel Mukherjee brings a novelist's imaginative sympathy to evoke what it feels like to depend on the rains. Supratik, the novel's protagonist, has become involved in the Naxalite movement, a violent revolutionary movement of Maoist inspiration, committed to rural revolution in India. As part of his work of political outreach, Supratik, a privileged city boy, spends time in the countryside, growing accustomed to its rhythms. He inhabits a rural Bengal that is impoverished, indebted, under the heel of landlords—and governed by the monsoon. As they prepare the seedbeds to transplant paddy saplings, Supratik's host, Kanu, gazes at the sky. His questions are perennial ones: "Would it arrive this year? his eyes seemed to be asking; would it be late? would it be enough? There was both anxiety and resignation in his face." And then, this time, the rains came. Supratik continues: "It was exactly as I remembered from childhood—sheets of water coming down for hours and hitting the ground with such force that you thought the road would dissolve—except here the ground, which is earth, does dissolve."[1]

On the surface there is little to connect this imaginative account of the lived experience of climate with the meteorological data being accumulated by international scientists at the time. Human lives and voices could not be fed into climate models, where readings of pressure and wind and moisture could. But they told different facets of the same story.

This chapter tells two intersecting stories—the connections between them were not fully evident at the time. The first is the story of monsoon science in the 1960s; the second is the political and economic history of India's mid-1960s droughts. The International Indian Ocean Expedition ran between 1959 and 1965. It used new technology to uncover the forces underpinning the monsoon; it re-situated South Asia in relation to the vastness of the Indian Ocean; it established new links between the countries along the ocean's rim, including a web of weather-monitoring stations. The scientific expedition raised the alarm that human activity was starting to make itself felt in the oceans—perhaps even that it was altering Earth's climate. At just the moment when Asian states were sloughing off the web of connections that linked them across borders and seas, satellites and aerial photographs and deep sea probes projected a view of Asia shaped by a vast, connected climate system—a system with very tangible consequences for the large development schemes that states around the Indian Ocean rim had embarked on.

In the same years the monsoon came urgently back into view in India, which was in the grip of drought for three pivotal years in the mid-1960s. India's enthusiasm for the Indian Ocean Expedition, in common with many other countries in the region, was driven primarily by short-term concerns with a looming food crisis, which threatened political unrest. The Indian Ocean Expedition promised a survey of the sea as a set of "material resources to be exploited."[2] Most of all, it promised more accurate weather forecasts. India's response to the crisis of the mid-1960s was to intensify its quest for water. The dam building of the 1950s had not gone far or fast enough. Old fears of the monsoon climate resurfaced, regardless of the advances in climatic understanding that the new science promised. The government adopted a package of agricultural reforms that included high-yielding crops, vast quantities of chemical fertilizer, and the more intensive exploitation of groundwater using electric pumps.

IN HIS 1981 PRESIDENTIAL ADDRESS TO THE AMERICAN HISTORIcal Association, on "The Challenge of Modern Historiography," historian of the Atlantic world Bernard Bailyn spoke of the relationship between what he called latent and manifest processes in history. The former he described as "events that contemporaries were not fully or clearly aware of . . . however much they might have been forced unwittingly to grapple with their consequences." He described the relationship between latent and manifest events like this:

> The events I am referring to were known, if at all, only vaguely by contemporaries or by previous historians to have *been* events; they are being discovered as particular happenings now for the first time. Taken together, they form a new landscape, a landscape like that of the ocean floor, assumed to have existed in some vague way by people struggling at the surface of the waves but never seen before as actual rocks, ravines, and cliffs. And like the newly discovered ocean floor— so rich, complex, and busy a world in itself—the world of latent events can be seen to be part of, directly involved with, the manifest history of the surface world itself.[3]

Bailyn's oceanographic metaphor is especially apt in this chapter, where it takes on literal as well as symbolic meaning. We can now, for the first time, integrate the discovery of the Indian Ocean's effects on the monsoon, the early signs of climate change, with the manifest political and economic transformations of India and other parts of Asia in the 1960s and 1970s.

The lessons of the new climate science—that Asia was intensely vulnerable, increasingly interconnected, bound by growing risks from the destabilization of its climate—went unheeded before a renewed quest to conquer nature. The 1960s and 1970s were the decades that pushed India and other parts of Asia more fully toward a crisis of water.

|

By the 1960s the Indian Ocean was largely invisible to states in South Asia who looked no further than the waters immediately off their coasts. Migrants had once traversed the sea with few restrictions, in a pattern of circular migration. Now they faced an obstacle-strewn space governed by passports and visa restrictions.[4] As India prepared for the International Conference on the Law of the Sea, to be held in Geneva under the auspices of the United Nations in early 1958, it was clear to Indian negotiators that the meeting would have what they called "far-reaching consequences." At stake was the renegotiation of the customary three-mile limit on the extent of each state's "territorial waters," a legal conception that came into widespread use around this time. Lawyers at India's Ministry of External Affairs pinpointed the core conflict: "The countries which support the three mile limit," they noted, "own over 80% of the world tonnage and are therefore interested in maintaining freedom of the seas." India, on the other hand, along with many developing countries, claimed a greater expanse of water of its coasts "on the ground of security or for economic reasons such as the preservation of exclusive fishing rights for their nationals." Among the pressing reasons for a change were technological developments that allowed for the exploitation of resources further offshore, not least fishing by large trawlers. India was equally concerned with the "conditions under which the waters of a bay can be regarded as internal waters," given the "close linking of the waters to the land domain" and the "utility of the bay to the economic needs of the country."[5] Although negotiations over the UN Convention on the Law of the Sea would continue through the early 1980s, the final agreement recognized the claims of countries pressing for an extended definition of territorial waters, to which was added a wider exclusive economic zone. The sea came, more and more, to resemble the land—as a form of territory.

To scientists, the Indian Ocean was "the largest unknown area on earth." Paul Tchernia, who worked in the physical oceanography laboratory of the National Museum of Natural History in Paris, described it as the "forlorn ocean." Returning from a voyage through the Indian Ocean en route to and from the Antarctic, he suggested that an international investigation of the Indian Ocean should be incorporated into the activities of the UN's International Geophysical Year in 1957–1958: a massive exercise in coordinated data gathering that transformed knowledge of Earth's physical processes. Tchernia's suggestion came too late to include the Indian Ocean in that giant program, but there was a convergence of interest in investigating the least well-studied among the world's oceans. The catalyst came from a meeting of the Special Committee on Oceanic Research, set up by the International Council of Scientific Unions. It met for the first time at the Woods Hole Oceanographic Institution in coastal Massachusetts in August 1958. Among its champions was the Scripps Institution of Oceanography's Roger Revelle (1909–1991), a pioneer in the study of global warming and the effects of carbon emissions on the oceans.[6]

Midcentury oceanographers were drawn to the Indian Ocean for the same reason that medieval traders could cross it—the seasonal reversal of the monsoon winds. This pattern of reversing winds made the Indian Ocean unique; this made it a "model of the world ocean," upon which scientists could test their "wind-driven models."[7] Many scientists who lived on the ocean's rim, especially those in government service, had more immediate interests. The ocean's fisheries held the potential to address concerns about food shortages in Asia and Africa; its mineral wealth had barely begun to be exploited. Unlocking the mechanism of the ocean's influence on climate could provide the key to food security and economic development.[8]

From the start there was tension between short-term and long-term aims of the project; between the search for quick results and the patient accumulation of data on which to build scientific understanding. Oceanographer Henry Stommel approached the wild enthusiasm for the new Indian Ocean Expedition with skepticism. He published a few editions of an anonymous newsletter he called *Indian Ocean Bubble*—named to invoke the eighteenth-century speculative craze known as the South Sea Bubble—with the implication that his colleagues' craze for the Indian Ocean, too, was built on speculation. Its circulation was limited to a short list of oceanographers. Its final editorial was honest to a fault. "I think there is only a very remote chance that the Expedition will help improve fisheries and alleviate the poverty of the people in many Indian Ocean countries," Stommel wrote. He found it "disheartening" to see "oceanography join the long line of pressure groups acting—under the guise of humanitarianism—to advance their own interests." Those "interests" were "in themselves legitimate, but essentially unrelated to the moral and 'socio-economic' issues which they pretend to serve."[9]

In the end the International Indian Ocean Expedition involved forty ships from thirteen countries. Its capacious agenda encompassed what the mission's official chronicle called "moral and 'socio-economic' issues" as well as the "interests" of oceanography in basic research. The list of countries involved does not map easily onto the geography of the Cold War. Many large states bordering the Indian Ocean were enthusiastic participants in the program, including hostile neighbors India and Pakistan, as well as Indonesia and Australia. The United States played a leading role, involving scientists from the Scripps and Woods Hole institutes of oceanography, as well as the US Weather Bureau and the navy. The British, too, were heavily involved, given that they still had a substantial colonial and strategic presence in the region in the early 1960s. The Soviet Union contributed the 6,500-ton *Vityaz*, the largest ship in the program. The Indian Ocean Expedition also marked the rebirth of German oceanography after the war, and it showed the resurgent scientific and technical prowess of Japan, which contributed two vessels, the *Kagoshima Maru* and the *Umitaka Maru*.[10]

The Indian ships on the expedition reflected—in their origins, their shape, their materials—different epochs of seafaring history. The *Kistna*, its "sleek lines betraying . . . naval origins," as one observer put it, was built as a naval frigate in 1943, a product of the Second World War's fillip to Indian industry.[11] Now armed with an Edo echo sounder with a range of six thousand fathoms, the ship was fitted for oceanographic research, but it came with a warning: "Austere living conditions; not fit for women scientists. No salt water bath fitted." The smallest vessel in the expedition was the *R.V. Conch,* which belonged to the University of Kerala. It represented a much older tradition of shipbuilding: it was a small ship built of hardwood, in the long tradition of coastal craft that had threaded together India's western coast for centuries. By contrast the trawler *R.V. Varuna* was brand-new, purpose-built in Norway in 1961 in connection with the Indo-Norwegian fisheries project. Despite its novelty, it came with the same "men only" warning as the naval frigate: "Women scientists cannot be housed."[12] From the earliest days of the spice traders, the Indian Ocean was crossed predominantly by men—some things were very slow to change, and the loss to Indian Ocean science has been considerable.

The expedition's research aims encompassed the study of ocean currents and littoral drift; an investigation of ocean chemistry, salinity, and temperature; the exploration of marine life, and especially fisheries; the study of wind and atmospheric conditions and rainfall. Much of the excitement came from the new technologies that allowed scientists to see the sea anew. Sonar technology allowed them to hear enough to map the Indian Ocean's sea floor with heightened accuracy—their images evoked an underwater continent as varied in its topography as the land above. Advances in satellite technology provided synoptic pictures of cloud cover and precipitation. Computers allowed scientists to process quantities of data beyond all precedent: Klaus Wyrtki of the University of Hawaii oversaw the production of an oceanographic atlas, which processed data from twelve thousand hydrographic stations stored on two hundred thousand computer cards.[13]

Among all of the Indian Ocean Expedition's endeavors, one observer wrote, "none shows more contrast between past and present than meteorology."[14] The Indian Ocean Expedition marked the most intensive investigation of the South Asian monsoon since the days of Gilbert Walker, now with a raft of new tools. Fascinating though it was to glimpse the ocean's floor, for many scientists the most urgent priority for the Indian Ocean Expedition was to provide a better picture of Asia's climate. Almost a century after the establishment of the Indian Meteorological Department, scientists wrote in 1962 that "inadequate knowledge of the large-scale influences on weather have always hampered weather forecasting." The need to understand the monsoon "has become even greater and more urgent," they argued, "in view of the large scale development plans of many of the countries in the field of agriculture, exploitation of water resources, flood control programmes, and programmes for ameliorating the consequences of weather extremes." Economic planning, they wrote, demanded "accurate advance information on the onset of the rains, its variations from day-to-day" and "the occurrence of spells of heavy rain and breaks."[15]

In his 1927 presidential address to the Royal Meteorological Society, Gilbert Walker had speculated that "variations in activity of the general oceanic circulation" would likely be "far reaching and important" in understanding the world's climate.[16] It was not until the 1960s, bolstered by data collected during the International Geophysical Year and the Indian Ocean Expedition, that his insight would be developed. Walker's pioneering work on the lateral connection between the climates of Indian Ocean and the Pacific—his Southern Oscillation—now acquired a vertical dimension. The Indian Ocean Expedition focused on understanding the exchange of energy between the ocean and the atmosphere, driven by the monsoon winds. Piece by piece, scientists sought to understand the large-scale monsoon circulation of the Indian Ocean. The reversal in the direction of the monsoon winds had been well known for centuries, but it was more difficult, one scientist wrote, to "determine the vertical limits—than the horizontal—of the monsoon influence." Changes on Earth's surface were linked with changes in the deep sea, and in the upper atmosphere.[17]

A crucial component of the Indian Ocean Expedition was the International Meteorological Centre that was set up in Bombay in 1963, at the Colaba observatory, which was first built as an astronomical observatory by the East India Company in 1826. Along with India, the project received support from Ceylon, Indonesia, Japan, the Malagasy Republic, Malaya, Mauritius, Pakistan, Thailand, the states of East Africa, the United States, and the United Kingdom. Its prized possession was an IBM 1620 "computor" (as the word was then spelled), financed by the United Nations Special Fund. The center's director was tropical meteorologist Colin Ramage. In 1958 Ramage moved from a position as deputy director of the Royal Observatory in Hong Kong—where he had studied the South China Sea's typhoons—to a professorship at the University of Hawaii at Manoa, where he directed a US Air Force–funded research station on meteorology. "Just as every viewer has his personal rainbow," wrote Colin Ramage, "so every meteorologist seems to possess a personal and singular understanding of what is meant by 'monsoon.'" The one point of agreement, Ramage noted, was that the Indian monsoon is the largest and most dramatic.[18] For all the international involvement, the core of the International Meteorological Centre's personnel came directly from the Indian Meteorological Department, which contributed one hundred staff.

Already during the UN's International Geophysical Year in 1957–1958, Indian meteorologists had made a signal contribution. Among them was Anna Mani. Born in 1918 in the princely state of Travancore, Anna Mani studied physics at Presidency College Madras and then worked at Nobel laureate C. V. Raman's laboratory at the Indian Institute of Science, Bangalore. In 1945, she received a scholarship to Imperial College—I heard many stories of Anna Mani from the Indian meteorologists I met, including a story, perhaps apocryphal, of how

she endured her voyage to Southampton as one of the few women on
a ship full of demobilized troops. Mani joined the meteorological department in 1948, and during the International Geophysical Year, she
took charge of a network of stations to measure solar radiation across
India.[19] Just the year before the International Meteorological Centre was established, India had augmented its research capacity with
the establishment of the Indian Institute of Tropical Meteorology in
Poona, which remains one of the country's preeminent institutions.
Under the Indian Ocean Expedition, those most intimately familiar
with the monsoon now contributed to research on a global scale.

In a pamphlet published by the World Meteorological Organization, Ramage described the Colaba center at work:

> Throughout the night, staff in the small, air-conditioned communica-
> tions room have been receiving broadcast coded weather reports from
> the Indian Ocean region in morse code and on teleprinters. Pictures of
> charts analysed a few minutes before in the meteorological centres at
> Nairobi, Moscow, Sangley Point and Canberra unroll from facsimile
> printers. Across the compound of Colaba Observatory in the Signal
> Office of the western Regional Meteorological Centre, other teleprint-
> ers disgorge figure-crammed sheets of paper containing detailed infor-
> mation on Indian weather, and on the weather over the whole eastern
> hemisphere north of the equator.[20]

In this scene the Indian Ocean came alive, where as a zone of trade
and as a political idea it was dead. The sea was connected in a new
way by weather maps and the flow of data through fascimile printers.
In the late nineteenth century, the expanded collection of data trans-
mitted through the telegraph allowed the first synoptic weather maps
of large regions to be drawn. The Colaba center's work reflected a
new geography as well as advances in technology. Overlaid on the
old British imperial networks of weather reporting were new centers
of knowledge and power, including Moscow and even Vladivostock.
This nocturnal hive of activity, in a small corner of south Bombay,

gave substance to the idea of the Indian Ocean as a vast weather system stretching beyond national boundaries. Not all of the exchange was unfettered. Thousands of reports came in, but "problems in radio transmission meant that the centre got less than half the observations made." Nevertheless, Ramage was reassured, "copies of all observations were sent by mail," furnishing a paper archive of minute observations of the Indian Ocean's climate, even if, by the time they arrived, "they were not much good for forecasting." More disappointing was the fate of a floating automatic weather station provided by the US government and anchored in the Bay of Bengal by the Indian navy: "After a few months, its radio quit and it was neither seen nor heard from again."[21]

The new quest to understand the monsoon relied on two breakthroughs in technology: aerial video and satellite photography. Five research aircraft were based in Bombay to support the International Meteorological Centre's work: one belonged to Woods Hole, the other four to the US Weather Bureau, including two large DC-6 airplanes. As in the nineteenth century, cyclones held a particular fascination for meteorologists—and now improved storm forecasting would benefit the growing numbers of people who lived in South Asia's coastal cities. The US Weather Bureau's two DC-6 aircraft flew the first aerial reconnaissance mission into a cyclone in the Indian Ocean (though there had been many such missions over the Atlantic): Ramage was in the scientific observer's seat on one of them. Ramage's plane flew through the storm at 20,000 feet; the other was way down at 1,500 feet above the ocean's surface. "I thought the aircraft was falling to pieces," Ramage wrote, "we dropped 300 feet in a single second"—a common complaint of nervous flyers in turbulence, but in Ramage's case it was likely accurate. Most awesome, for this lifelong student of tropical storms, was the experience of flying into the eye of a monsoon depression. He described flying "into an amphitheatre of multi-layered nimbo-stratus cloud. In the centre only thin milky cloud above us and almost none below, and five minutes later we were once again in rain clouds."[22]

When Ramage and Indian meteorologist C. R. Raman (brother of Nobel laureate in physics C. V. Raman) produced their atlas of Indian Ocean meteorology, they were able to draw 144 charts based on 194,000 shipboard observations and 750,000 balloon ascents into the upper atmosphere. The ability to visualize the weather in new ways was especially compelling. Scientists installed cameras on commercial and military aircraft flying through the region to take time-lapse films of the clouds they encountered in flight over the Indian Ocean. Ramage described how time-lapse cameras had previously been used "spectacularly in science films to compress into a few seconds the blooming cycles of flowers"; now they were placed on aircraft to photograph clouds at the rate of one frame every three seconds. On a six-hour flight, they would record every cloud on thirty meters of 16 mm film. Watching the videos later, Ramage wrote, "The viewer gets the rather exciting impression of flying at about 50 times the speed of the aircraft."[23]

An even more promising development was underway by the end of the expedition—daily satellite photographs of the Indian Ocean. "We now have for the first time," Ramage enthused, "the opportunity to attempt a complete description of the whole atmospheric distribution over the Indian Ocean." That "complete description" is what Ramage and Raman attempted, chart by chart, in their meticulous work of climatic reconstruction. The most exciting prospect lay just over the horizon of possibility. The force of the monsoon came from the exchange of energy between air and sea—it was now possible to study this complex process. The promise of satellite photography was that it might "elucidate the role of the monsoons in the total atmospheric circulation."[24]

Despite his optimism, Ramage delivered a modest assessment of progress. The goal of Henry Blanford and Gilbert Walker—a long-range forecast of the monsoon—remained elusive. Even incremental improvements in forecasting, Ramage thought, would help in "aiding flood prevention and control and in enabling irrigation engineers to make the best possible use of stored water," as well as helping the fishers of the Indian Ocean rim to take advantage of lulls in the monsoon, up to a week long, to take to sea. But the prospect of an accurate long-range forecast, Ramage lamented, was "as remote as ever." The vast accumulation of data had not altered a truth well known—that "the atmosphere is turbulent and chaotic." There was no substitute for patient observation. The best meteorologists could do was to keep doing what was embedded in their practice: to use "long climatological records and detailed statistics to come up with a sort of odds on what the next season's rainfall will be." His conclusion was sober:

> The apparently rhythmic nature of rain-and-break, rain-and-break, during the summer monsoon encourages us to delve more deeply into the underlying causes of the rhythm and in particular the causes for interruptions or changes in the rhythm. Finding the rhythm of a total season, however, seems almost certainly beyond our immediate grasp.[25]

MORE OMINOUS SIGNS EMERGED FROM THE INDIAN OCEAN Expedition. Two years before that expedition began, Revelle had written, with his colleague the geochemist Hans Seuss, that human beings were conducting, unwittingly, a "large scale geophysical experiment" with the world's climate. "Within a few centuries," Revelle and Seuss wrote, "we are returning to the atmosphere and oceans the concentrated organic carbon stored in sedimentary rocks over hundreds of millions of years."[26] One of Revelle's students, Charles Keeling, was the first to start systematic measurements of atmospheric carbon the following year, in 1958. Revelle and colleagues had long-range goals for their study of the Indian Ocean: they wanted to see how far the Indian Ocean was a "dump for the waste products of industrial civilization." And they sought to determine "the role of the ocean in climatic change, especially in absorbing the carbon dioxide spewed into the atmosphere when fossil fuels are burned."[27] We have forgotten how important the Indian Ocean was to documenting anthropogenic climate change, prompting early stirrings of alarm. The data from the Indian Ocean voyages suggested that the sea and the atmosphere were being affected by human activity on land. But these "long range" problems were then distant from the level of human experience. The time horizons of oceanic research were incommensurable with those of planning for food security. Because the long-range monsoon forecast remained elusive, because understandings of climate grew more complex, it was easier to focus on what could be contained and controlled—one river valley at a time.

The Indian Ocean Expedition generated a picture of the South Asian monsoon that was an integral part of the global climate system. South Asia's climate was part of a large-scale interchange of energy and moisture between the ocean and the atmosphere. This expansion in the scale and complexity of understanding sat uneasily with the confidence, so prevalent in Asia among dam builders and planners, that nature could be anticipated, or its effects engineered away. A 1968 pamphlet on the monsoon—written by P. K. Das, director of the Indian Meteorological Department, and published as part of a National Book Trust series, "India: The Land and People"—contrasted the tendency of Indian farmers and poets to see the monsoon in vast, even cosmic, terms with "more rational techniques, such as the scientific control of river valleys." In a sense, with its emphasis on turbulence and chaos and complexity, the new monsoon science resonated more easily with what Das saw as the superstitious view than with a view of the monsoon as simply a variable to be controlled.[28]

II

The final year of the International Indian Ocean Expedition coincided with the worst failure of the South Asian monsoon in decades. For two successive years, in 1965 and 1966, large parts of India suffered from drought. The drought coincided with India's first major political transition after independence, following the death of

Jawaharlal Nehru at the end of May 1964, and it pushed forward a change in economic strategy that was already underway. The two years after Nehru's death made clear the limits of India's progress toward self-sufficiency in food.

Nehru was succeeded by the diminutive and mild-mannered Lal Bahadur Shastri, a party stalwart from the Hindi heartland of North India. When Shastri took office, alarm about India's food situation was widespread. There had been many flare-ups of protest in the cities, dubbed "bread riots" in the media. The Congress party's distinctive style of accommodative politics came under strain after the disaster of the China war. Groups that had long formed part of the Congress coalition—middle- and upper-caste landowners, urban workers, industrialists—were no longer content to defer to the urban elite, no longer willing to keep a lid on their conflicts with each other. Some of them found a voice outside the Congress umbrella. Under attack from left and right, Shastri, who was easy to underestimate, undertook a series of quiet but decisive reforms. Political scientist Francine Frankel, who was doing research in India at the time, describes the result: "A series of undramatic initiatives in economic policy that went virtually unnoticed at the time cumulatively altered the entire approach to India's development strategy."[29]

The core of Nehru's approach had been a push for import-substituting industrialization. India in the 1950s had developed as a mixed economy, but one in which the public sector played a leading role, especially at the "commanding heights" of the economy. This was the strategy championed in the 1950s by the planning commission, to which Nehru gave considerable autonomy under the leadership of master statistician (and sometime meteorologist) P. C. Mahalanobis, dubbed "the Professor." The Indian countryside was given two roles to play in this "drama," as Mahalanobis insisted that the second five-year plan must be: The first was to ensure food security to a country still scarred by the memory of colonial famines, with the ultimate aim being self-sufficiency in food grains. The second was to generate foreign exchange through the export of nonfood crops to pay for the imported machinery that India would need until its own factories could make them. Jute and cotton were two of India's most valuable exports.

But at odds with the emphasis on self-reliance came an increasing reliance, through the 1950s, on food aid from the United States. From the time of its institution in 1954, Public Law 480, or PL-480—known widely as the "food for peace" scheme—disposed of the large agricultural surpluses of the American Midwest in the postcolonial world, on preferential terms. India was by far the largest recipient of this aid. Indian imports of American wheat grew from two hundred thousand tons in 1954 to more than 4 million tons by 1960. Given the ups and downs of the Indo-American relationship in the 1950s, this struck many Indian politicians as an uncomfortable level of dependence on an unreliable patron. The importance of American food aid was only one indication that the Indian government's agricultural strategy had faltered. Looking back, in the late 1960s, an official report acknowledged the problem. "All the efforts at achieving self-sufficiency in foodgrain production during the three Plan periods did not fully succeed," the Indian ministry of agriculture acknowledged; instead, a sharp drop in agricultural production in the early 1960s came as a "great shock to everyone concerned with agriculture." After 1961, per capita income in India did not increase, and by the mid-1960s the availability of food per capita was lower than it was in 1956.[30]

The man charged with addressing this "shock" was C. Subramaniam. He was born in 1910 to a farming family in Coimbatore district in Madras, a prosperous region of irrigated export agriculture at the edge of the Western Ghat mountains and a center of India's textile industry. Subramaniam was a protégé of the veteran Madras leader of the Congress party, C. Rajagopalachari. He joined the Indian freedom movement as a young man; he was one of an army of Congress party workers imprisoned by the British during the Quit India movement in 1942. He spent the 1950s in the state government of Madras, until Nehru appointed him to the coveted industries portfolio in his cabinet. For Shastri to move him to agriculture, something of a Cinderella ministry, seemed a step down. But Subramaniam embraced the challenge. He came to represent a strand of economic thinking in India that had always run alongside, sometimes in tension with, the planning commission's "industry first" approach. Subramaniam believed, contrary to the planning commission, that the Indian countryside was the key to security and progress. Subramaniam drew on ideas that had been in circulation from the late nineteenth century. They were prominent in the writings of India's early "economic nationalists," and found new expression in the 1920s and 1930s in detailed studies of agricultural economics.[31]

In the circumstances of the early 1960s, a rediscovery of rural India's importance converged with a line of thought that American development experts pressed upon India. From the late 1950s the World Bank and many American government observers began to urge that India should pay more attention to agriculture, even at the expense of scaling back its grand industrial vision. Specifically, they advocated for markets to play a greater role in Indian agricultural policy, which would in turn spur investment in new technologies. They were skeptical of the Nehru government's emphasis on agricultural cooperatives; they argued, instead, that India could boost its food production most rapidly by providing incentives to farmers with capital, those who already benefited from larger landholdings and irrigation facilities, even if this came at the price of higher levels of inequality in the countryside. Wielding the powers of persuasion and veiled threat that they gained from India's dependence on American food aid, these outside experts gained a sympathetic hearing from within the Indian government, which had its own "America lobby" as well as pro-Soviet faction.[32]

From the outset Subramaniam believed that new technologies were the key to the transformation of India. One of his first initiatives was to strengthen the moribund Indian Council for Agricultural Research, and to bolster agricultural education in India. Early in his tenure as agriculture minister, Subramaniam was impressed by reports of the stunning results shown by Rockefeller-sponsored experiments in Mexico with high-yielding varieties of maize and wheat, and by experiments with new hybrid strains of rice in Taiwan and the Philippines. Could they work in India? A team of scientists led by the Canadian plant pathologist R. Glenn Anderson had already initiated a series of experimental stations in India with pilot projects in Delhi, Ludhiana, Pusa, and Kanpur; when one hundred kilograms of seed, flown in from Mexico, arrived in India in 1964, they were ready to test them in Indian conditions. Here lay the roots of what would come to be known as the "Green Revolution," which would transform Indian and global agriculture in the final third of the twentieth century.[33]

But first Subramaniam had to prevail over his cabinet colleagues who remained committed to the Congress's stated goal of moving toward a "socialist pattern of society"—or, at least, to rapid industrialization first and foremost. T. T. Krishnamachari was the planning commission's most vocal proponent of focusing on heavy industry. His concern was primarily with keeping food prices down for urban workers. To achieve this, he argued for a national food distribution system based on a system of price controls and monopoly food procurement by the state. Subramaniam pointed out that this infrastructure of food control, which had its roots in the wartime economy, was "uneconomical"—government prices for compulsory procurement were so low that they gave farmers no incentive to invest in new technology. The two sides crossed swords over inequality. Subramaniam recalled in his memoirs that his opponent argued against high-yielding varieties because "this strategy would lead to greater social tension within the rural areas, because benefits would be unequally distributed." Subramaniam's response was to ask his critics "what other option we had." To those who argued that he was caving to American pressure, Subramaniam countered that only a new approach to agriculture would save India from subjection, for he feared that "once we became dependent on these imported foodgrains other political strings would be attached to them."[34] The turn to high-yielding varieties made Indian agriculture dependent, instead, on large imports of chemical fertilizer—and more dependent than ever on new sources of water.

The crux of Subramaniam's strategy was to "concentrate modern inputs in irrigated areas."[35] This is a prosaic way to describe a fundamental change. From the nineteenth century India's geography of water had shaped plans for the country's future. Now the difference between irrigated and rain-fed lands would be accepted as a necessary inequality—even a matter of strategy.

SUBRAMANIAM'S APPROACH TO AGRICULTURE GAINED A FILLIP from the overwhelming sense of economic and political crisis surrounding India in 1965. By September 1965 it was clear that the year's summer rainfall was far short of normal: aggregate agricultural production was 17 percent down on the previous year. In a speech to persuade the chief ministers of India's states of his policy, Subramaniam spoke of a "race against time." Facing down criticism from the Communist H. N. Mukherjee in Parliament, who accused Subramaniam of forcing through his strategy under American pressure, the agriculture minister accused his questioner of exploiting "the psychological hour when the monsoon has failed," and preying on the "fears" of the country.[36] India's vulnerability to the monsoons, it seemed, was as deep as ever. "How helplessly we are at the mercy of the elements," a newspaper editorial lamented in 1965, arguing that all India had to show for the previous decade of development efforts were some "shallow and tentative improvements in irrigation."[37]

Economic crisis merged with political turmoil. A series of military skirmishes on the border escalated into war with Pakistan, from August 5 to September 22, 1965. The catalyst was Pakistani military infiltration into Kashmir, undertaken in the hope of sparking a rebellion against Indian rule. Both sides claimed victory after a UN-brokered cease-fire; in contrast with the China war, however, the

Shastri government's military campaign was greeted as a success at home. Shastri urged struggle on both fronts with his resonant phrase: "Jai Jawan! Jai Kisan!"—victory to the soldier, and victory to the farmer. An immediate consequence of the war was the abrupt cessation of American aid both to India and to Pakistan.

US president Lyndon Baines Johnson had already assumed direct charge of PL-480 shipments; with characteristic bluntness he called it the "short tether." He would ship only enough grain to India to meet immediate needs: a direct and unabashed form of political leverage, which angered the Indian government. Subramaniam played an essential role in negotiations over restarting the shipments after the war. He had developed a reputation in Washington as an Indian leader favorable to the United States, and whose policy views tallied with American advisors' ideas. Subramaniam forged a rapport with Orville Freeman, Johnson's secretary of agriculture. The two men met at a UN Food and Agriculture Organization meeting in Rome in November 1965 and signed a secret agreement to accelerate India's agrarian reforms in exchange for expanded American food aid. Taking over as prime minister after Shastri's sudden death in Tashkent, where he had gone to sign the peace treaty with Pakistan, Nehru's daughter Indira Gandhi visited the United States in 1966 to cement this deal. She received a warm reception from LBJ, but she felt deep humiliation at having to go to the United States as a supplicant. "I don't ever want us to beg for food again," Indira Gandhi told an aide in December 1966.[38]

When the summer monsoon failed for a second successive year, in 1966, India's food situation worsened. The state of Bihar, one of India's poorest, was most directly affected. Facing Republican congressmen who were hostile to aid and hostile to India, the Johnson administration emphasized the scale of the food emergency in India— evoking the specter of famine for the first time since independence. In order to push through a bill boosting aid to India, Johnson told Freeman that he wanted the American public to know "that people were being hauled away dead in trucks, and that they needed food."[39] By contrast Indira Gandhi's government chose not to declare a famine

Indira Gandhi in January 1967, soon after she became India's prime minister. CRED-IT: Express Newspapers/Getty Images

in Bihar for fear of the domestic political fallout. Instead, the central government dismissed early reports of starvation from Bihar as hyperbole—just as the British imperial government had been wont to do. Famine was too much a symbol of a dark past, its conquest too vital to the political legitimacy of the Indian state, to concede this, the first famine since India's independence. But the scale of suffering in Bihar threatened to explode and on April 20, 1967, the government of India declared the existence of famine in Palamau and Hazaribag districts; five more districts were added in time, along with others that suffered "scarcity." The nineteenth-century Famine Codes, revised incrementally over the years, came into effect. The full force of the state swung into action to counter food shortages in Bihar. PL-480 shipments from the United States were vital, distributed in twenty thousand fair-price shops. In keeping with traditional practice, the government initiated public works on a large scale to provide employment to augment local incomes and to encourage food imports from other regions of India. Under the leadership of the socialist Jayaprakash

Narain, the Bihar Relief Committee mobilized an army of volunteers, as well as donations. A year before the Biafra crisis of 1968, often held up as a watershed in the development of a global humanitarian consciousness, the Bihar crisis reached the wallets, and the television screens, of a public far away.[40]

By most measures, the Indian and American response to dearth in Bihar was a success. Despite major food shortages, far fewer people died than during the nineteenth-century famines; the official death toll was in the region of 2,300 people. Even if this was an underestimate, the contrast with India's experience of famine under colonial rule was stark.

But the story has a strange and unexpected coda. Housed in the Johnson Presidential Library in Austin, Texas, is an innocuously named box of archival files: "India Memos and Misc., 1 of 2, volume 8." I learned about its contents in a fascinating article by two historians of Cold War science, Ronald Doel and Kristine Harper. Doel and Harper argue that the environmental sciences were vital to the Johnson administration's foreign policy, and to its projection of American power overseas.[41] From the 1950s, the deepening American involvement in Vietnam had generated a deepening interest in the hydrology of the Mekong River. As the US military intervened more intensively in Vietnam after 1962, mastering nature became strategically vital. American medics experimented with new drugs for the control of malaria, a prerequisite for jungle warfare: they devised mefloquine. Some went further: they had visions of intervening to alter patterns of rainfall to disrupt the agricultural base of North Vietnam. The United States had seen a long series of attempts at weather control in the twentieth century, a history characterized more by outlandish schemes than by any measurable success.[42] Now it became a plank of military and diplomatic strategy. The Bihar drought occurred just when secret American plans for weather control in Vietnam were in testing. President Johnson connected the two, as he became the most unlikely of experts on the South Asian monsoon. Johnson would write in his memoirs that, as he looked at weather charts before approving each PL-480 shipment to India, he came to know "exactly where the rain fell and where it failed to fall in India."[43]

In January 1967, Pierre St. Amand arrived in Delhi with others from the Naval Ordnance Test Station on a highly classified mission; indeed, only a decade ago did the work of Doel and Harper bring it to light. Nicknamed Project Gromet, the scheme—with the secret approval of Indira Gandhi's government—aimed to inject silver iodide into "large, high-altitude cold clouds" to force precipitation. Official acknowledgment of the program came in the form of a sly and prosaic statement: "Scientists from the United States and India are cooperating in a joint agro-meteorological research project, localized in Eastern U.P. [Uttar Pradesh] and Bihar to study the cloud physics and rain producing mechanisms over these areas of India which have incurred several droughts during the last few years." The problem was that, in January, the skies over Bihar were virtually cloudless. They expanded their quest toward Punjab. US ambassador Chester Bowles, an Indophile, wrote in secret that "both we and the Indians want to demonstrate that if we can [force precipitation] India's food and agriculture need not be entirely at the mercy of weather vagaries." Soon after that, the archival trail runs cold; "GROMET quietly died," its historians conclude.[44]

Doel and Harper were interested in Project Gromet primarily for what it tells us about US foreign policy under Johnson. Seen in the light of India's long history of struggle with the monsoon, it acquires other layers of meaning. In a sense, Project Gromet was the antithesis of the Indian Ocean Expedition. Where the new science emphasized the complexity of the monsoon climate, rooting it in teleconnected land-ocean-atmosphere interactions on a planetary scale, the attempt to make rain in Bihar epitomized the logic of control and containment. Even then, the architects of the plan feared the consequences of engineered rain in India having an unwanted impact across the border in Pakistan.

THE YEAR 1967 WAS A TURNING POINT IN INDIA'S POLITICAL HIS-
tory. Having ruled India with a comfortable parliamentary majority
since independence, the Congress party took a drubbing at the polls
in India's third general election. While they remained in power in
New Delhi, their majority was reduced—more significantly, they lost
control of state governments in states including Tamil Nadu (to the
regional party born of the anti-caste movement, the Dravida Mun-
netra Kazhagam), Kerala (to the Communists), and West Bengal (to
a coalition including the Communist Party). A generation after independence, the mantle of Mahatma Gandhi and Jawaharlal Nehru
was no longer enough to muster support for the party that had led
the nationalist movement. The coalition of social forces that underpinned Congress dominance was fragile; in many places it unraveled.
A confidential assessment by the British High Commission in Delhi
reported to London that the election results reflected "impatience
with the chronic failure to deal with rising prices, low wages [and]
food shortages, amounting to famine in certain areas."[45]

Occasionally, conventional accounts of this political moment include climate among their explanations for the change. India's political
and economic transformation in the 1960s, writes political scientist
Ashutosh Varshney, owes much to the "serendipity of the monsoon."[46]
In recent years, it has become common once again for historians to
see climate as a force underpinning political events.[47] But to think of
India's political transformation as "caused" by the failure of the monsoon would be unduly simplistic. The outcome of the 1967 elections
reflected the hopes of India's voters; it was the outcome of new languages of political mobilization deployed by India's parties; it distilled
new struggles for power and justice and recognition unleashed by
mass democracy, which could no longer be contained within a dominant party system. A richer picture emerges if we see climate not as an
external force determining human outcomes, but rather as a source of
all-too-human fears and anxieties. Only in the context of century-long
fears about India's monsoon climate—the deep historical association

of monsoon failure with famine—can we understand why so many ex-perienced the drought of the mid-1960s as evidence of political failure.

III

Indira Gandhi was one of few heads of state to attend the UN's first conference on the human environment, held in Stockholm in June 1972. In her rousing speech to the plenary session, she set out ecological problems that were already a matter of public discussion in India:

> We share your concern at the rapid deterioration of flora and fauna. Some of our own wildlife has been wiped out, miles of forests with beautiful old trees, mute witnesses of history, have been destroyed. Even though our industrial development is in its infancy, and at its most difficult stage, we are taking various steps to deal with incipient environmental imbalances. The more so because of our concern for the human being—a species which is also imperiled. In poverty he is threatened by malnutrition and disease, in weakness by war, in richness by the pollution brought about by his own prosperity.[48]

But her diagnosis of the root cause differed from that of many of the conference's promoters, whose vision was consumed by dark Malthusian nightmares in the Third World, epitomized by biologist Paul Ehrlich's *Population Bomb,* published in 1968. The opening lines of Ehrlich's book described a "stinking hot night" in Delhi. "As we crawled through the city, we entered a crowded slum area . . . the streets seemed alive with people," he wrote: "People eating, people washing, people sleeping. People visiting, arguing, and screaming . . . People, people, people, people."[49]

In response, Mrs. Gandhi set out a position that saw environmental degradation as primarily a problem of poverty: a problem of distribution, not of numbers. She reminded her audience that "we inhabit a divided world." She attributed historical responsibility for the despoliation of Earth where it rightly belonged: with the wealthy countries of the world. "Many of the advanced countries of today have reached their present affluence by their domination over other races and countries," she said, and through "the exploitation of their own natural resources." The rich world "got a head start through sheer ruthlessness, undisturbed by feelings of compassion or by abstract theories of freedom, equality or justice." But now the poor countries were being told that they could not do the same. "The riches and the labour of the colonized countries played no small part in the industrialization and prosperity of the West," she reminded her audience, but in the 1970s, "as we struggle to create a better life for our people, it is in vastly different circumstances, for obviously in today's eagle-eyed watchfulness we cannot indulge in such practices even for a worthwhile purpose." The development of a middle class in India, or even just the provision of minimally decent standards of living to its poorest citizens, took place with a growing awareness of finite resources. "We do not wish to impoverish the environment any further," she insisted, "and yet we cannot for a moment forget the grim poverty of large numbers of people." Her most resonant phrase, and the one for which her speech is remembered, was in the form of a question: "Are not poverty and need the greatest polluters?"

She raised the stakes, asking: "How can we speak to those who live in villages and in slums about keeping the oceans, the rivers and the air clean when their own lives are contaminated at the source?" She resisted the binary choice of development *or* environmental protection. "The environment cannot be improved in conditions of poverty," she declared, and "nor can poverty be eradicated without the use of science and technology." In "science and technology" lay India's great hope.

Gandhi concluded by describing to the audience how she saw India's quest since independence. "For the last quarter of a century," she said, "we have been engaged in an enterprise unparalleled in human history—the provision of basic needs to one-sixth of mankind within the span of one or two generations."[50] In this evocation of speed, urgency, and scale lay Gandhi's recognition of the demographic and material transformation that was sweeping the world. The edge of Malthusian panic remained, despite Indira Gandhi's eloquent rebuttal at Stockholm. In time just such a sense, that population growth was an exorable and destabilizing force, fed her own fears of conspiracy. In the context of labor unrest and judicial challenges to the legitimacy of her election victory, they underpinned the siege mentality that led her to declare a state of emergency in 1975, suspending India's democratic constitution for the first and (so far) the only time, using a colonial-era provision. It would lead Indira Gandhi's government to enact a brutal population control program involving gross abuses, including forced sterilizations.[51]

In Indira Gandhi's vision, many disparate concerns came together to constitute an overarching problem of "the environment"—population growth and the finitude of natural resources; concerns about the impact of rapid development on human health; concerns about species extinction and disappearing habitats. The international conference coincided with the earliest attempts to confront these challenges at home in India. Mrs. Gandhi's government sponsored the Water Act of 1974, which was among the first attempts to deal with an environmental issue on a national basis in India. The act created pollution control boards at both the state and the national levels; the boards were given the authority to determine permissible levels of pollution, setting limits on the composition and quantity of effluent that factories, for instance, could discharge into water bodies. The legislation took an expansive view of water, covering "streams, inland waters, subterranean waters, and sea or tidal waters." But the law proved, and it continues to prove, difficult to implement. Pollution control bodies did not have the will or the power to prosecute powerful local industrialists. The Water Act faced many challenges in the courts, often on the grounds that it violated the constitutional right to carry on a trade or business. For instance, in 1981 lawyers for a firm called Aggarwal Textile Industries, challenging a ruling from Rajasthan state's pollution control board, argued that "the problem

of prevention of water pollution is a problem of vast magnitude
and . . . it would be beyond the means of an individual to prevent or
control the pollution resulting from an industry set up by him." In a
significant number of cases, offenders were allowed to continue their
polluting activities.[52]

DESPITE DAWNING AWARENESS OF THE QUESTION OF SUSTAINABIL-
ity, the crux of India's response to the crisis of food and population lay
in a massive increase in the use of water. The early twentieth century's
faith in the possibility of expansion without limit was reinvigorated.
The precondition for the growth of the Green Revolution in India was
an expansion in irrigation. And the water available from surface irrigation works, however ambitious in scale, was insufficient. Cultivators
in arid parts of India had known for centuries, and British administrators recognized in the nineteenth century, that South Asia's groundwater resources provided the most immediate insurance against drought.
Many regions of South Asia had ancient and elaborate systems of well
irrigation, though this infrastructure had fallen into disrepair in many
places by the nineteenth century. The advantages of groundwater resources are manifest in South Asia: groundwater is locally available
on demand and requires far less infrastructure than surface irrigation
works; groundwater is spared the large-scale water loss from evaporation that reservoirs experience. Groundwater resources are more
resilient to episodic monsoon failure. Until the 1960s, groundwater
could not be mobilized on a large enough scale to meet India's food
requirements. The arrival of the tubewell changed that decisively.

Tubewells are a humble technology, unlikely harbingers of a hydrological revolution. A tubewell is a well driven by an electric pump,
consisting of a long stainless steel tube that is bored into an underground aquifer. Historian and architect Anthony Acciavatti argues that
they represent an "inversion" of the monumental water technologies:
dams and canals. In contrast with dams, the tubewell has a "minimal

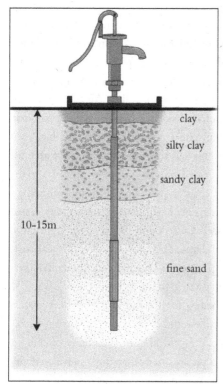

clay

silty clay

sandy clay

10–15m

fine sand

An electric tubewell, of the kind that proliferated in India in the 1960s. CREDIT: Illustration by Matilde Grimaldi

footprint and maximum draft of water, it creates undreamed of independence and three-dimensional chaos."[53]

The Indian government encouraged the adoption of high-yielding varieties of wheat and rice by subsidizing the capital costs of infrastructure for the intensive exploitation of groundwater. State electricity boards reduced the cost of electricity. By the 1970s, unable to bear the cost of monitoring energy use by millions of farmers across widely dispersed areas, state electricity boards opted for flat tariffs. This gave large farmers an incentive to use as much energy as possible to extract water from underground. As a result, agriculture's share of total energy use in India grew from 10 percent in 1970–1971 to 30 percent by 1995, even as state electricity boards accumulated huge losses. By 2009, groundwater accounted for 60 percent of India's irrigated area, and surface irrigation for only 30 percent.[54] Just as they shared the waters of the Indus, India and Pakistan came to share a new dependence on groundwater. Embracing the same package of hybrid seeds and intensive fertilizer use, Pakistan's food security became even more reliant on irrigation than India's. All the while, and despite the declining importance of surface irrigation, the profusion of large dams continued. Dam construction was unstoppable, even as underground water now supplied the greater share of water for irrigation. Large dams had acquired enormous symbolic power, to the point where they epitomized the conquest of nature by technology. Three decades after India's independence there were also many vested interests in the engineering and construction industries committed to the continued proliferation of dams. Their social and ecological costs multiplied through the 1970s; as we shall see in the next chapter, their costs provoked widespread resistance in the 1980s.

Utopian technological schemes for the capture of India's waters flourished in the 1970s, alongside—and as though unaware of—growing understanding of the scale, power, and unpredictability of climate. Although Roger Revelle was a pioneer of oceanography and an architect of the 1960s' Indian Ocean Expedition, he turned in the 1970s to more practical matters. Revelle moved from Scripps to Harvard to found the Harvard Center for Population and Development Studies. In 1975 he wrote an essay with his Indian colleague V. Lakshminarayana on what they called the "Ganges Water Machine." They expressed concern that "deeply embedded cultural, social, and economic problems inhibit modernization of agriculture and fuller utilization of water resources" in India. They envisaged that "the introduction of technological changes on the required scale might break the chains of tradition and injustice that now bind the people in misery and poverty." They had in mind a technological mirroring of the vast, interconnected hydraulic system that linked the monsoon rains, the Himalayan rivers, and the waters underground—an expanded network of bunds, dams, and, above all, the massive expansion of groundwater pumps. The same year, K. L. Rao, a veteran of Indian irrigation, published an even grander plan. He returned to the dream of Sir Arthur Cotton, irrigation pioneer of the nineteenth century, in proposing a large scheme to transfer water—through a network of canals—from the wettest to the driest parts of India.[55]

India's experience of water-driven growth in the 1970s found echoes across Asia. The 1970s also saw rapid growth in Chinese agriculture, as China developed its own path toward a green revolution. The unprecedented expansion in food production in China in the 1970s built upon the extension of agricultural research stations right down to the level of local communes. Like the Indian government, the Chinese state viewed the rapid growth in population and the pressure on arable land with alarm, reversing the pronatalism that characterized the first two decades after the revolution. As in India, Chinese farmers' adoption of high-yielding seed varieties depended on large quantities of chemical fertilizer. In the 1970s, the Chinese fertilizer industry expanded through a dispersed network of small-scale factories. High-yield dwarf rice varieties spread especially rapidly in China in the 1970s, boosting harvests. And in China, as in India, the agricultural growth of the 1970s depended on mining underground water. Electric pumps played almost as significant a role in expanding irrigation in China as they did in India. In 1965, there were approximately half a million mechanized irrigation and drainage devices in China; by 1978, there were more than 5 million.[56] But in other ways, the Chinese approach to growing more food diverged from India's. In keeping with the Maoist emphasis on mass political mobilization, China's agricultural strategy was more broadly based than India's. Historian Sigrid Schmalzer describes it as a "patchwork of methodologies," in which mechanization coexisted with labor-intensive terracing, chemical fertilizer with traditional practices of night-soil collection and the application of pig manure.[57] The water- and fertilizer-fueled growth of the 1970s laid the groundwork for China's further agricultural expansion in the 1980s; but with the end of agrarian collectivization and the arrival of market reforms, rural inequalities, too, grew wider.

IV

The water inequalities that India has always faced deepened in the 1960s and 1970s; they were accentuated by the uneven spread of tubewells. From the late 1960s, as the Green Revolution took off, the drier regions of India's northwest and southeast emerged as the centers of agricultural growth—a result of groundwater exploitation, fueled by electrification to allow the use of high-yielding seeds. The water-rich areas of India's northeast, by contrast, continued to rely on rainfall, utilizing relatively inefficient diesel pumps for shallow groundwater irrigation; they remained at risk of regular waterlogging, but lacked the infrastructure to use the surplus water for storage or groundwater recharge.[58] Where large dams promised to create energy through hydroelectric power, pumps used it in large quantities to mine water.

These inequalities were manifest between 1970 and 1973, when parts of western and central India experienced three successive years of drought. The western state of Maharashtra was worst affected. In the World Bank's archives in Washington is a fifty-page typescript account of the Maharashtra drought; scrawled at the top is a handwritten instruction: "Circulate." The author was agricultural economist Wolf Ladejinsky (1899–1975). He was born to a Ukrainian Jewish family who fled the Russian civil war to the United States in 1922. Ladejinsky studied at Columbia University and joined the US Department of Agriculture's foreign service, coming to specialize in Asia's agrarian problems. He served in the American occupation of Japan, where he played a key role in overseeing land reform, as he then did in Taiwan. Ladejinsky came under suspicion in the McCarthy era, but President Eisenhower defended him and appointed him to direct land reform in South Vietnam in the late 1950s, just as American involvement there was deepening. Through the 1960s, Ladejinsky continued to focus on the problems of rural Asia, working with the World Bank. He was anti-communist but saw the importance of land reform in societies where landholdings were highly concentrated in a few hands.[59]

Ladejinsky had worked in India on numerous occasions, and the World Bank sent him back in 1972 to investigate a drought that threatened hunger in Maharashtra. "This is an occasion when the writer intends to keep his emotions in leash," he promised at the out-set of his note, fearing that his "credibility may suffer from drama-tizing the incontrovertible cruelty of nature—no rain and no crops." He observed that "historically the struggle of the Maharashtra farmer has been one of quest for water." In an echo of British commentary in the late nineteenth century, he personified the monsoon as a force, as when he referred to "the monsoon playing truant." He described his journey out to the countryside from Poona; it was not long before he found himself traveling through a landscape that "leaves one shaken about the perversity of nature." Lack of drinking water was a prob-lem everywhere. More than shortages of food, it was a lack of water that, Ladejinsky saw, led farmers to uproot themselves and migrate in search of work. He described the sight of water tankers surrounded by people at famine relief camps—tankers that had been donated by oil companies as an act of charity. Words mattered, Ladejinsky ar-gued: neither the state nor the central government wished to invoke the term "famine"—seen as a relic of a dark colonial past—but their use of the mild term "scarcity" masked the severity of the crisis.[60]

The drought in Maharashtra showed how little the hydrologic revolution of the 1950s and 1960s had touched many parts of rural India. In his detailed study of the drought, economist Jean Drèze noted that only 8 percent of land in the region was irrigated. For those on rain-fed lands, Drèze wrote, "the meagre harvest of coarse grains remain a gamble on the monsoon and the land offers a spec-tacle of desolation and dust during the slack season." The drought led to a 14 percent drop in food availability; the threat of starvation was very real. It was averted by concerted government response, arguably one of the most effective in the history of independent In-dia. The Food Corporation of India organized the transportation of wheat from other parts of the country. It was sold at subsidized prices through thirty thousand fair-price shops distributed across the state. At the same time, a large program of public works generated employment; up to 2 million people a day attended these works, building roads and bridges and digging wells. This boosted local incomes and in turn pushed up food prices in Maharashtra, attracting supplies from beyond the state—often illegally, since the government had barred the interstate trade in grain during the crisis. Those illicit supplies, even at inflated prices, helped to compensate for the shortfall.[61]

In the end it was not big technology but rather the unheralded public distribution system of India that averted catastrophe in Maharashtra. The drought showed how patchy and uneven the reach of water engineering was. It showed the importance of public policy and prompt intervention. But these were not the lessons learned. The drought did nothing to dent confidence in the idea that all India needed was irrigation, now from deeper and deeper underground.

V

Scientists' understanding of the monsoon advanced in the 1960s and 1970s, spurred by the data collected by the Indian Ocean Expedition and by the International Geophysical Year that had preceded it, in 1957–1958. At the heart of monsoon science now were two phenomena. The first was "moist processes"—most simply, the release of latent heat and the effect of clouds on radiation. The second was the coupling of ocean and atmosphere. There was a new awareness that the monsoon system formed, as one meteorologist put it, a "complex of seemingly disparate parts: two fluids, the mobile air and the changing ocean below." Increasingly sophisticated computer models could turn each of these processes on and off in an attempt to isolate and investigate different variables.[62]

The most important breakthrough came with the work of Jacob Bjerknes, a Norwegian meteorologist at UCLA—son of Vilhelm Bjerknes. Father and son were both part of the team that had, in the 1910s, discovered and named the phenomenon of polar fronts

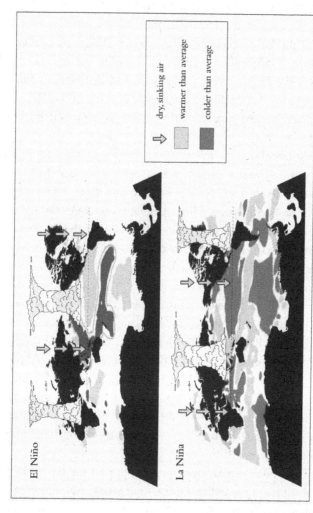

In the 1960s, Jacob Bjerknes discovered the El Niño Southern Oscillation (ENSO)—the illustrations show its "warm" (El Niño) and "cool" (La Niña) phases. CREDIT: Illustration by Matilde Grimaldi

from their observatory in Bergen. Now, using data generated by the expeditions of the International Geophysical Year, Jacob Bjerknes determined the mechanism driving a phenomenon that Gilbert Walker had first observed in the 1920s, during and just after his time as director of the Indian meteorological service. Walker had called it the Southern Oscillation, an oscillating contrast in atmospheric pressure across the Pacific Ocean, as measured in Darwin and Tahiti. But Walker had been unable to determine the cause of this swing in pressure; Bjerknes discovered that the answer lay in the waters of the Pacific Ocean.

Bjerknes found that the key to the southern oscillation lay in the periodic warming of sea surface temperatures in the eastern Pacific Ocean, and he dubbed it El Niño in keeping with the term that fishers had given the phenomenon. This warming reverberates throughout the world's climate. Most of the time, the waters of the western Pacific, off Indonesia, are warmer than in the eastern Pacific—this drives the easterly "trade winds," but Bjerknes saw that these were a surface manifestation of an overturning circulation in the upper atmosphere at higher latitudes. He named this the Walker Circulation in honor of Sir Gilbert. During an El Niño episode the contrast narrows as the waters of the eastern Pacific warm up; in response, the Walker Circulation weakens, since it is driven by that difference in temperature and pressure. Less intensive surface winds reduce the ocean's churn, leading to less of the colder water from the depths welling up to the surface; this sustains the abnormal warmth in the eastern Pacific, and so flattens the usual temperature contrast across the ocean—attenuating further the Walker Circulation.[63]

This disruption in circulation has consequences for rainfall in the western Pacific and the Indian Ocean—and even the North Atlantic. El Niño years tend to be associated with weak monsoons in Asia, and with excessive rainfall in South America. Bjerknes dubbed the overall oceanic-atmospheric system the El Niño Southern Oscillation (known as ENSO), of which El Niño was the phase of ocean surface warming and La Niña (girl child) of cooling; years that exhibit neither extreme are known as "neutral." La Niña has the opposite effect of El Niño, strengthening the temperature contrast between the eastern and western Pacific, strengthening the Walker Circulation, and bringing more rain than usual to Asian shores.

The discovery of ENSO marked a breakthrough for understanding the Asian monsoon. Once it had been identified, historical climatologists showed that many of the worst droughts in Asian history—including the droughts that brought famine in the 1870s and the 1890s, and also the Maharashtra drought of 1972–1973, discussed earlier in this chapter—coincided with El Niño events. But the causal relationship between ENSO and the monsoon is complex. There is some evidence to suggest that an especially strong or weak monsoon might foreshadow rather than follow the corresponding phase in the ENSO cycle. Tropical meteorologist Peter Webster argues that there may be truth in Charles Normand's suggestion—made in the 1950s after his retirement as head of India's meteorological service and before Bjerknes had discovered El Niño—that India's weather was more use in predicting what was in store for other parts of the world than it was itself amenable to prediction.[64]

Knowledge of ENSO raised new questions about the periodicity of drought in Asia—a question that, as we have seen, had provoked much discussion in the 1870s. It reinforced the sense, drawn from the Indian Ocean Expedition, that Asia's climate was fiendishly complex, associated with many other parts of the planet's climate. ENSO is quasiperiodic; it recurs but the intervals between events vary and are not easy to predict. The early 1970s also brought new knowledge of internal climatic variability on shorter timescales. In 1971, Roland Madden and Paul Julian, based at the US National Center for Atmospheric Research, discovered what came to be known as the Madden-Julian Oscillation (MJO): an oscillation in surface pressure and wind direction over large areas, with consequences on a planetary scale.[65] The MJO has a clear periodicity; it is, as meteorologist Adam Sobel has put it, "a signal that emerges above the meteorological noise." Migrating from west to east, from the Indian Ocean to the Pacific, the

MJO lasts between thirty and sixty days, and its intensity varies from year to year. In its "active" phase, the MJO brings heavy rain, and a heightened chance of tropical cyclones; in its "suppressed" phase, it interrupts the monsoon flow, even reversing the wind direction, bringing clear skies. The MJO is associated particularly with the northern winter; but scientists also discovered another intraseasonal oscillation in the northern summer months, which propagates northward rather than eastward. Known as the Boreal Summer Intraseasonal Oscillation, its connection with or independence from the MJO has been the subject of debate, but it, too, is thought to play a vital role in the fluctuations of rainfall over Asia each summer.[66] In the 1980s further research was done to uncover the mechanisms at work behind these intraseasonal oscillations, though some uncertainty remains.[67] These intraseasonal oscillations might well explain the alternation between active and break periods in any monsoon season, which has such vital and direct effects on agriculture.

Advances in technology and understanding did little to revise Colin Ramage's verdict, at the end of the Indian Ocean Expedition, that little headway had been made in forecasting the monsoon in a practical sense. Progress had been made in understanding the system on a large scale. But what mattered most to Asian farmers were the rhythms of rainfall within a given monsoon season—the relationship between what meteorologists call "active" and "break" periods of the monsoon. So finely attuned is Asian agriculture to the monsoon that the most devastating effects on cultivation often come from unexpected breaks in the midst of the summer rains, even when rainfall overall is plentiful—the skies brighten suddenly, and crops do not receive the water they need to thrive at a critical phase in their life cycle.

A further push to crack the monsoon's code came in 1979, as part of a worldwide effort called the Global Weather Experiment—the Indian Ocean component of that came to be known as the Monsoon Experiment, or MONEX. The scale of the operation was vast, even larger than the Indian Ocean Expedition of the 1960s, and more fully equipped with satellite technology. It encompassed 3,400 land stations, 800 upper air observatories, 9 weather ships, 7,000 merchant ships, and 1,000 commercial aircraft drawn in to record observations, 100 dedicated research aircraft, 50 research ships, 5 weather satellites, and 300 balloons. Despite this scale, despite the dazzling advances in equipment, Webster noted that much older ways of knowing the monsoon—the instinctive knowledge of mariners—were still in evidence in 1979; he saw that *dhows,* the traditional sailing vessels of the northwestern Indian Ocean that had for centuries harnessed the monsoon currents, were still widely used.[68]

———

COLIN RAMAGE, DIRECTOR OF THE METEOROLOGICAL COMPONENT of the Indian Ocean Expedition, returned to India in the 1970s. In his spare time while there on assignment, he turned amateur historian and wrote a short and provocative essay on how his predecessor John Eliot—the second director of Indian meteorology, after Henry Blanford—had failed spectacularly to forecast the crushing drought of 1899–1900. Ramage delved into the archives of the *Times of India* and wrote a powerful account of the famine that followed. "The government refused with religious fervor to modify the holy writ of laissez-faire," he declared. Like many critics of imperial policy at the time, he saw that the railways had done as much to worsen as to alleviate the famine, by making it easier for speculators to ship grain out to areas where purchasing power was higher. He praised the Indian government's efforts to protect its citizens against the threat of famine, but noted that, despite best efforts, India remained dependent on food imports, as it had since the 1920s. His conclusion was ominous. "Can we be sure that such a devastating famine will not recur?" he asked; not since Indian independence had there been a drought as severe as the drought of 1899. He ended his essay on that note, leaving implicit the underlying question: what would happen if another drought of that magnitude were to materialize?[69]

But another threat was now on the horizon. In 1979, the same year as MONEX, the World Meteorological Organization held its first World Climate Conference. The conference declaration recognized the need to "foresee and prevent potential man-made changes in climate that might be adverse to the well-being of humanity."[70] From the 1980s, the combination of climate change and other environmental threats compounded the water-related risks faced by billions of people in Asia.

NINE

STORMY HORIZONS

I N THE 1980S THE FULL PROPORTIONS OF ASIA'S WATER CRISIS BECAME manifest—including its vertical dimension. In that decade, satellite images and remote sensing data revealed to scientists how fully human activities had transformed the physical environment, reaching from the underground waters to the upper atmosphere. Much of the impact was visible without the aid of sophisticated technology. It became viscerally clear in the quality of the air that people breathed and the water that they drank, in the strangulation of the rivers they lived by and lived from. In 1984, the chemical pollutants that coursed through the holy Ganges reached such concentrations that a stretch of water caught fire: it became a river of flames.

Asia's waters were subject to unprecedented demands that came from the convergence of two large processes. The first, which began in the 1950s, was population growth. India's population grew from just under 370 million people in 1950 to 684.8 million in 1980, an

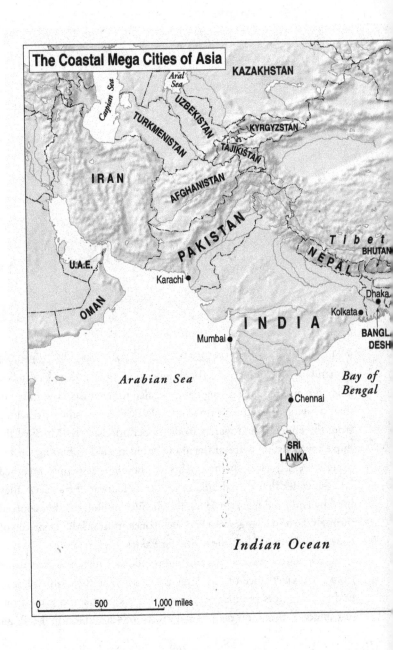

The Coastal Mega Cities of Asia

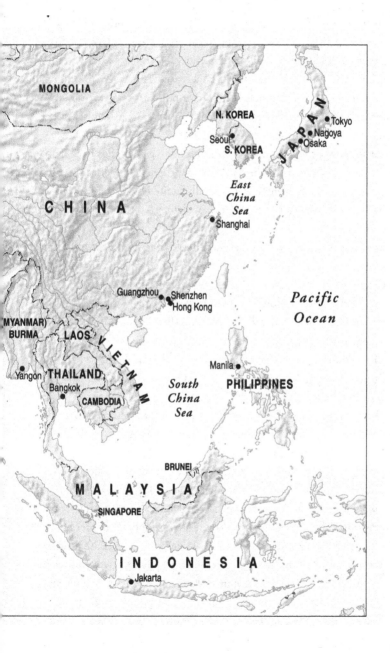

increase of 185 percent; through the 1970s alone, India added 131 million new citizens. China's population grew more slowly, but from a larger base: from 562 million people in 1950 to just under 988 million by 1980, and with an absolute increase of close to 166 million people in the 1970s. Although the rate of population growth in both India and China had slowed considerably by the 1980s, not least because of coercive population "control" measures in both countries, the cumulative impact of the previous decades' growth has been manifest. Belying the fears of the Malthusian prophets, the effects of this expansion on the biosphere were initially limited by very low levels of income per capita—and limited, too, by deliberate efforts by both the Indian and Chinese governments to hold back consumption to generate savings for future investment in industrial development. But then arrived the second major transformation, which began in the 1980s: the rapid economic growth of Asia's two largest countries, China and India, both of them following a path forged by other countries in East and Southeast Asia a decade or more earlier, but on an altogether different scale.

The bonfire of socialist austerity began first in China, where the reforms of Deng Xiaoping enshrined the notion that "to get rich is glorious." From 1978 to 2012, the Chinese economy grew at an average annual rate of 9.4 percent, "the fastest sustained expansion by a major economy in history."[1] The Indian economy was slower to accelerate, but by the 1980s average annual growth was around 5 percent. Following an emergency loan from the International Monetary Fund to meet a critical shortage of foreign exchange, the Indian economy underwent a process of liberalization after 1991, orchestrated by the economist Manmohan Singh. This involved a dismantling of elaborate regulations governing private investment and trade, dubbed the "License-Permit Raj." High growth followed, picking up in the late 1990s; but it was accompanied by galloping inequality. India has remained home to more poor people, in absolute terms, than any country on Earth.[2] In India, more than in China, the ecological threats generated by new prosperity intensified the more familiar, water- and weather-related risks of extreme poverty. Unlike China, India's population has remained predominantly rural, and will continue to be so by the middle of the twenty-first century. The destabilization of Asia's water ecology, which accelerated in the 1980s, put more people at risk in India and in neighboring Bangladesh than anywhere else.

This chapter shows how, starting in the 1980s, Asia's waters submitted to a concatenation of demands from industry, from agriculture, and from the needs of booming cities. The mining of groundwater exceeded the capacity of the hydraulic cycle to replenish aquifers. A hunger for energy led to a renewed interest in hydroelectric power. States and private investors eyed the upper reaches of the great rivers. From the 1980s hydraulic projects converged upon the Himalayas. The most promising lowland dam sites were exhausted by the 1970s; the steep drops of the mountain rivers made them ideal for power generation. As a cluster of competing projects lined up along the rivers' descent from the 1980s, the potential for conflict grew. States acquired the capacity to deny water to others downstream; not so much the technical capacity, since dam technology had changed relatively little from the 1950s, but rather the financial and infrastructural capacity—and above all, the will. New demands on resources, and new demands for water, came from the revival of trade between South, Southeast, and East Asia, which had ebbed in the 1950s and 1960s.

The final ingredient in this cocktail of ecological destabilization came with the accelerating effects of climate change. Already in 1982, environmental activists in India invoked what they called a "rather futuristic problem"—the "possibility of global climatic change taking place by the end of the century because of increasing carbon dioxide in the atmosphere." They raised an ominous prospect: "It is quite possible . . . that agriculture as practiced for centuries in India may have to change and crop outputs may become a matter of even greater uncertainty than today."[3] Since then, the scientific consensus on the reality of anthropogenic climate change has been overwhelming.[4] Climate change is no longer a "futuristic" problem—its effects are here, now. And its effects menace the coastal rim that stretches from India to China.

Climate change affects water in every form: it affects the rain clouds and the Himalayan glaciers, the flow of rivers and the shape of coastlines, the level of the ocean and the intensity of cyclones.[5] Climate change is irreducibly historical. As historian and Marxist theorist Andreas Malm observes, "The storm of climate change draws its force from countless acts of combustion over, to be exact, the past two centuries."[6] But the current crisis is a product of history in another sense too. The acute impact of climate change on Asia, and on South Asia in particular, will play out across a landscape shaped by the past—shaped by the cumulative effects of social inequality, shaped by the borders of the mid-twentieth century, shaped by infrastructures of water control. And it will be shaped by the legacy of ideas from the past, including ideas about climate and the economy.

I

Water was a core ingredient in Asia's experience of what economist Angus Deaton calls the "great escape" from scarcity.[7] The intensification of agriculture driven by the Green Revolution—a package of high-yielding seed varieties and extensive fertilizer use, sustained by vast quantities of water—augmented food production to an extent that would have been unthinkable even one generation earlier. Between 1970 and 2014, India's production of cereals grew by 238 percent, compared with a 182 percent expansion in population over the same period. This took place with only a marginal increase in the quantity of land given over to food crops. In China the expansion was more dramatic still: a 420 percent increase in cereal output with no increase in land area under cultivation.[8] Just a decade after the desperate recourse to American food aid during the monsoon failures of the 1960s, India became a food surplus country.

Intangible though it was, an unshakable sense took hold among the Indian elite that the threat of an uncertain monsoon had receded. It was a sense expressed by writer and newspaper editor Khushwant Singh in a 1987 essay on the monsoon in Indian literature. Singh ranged widely across Indian epics and poetry to show how deeply the monsoon had shaped Indian cultural sensibilities over hundreds of years. But he concluded that, in recent decades, "India has taken enormous strides toward freeing herself from dependence on the vagaries of the monsoons." Technology led the charge: India had "raised enormous dams, laid thousands of miles of irrigation canals, and dug innumerable electrically operated tube-wells to supply water to her farms." A sense of security brought disenchantment. "There is no longer the same agony waiting through long summer months of searing heat to catch a glimpse of the first clouds," he argued. The monsoon had vanished from Indian literature; it "no longer stirs the imagination of the poet or the novelist with the same intensity it used to."[9]

Those closer to rural India had a different view. The same year as Singh, the modernist artist Jyoti Bhatt, trained in the influential Baroda school in Gujarat and immersed in local artistic traditions, wrote that for all of the improvements in weather forecasting, the ability to predict the character of a whole monsoon season remained elusive. In folk culture, if not in high poetry, the monsoon's mysteries lived on. Bhatt described an annual festival in the arid lands of Kutch and Saurashtra, in Gujarat, celebrating Bhadali—the daughter of a shepherd and a gifted diviner of rain. The festival was bound up with anxious expectation. Villagers in Gujarat, he wrote, "keep observing and interpreting various omens, signs, and factual symptoms around them." They relied on the "collected experience of many generations" to decide when to plant their crops each year. Bhatt was agnostic about how far these rituals helped farmers, but at the very least he saw that they provided more excitement and drama than "watching a Door Darshan [the state broadcaster] weather forecast based on data received from Insat, on a small TV screen."[10]

Also in 1987, but on a larger scale and in the language of economics rather than poetry, Harry Oshima revisited the old region of "monsoon Asia." Hawaii-born Oshima (1918–1998) wrote his dissertation on the national income statistics of Asia's new states; he worked for the United Nations in the 1950s, and served as the Rockefeller Foundation's representative in the Philippines in the 1970s.[11] Oshima found that monsoon Asia's coherence had been shattered by a transformation in the relationship between water and productivity. Oshima began with a timeless vision: across the coastal and deltaic sweep from South Asia to East Asia, the intense seasonality of rainfall created common patterns of agriculture—labor-intensive paddy cultivation, high population density, a preponderance of small farms. He wrote, too, of the "philosophy" of the "monsoon economy"—an ethos of "harmony . . . compromise, moderation, diligence, and cooperation." In writing this, Oshima echoed the language of an earlier era, which drew a straight line from climate to culture. But the period since 1970 had broken deep historical patterns. There had been an unexpected differentiation in income levels across the region, which was now, in Oshima's view, "crystallizing with a few modifications into the three basic regions of . . . East, Southeast, and South Asia." Oshima was least sanguine about South Asia's prospects; pessimism about India was widespread among economists at the time. South Asia was effectively now the residue of "monsoon Asia"; everywhere else, industrial growth and intensive irrigation had powered an escape from the monsoon.[12] But even in India, it was clear by the 1980s that something fundamental had changed.

In India, the revolution in food production depended, above all else, on groundwater. As we have seen, the first experiments with using motorized pumps to extract groundwater in India date from the late nineteenth century, but until the 1960s, their use was negligible. The greatest growth came in the use of private tubewells: there were half a million in use across India in 1968; that number had grown to 5 million by 1994. As the exploitation of groundwater increased, so too did the depth that tubewells had to reach. Investment in tubewells has been almost entirely private, in contrast with dams and other surface irrigation works that have been publicly funded. But under Indira Gandhi's government in the 1970s, landowners were encouraged to utilize groundwater and install tubewells through the provision of subsidized or even free electricity; state electricity boards were left to set prices, and many of them incurred heavy losses. The use of groundwater proceeded with no regulation. Large farmers, with the capital to invest in technology and with the large landholdings to benefit from irrigation, dug deeper than their neighbors, capturing groundwater for their private use and even selling it on to others. Cheap electricity provided an incentive for farmers to extract as much groundwater as they could, with little thought for replenishing the aquifers. Tushaar Shah, a leading expert on groundwater policy in India, has described it as "an atomistically managed water-scavenging irrigation regime involving tens of millions of pump owners who divert surface and groundwater at will." In all, nearly three-quarters of the expansion in India's irrigated cropland since independence has come from groundwater, and much of the expansion came in the 1970s and 1980s.[13]

The effects of this boom in water mining were clearest in Punjab. Already by the 1910s, Punjab was India's most prosperous agricultural region; the elaborate system of canals built by the British made its arid lands productive. By the early twenty-first century, Punjab produced 20 percent of India's wheat and 42 percent of its rice, on just 1.5 percent of the country's land area. Punjab possessed only around eleven thousand tubewells in the late 1960s, on the eve of the Green Revolution—that number would grow more than 100-fold to 1.3 million over the next forty years. Groundwater provides two-thirds of Punjab's water supply. But the water table has declined perilously since the late 1970s. The intensive use of pesticides in farming has contaminated water sources, and this is widely acknowledged to be responsible for a substantial increase in the incidence in cancers in the area. In the western Indian state of Gujarat, another region where agriculture is dependent on groundwater, the water table has dropped by 1.4 meters each year from the late 1970s through the end of the 1980s, and at an even faster rate since then.[14]

If the monsoon no longer inspired India's poets, as Khushwant Singh observed in the 1980s, the infrastructure of groundwater extraction has become an unavoidable feature of the landscape in ways that have left their mark on South Asian literature. In a powerful short story published early this century, Pakistani-American writer Daniyal Mueenuddin evokes the landscape of that part of Punjab that formed part of Pakistan after Partition—the agrarian heart of a country even more dependent on irrigation than India. The protagonist is Nawabdin, the village electrician; his special talent was "a technique for cheating the electric company by slowing down the revolutions of electric meters." This mattered deeply, because electricity was the lifeblood of agriculture—"In this Pakistani desert, behind Multan, where the tube wells ran day and night, Nawab's discovery eclipsed the philosopher's stone." In that simple detail, as Mueenuddin sets the scene, we glimpse a vast agrarian transformation.[15]

—

INDIA AND CHINA HAVE MUCH IN COMMON IN THEIR RELIANCE ON groundwater to secure an increase in food production, in their vulnerability to the depletion of water sources, in the economic geography of their water use, and in the sheer scale of change they have experienced since the 1980s. But China has grown much faster than India, and India has been even more vulnerable than China to water- and climate-related risks, as a result of its greater dependence on agriculture, its higher levels of poverty, and, to return to a theme that has recurred throughout *Unruly Waters,* because of the particular characteristics of the monsoon.

If China's use of groundwater since the 1970s has not been quite as prodigious as India's, it is not far behind. Underground aquifers provide water to 40 percent of China's farmland, and drinking water to 70 percent of the population of China's arid north and northwest. Across the North China Plain, groundwater levels have dropped by approximately 1 meter a year since 1974, a rate of depletion comparable with that of Punjab. Like India, China's groundwater is contaminated. A study undertaken by the Chinese government in the 2000s showed that 90 percent of China's groundwater was polluted, and 60 percent severely polluted with heavy metals and fertilizer and chemical waste.[16]

In the broad sweep of history, China and India have undergone comparable shifts in their economic geography—in both cases, groundwater and other sources of irrigation were the driving force of change. Historian David Pietz points out that China, in the second half of the twentieth century, underwent a "reversal of food production patterns that pertained for most of the imperial period." The dry North China Plain now produces 60 percent of China's wheat and 40 percent of its corn on 22 percent of its land, and just 4 percent of its water resources. This has led to the transfer of what hydrologists call "virtual water"; that is to say the water that is embedded in crops, from water-scarce to water-abundant areas.[17] This is a story that parallels, in nature and in timing, the emergence of arid Punjab as India's agricultural powerhouse. As the economist Harry Oshima noted in the 1980s, even before the scale of China's and India's transformation was evident, the old geography of monsoon Asia had been shattered. It had been shattered, above all, by new sources of water. At any point until the late nineteenth century, it would have been self-evident that agrarian wealth in Asia lay in areas of abundant rain—the essence of monsoon Asia was the intensity of cultivation, especially rice cultivation, that the monsoon climate allowed. In a remarkably short space of time—the forty or fifty years after 1960—this pattern had been reversed by technology, and by fossil fuels.

The terrible paradox is that this stunning expansion in food production was achieved in a way that cannot be sustained. Groundwater resources are under acute strain in the regions of Asia that most depend on them. A study using data from the NASA Gravity Recovery and Climate Experiment satellites showed that, between 2002 and 2008, groundwater depletion in northwestern India—the heartland of the Green Revolution—amounted to 109 cubic kilometers of water, an amount that exceeds the storage capacity of India's largest reservoir. Freshwater availability per capita in India is projected to fall to 1,335 cubic meters by 2025, in comparison with a global average of 6,000 cubic meters. Groundwater has been the cornerstone of India's and China's food security since the 1970s—but for how long?[18]

THE SUSTAINABILITY OF INDIA'S GROUNDWATER BOOM IS ONLY ONE aspect of a deeper crisis of water. It was clear from the earliest years of the Green Revolution that one consequence of the new approach to agriculture was deepening rural inequality. In his commentary on the Maharashtra drought of 1970 to 1973, economist Wolf Ladejinsky had seen how sharp the contrast was between irrigated and nonirrigated lands. Long-standing fault lines between wet and dry, rain-fed and groundwater-supplied lands grew deeper. Access to water was both a cause and a symptom of inequality.

In the 1980s, recognition of the extent of water inequality energized an intellectual and political movement that called into question the fundamental pillars of India's development strategy. Disagreements over economic policy were common enough in the 1950s and 1960s. India's policymakers included committed planners as well as those in favor of free markets. But they disagreed about the means, and not the ends of development. By the end of the 1960s, India faced a radical alternative, in the shape of a Maoist insurgency that began in West Bengal and soon spread to other parts of the country. The insurgents, led by an urban elite committed to the romance of revolution, believed that only the violent dispossession of India's landowning class could bring about substantive change. Paradoxically, they drew inspiration from China at just the moment when Chinese agriculture changed course, embracing its own version of the Green Revolution. Others looked to India's past, to the history of water, for inspiration as they considered alternative economic models.

Mahatma Gandhi was a clear source of inspiration for many of those who, in the 1980s, challenged the assumptions of the Indian state. Though their influence on economic policy was muted, Gandhians continued after independence to urge upon India a different model of development—more rooted in rural communities, less wedded to monumental technology. They called for a holistic approach to development that emphasized both social and ecological equilibrium. The essence of their philosophy was encapsulated in Gandhi's 1946 pronouncement that "the blood of the villages is the cement with which the edifice of the cities is built. I want the blood that is today inflating the arteries of the cities to run once again in the blood vessels of the villagers." The 1970s saw the rise of the Chipko movement that brought together concerns with environmental degradation in Himalayan forests with the assertion of forest peoples' rights to the resources on which their livelihoods depended. The movement was explicitly Gandhian in inspiration, and women played a leading role within it.[19]

That spirit infused a new approach to India's water problems in the 1980s, an approach that looked back to a golden age of local, sustainable water management, embedded in the ancient practices of rural India. Just as Gandhi evoked a largely mythic notion of India as a collection of village republics—an idea that he drew primarily from Western writers—environmental activists in the 1980s harked back to an ecologically responsible, traditional India. It mattered little that this vision bore little resemblance to the picture that was emerging from historical and archaeological research. Archaeologist Kathleen Morrison puts it this way: "nostalgic" environmentalists evoked "a mode of life that I have simply been unable to reconstruct even as my work has expanded to incorporate three thousand years of agrarian history."[20]

In those same decades, research on the history and diversity of India's water practices painted a more complex picture. Water management was often tied to the exercise of royal power. Irrigation works depended on coerced labor. Access to common property was governed by the exclusions of caste—and those commons were under pressure even before the nineteenth century. The valorization of communitarian solutions could often serve to legitimize inequality and oppression. Many of the architects of independent India, including Jawaharlal Nehru, had seen this clearly. B. R. Ambedkar, leader of India's Dalits and architect of the Indian constitution, was no rural romantic: "What is the village but a sink of localism, a den of ignorance, narrow-mindedness, and communalism," he had asked in the Constituent Assembly in 1948. The divergence between these different visions of India's past reminds us that water has a public as well as a scholarly history—throughout the 1980s, idealized narratives about water management in the past had rhetorical and strategic value for the debates of the present, and they informed contending visions of the future.[21]

Even if it was little more than a useful fiction, the idea of a return to a more ecologically harmonious past motivated many strands of the Indian environmental movement, which emerged in earnest in the early 1980s. The movement's foundational text was the *First Citizen's Report on the State of India's Environment,* written by Anil Agarwal and his colleagues at Delhi's Centre for Science and Environment, which Agarwal had founded in 1980. Agarwal was by then one of India's most influential environmentalists. Born in the northern industrial city of Kanpur in 1947, to a landowning family, Agarwal studied mechanical engineering at the Indian Institute of Technology there. His work as the science correspondent of the *Hindustan Times* in the 1970s brought his writing to international attention. The citizen's report paid close attention to the water crisis facing rural India. A few years later, in 1985, a report by the center urged the importance of recovering and repairing India's ancient practices of harvesting the waters of the monsoon. The report's authors, Agarwal and his protégé Sunita Narain, went further; in their view, nothing less than a revitalization of rural India would reverse India's slide toward ecological degradation and social crisis. They described India's traditional villages as "integrated agro-sylvo-pastoral entities," dependent on common property resources: the rivers and the lakes and the forests. They claimed that the Indian state's approach to development—top-down, reliant on big technology—had "torn asunder this integrated character of the villages." In this was more than a trace of the holistic "rural sociology" of Radhakamal Mukerjee in the 1920s. Agarwal and Narain argued that "the process of state control over natural resources that started with colonialism must be rolled back." Their prescribed solution was the final decolonization of rural India, a reversal of the process that began in the mid-nineteenth century with the British search for India's water wealth.[22]

A similar commitment to elevating traditional practices and indigenous knowledge underpins the most wide-ranging and influential condemnation of the Green Revolution, written by environmental activist Vandana Shiva. Shiva emerged as a distinctive voice in Indian debates in the 1980s, and in the 1990s she would go on to become internationally influential within the antiglobalization movement. Trained as both a physicist and a philosopher, Shiva started the Research Foundation in Science, Technology and Ecology in the Himalayan town of Dehra Dun in 1982. In the opening pages of her book, *The Violence of the Green Revolution,* published in 1991, Shiva looked back on the 1980s, and described that as the decade when Asian societies came under the grip of what she described as "an ecological crisis and the threat to life support systems posed by the destruction of natural resources." Taking aim at the idea that the Green Revolution had brought about an agricultural "miracle," Shiva highlighted its costs. Many of these were well known, but Shiva's forceful prose gave them new prominence. She called Punjab a "tragedy," a cautionary tale of the folly of "breaking out of nature's limits and variabilities." She argued that the use of high-yielding seeds, pesticides, and ever-more water had left Punjab with "diseased soils, pest-infested crops, water-logged deserts, and indebted and discontented farmers." She challenged the Indian state's obsession with technological solutions to social and ecological problems; echoing Agarwal and Narain, she implied that India had not rid itself of its colonial legacy. Juxtaposing the quest for water with her emphasis on conservation, she posed it as a struggle between "diversity, decentralization and democracy," on one side, against "uniformity, centralization, and militarization" on the other.[23]

Shiva's book epitomized a new sort of environmental thinking in India. But it also reflected a new set of intellectual and political connections that crossed Asia's borders in the 1980s. Her book was published by the Third World Network, which was based in Penang, Malaysia. Formed in 1984, the Third World Network was an offshoot of the Consumers' Association of Penang—which, founded in 1970, was one of Asia's earliest pressure groups devoted to a broad range of causes ranging from fair prices and housing to food safety. It was the group's report on the *State of Malaysia's Environment* that had inspired Anil Agarwal to undertake a similar exercise in India after attending a conference in Penang. The Third World Network marked the incorporation of environmental concerns fully under the umbrella of issues on which Asian activists made common cause. The network—which reached beyond Asia to encompass groups in Africa and Latin America, with many allies among activists and nongovernmental organizations in Europe and North America—brought together a commitment to social and economic justice with a new concern about sustainability. The Third World Network helped to bring Shiva's work to a wide audience among activists in Asia and beyond. For her part, Shiva applied her analysis to Asian societies writ large, and not just to Punjab. Even if the 1970s' movement for a New International Economic Order at the United Nations had proved short-lived, marginalized by the Anglo-American turn toward privatization in the 1980s under Reagan and Thatcher, ethical claims on behalf of what we now call the Global South lived on. The sense of a unified Third World fighting against the legacies of colonialism as well as new forms of imperial power began to crumble as parts of Asia began to experience very rapid economic growth, but it continued to influence movements for environmental justice that focused on the ways poverty heightened environmental vulnerability and inequality worsened environmental harm.[24]

The networks of activism that linked environmentalists across Asia's borders turned, in the early 1990s, to the problem of climate change. In a 1991 text that remains influential to this day, one of the earliest and most eloquent expressions of the argument for global environmental justice, Anil Agarwal and Sunita Narain wrote about the problem of *Global Warming in an Unequal World.* Their opening sentence is powerful and stark: "The idea that developing countries like India and China must share the blame for heating up the earth and destabilizing its climate . . . is an excellent example of *environmental colonialism.*" They pointed out that historical responsibility for the accumulation of carbon in the atmosphere lay entirely with the advanced industrial countries of the world; they highlighted the hypocrisy of those countries now telling India and China to cut their emissions, when in per capita terms, India's or China's emissions were miniscule. Their conclusion was that "the Third World today needs far-sighted political leadership" to resist the calls by Western political leaders and environmentalists to "manage the world as one entity," which could only be a mask for exploitation as long as the world remained so unequal and so divided.[25]

The pamphlet was published just on the eve of India's economic liberalization: a series of market reforms that followed an emergency IMF loan, secured when India faced a crisis of foreign exchange. In its language, it belongs firmly in the era that was closing, though few saw it at the time. The idea of the Third World, invoked repeatedly, had already started to come unstuck; with the collapse of the Soviet Union, it dissolved. What Agarwal and Narain could not have anticipated was just how rapidly the Indian environment, with the Indian economy, would be transformed by a new openness to global capitalism.

China's own economic transformation was well underway in 1991, but its colossal scale and its colossal implications for the world were only slowly becoming evident. With good reason, the 1991 pamphlet called for a united front against the powerful nations in international negotiations over climate and emissions. But it was blind to the proliferation of cross-border challenges confronting Asia—in the realm of water above all.

THE EXCHANGE OF IDEAS ABOUT SHARED ENVIRONMENTAL CONcerns took place alongside a focus on deeply local problems. As the crisis of rural India became more visible, it appeared to be rooted in climatic and social characteristics that were distinctive to the Indian subcontinent, familiar to observers going back to the nineteenth century—the deep and particular unevenness of water's distribution, and the pervasive social and caste inequalities that limited people's access to water.

Journalist Palagummi Sainath (b. 1957) spent the 1980s working on the Bombay tabloid *Blitz*. In 1993, he received a *Times of India* fellowship that he chose to spend traveling through India's poorest districts. He traveled one hundred thousand kilometers over a few years, more than five thousand of them on foot. The *Times* published his dispatches in installments, at a time when ever-less reporting from impoverished rural India reached a metropolitan audience that was now in the grip of economic expansion. Sainath wanted to move beyond what he saw as the media's focus on "the spectacular" and to highlight "the long-term trends that spell chaos [but that] don't make good copy." Sainath's articles were collected and published as a book in 1996 with the deeply ironic title *Everybody Loves a Good Drought*. In his many articles on water, Sainath drew attention to the opportunities for profit that water scarcity brought to a new cabal of "water lords." Sainath observed something familiar from earlier times—absolute scarcity of water was not always the problem, its distribution was. "Simply put," he wrote, "we have several districts in India that have an abundance of rainfall—but where one section, the poor, can suffer acute drought." With the insight that came from his immersion in rural India, Sainath distinguished between "agricultural drought" and "meteorological drought," arguing that the latter was not necessary for the former to bite—there were droughts that were "real," and droughts that were "rigged."[26]

The Indian countryside reeled from a double burden. Many farmers, those inhabiting the 60 percent of India's farmland without irrigation, suffered under the age-old burden of their dependence on an uncertain monsoon. But high-intensity farming brought its own burdens. When the history of late-twentieth-century India is written from a perspective of greater distance, alongside vertiginous economic transformation there will be a less visible, shameful, story: the story of an epidemic of farmer suicides on a scale that may be without parallel in the world. Sainath was among the first to bring this silent crisis to public attention. Starting in the late 1990s, an estimated seventeen thousand farmers each year have taken their lives—at least two hundred thousand deaths from suicide between 1997 and 2010. At the root of the intolerable pressure that many of India's farmers labor under are their growing debts—debts for purchases of seed and fertilizer and pesticide and fuel for groundwater pumps.[27]

Meanwhile access to water continues to be an indicator of the most fundamental social inequalities. In a comprehensive survey of the practice of untouchability in rural India, undertaken at the start of the twenty-first century, Delhi sociologist Gyansham Shah and colleagues found that Dalits in rural India regularly face exclusion from access to basic public services—and of these, the authors found, the most important by far was the denial of access to water. No fewer than 48 percent of villages surveyed reported such denial. Pervasive upper-caste beliefs about the polluting effects of Dalits having contact with water sources leads to systematic discrimination, enforced by violence. The practices that Shah and colleagues documented ranged from absolute exclusion from tubewells and tanks to Dalits being forced to wait until everyone else had taken their water before being allowed limited access. Ninety years after Ambedkar's march on the tank at Mahad, unequal access to water remains pervasive in India. And caste discrimination explains why such inequalities are sharper in India than anywhere else in Asia.[28]

Sainath's dispatches from rural India in the 1990s date from a time when climate change was not foremost among India's concerns. To revisit his urgent reportage two decades later, when the signs of climate change are everywhere, reminds us that, in India and elsewhere in Asia, climate change comes on top of a mountain of intersecting ecological and economic crises that has been building since the 1980s.

II

If the ocean underground began to recede as a result of the unsustainable use of groundwater in the 1980s, Asia's running waters—its rivers—were more visibly scarred. The rivers were the conduit between the countryside and the insistent demands of growing cities. Since the late nineteenth century, the engineering of rivers—damming, diverting, impounding them—has governed efforts to redistribute water. The twentieth-century quest for the "white gold" of hydroelectric power intensified that quest. In India and China alike, the abuse of rivers provoked a new environmental consciousness in the 1980s. The dreams of the 1950s and 1960s gave way to an unfolding nightmare. The circumstances under which the Indian and Chinese environmental movements emerged were very different from those of their counterparts in North America, Europe, and Japan earlier in the century. In Asia, rapid growth followed, rather than preceded, awareness of scarcity and natural limits. And a sense of fragility before the power of nature, a sense that hard-won gains were under threat, led authorities in India and China to the defiant, even violent defense of large technological solutions to the problem of water.

In India, the crisis of river pollution was clearly visible by the 1980s. In their first report on India's environment, Anil Agarwal and his colleagues at the Centre for Science and Environment wrote that "river pollution in India has reached a crisis point. A list of India's polluted rivers reads like a roll of the dead." They described the holy Ganges as a "network of cesspools," and came up with a grim list of industries responsible for the damage: "DDT factories, tanneries, paper and pulp mills, petrochemical and fertilizer complexes, rubber factories . . . "[29] A few years later, Darryl D'Monte, a pioneer of Indian environmental journalism and a contributor to the Centre for Science and Environment's report, declared that the "destruction of life support systems along the Himalayas" constituted "the world's single biggest ecological crisis."[30]

In 1985, a campaigning lawyer, M. C. Mehta, took up the river's cause. Mehta, born in a small village in Jammu and Kashmir state, worked as a public interest litigation lawyer in India's supreme court. A 1984 visit to the Taj Mahal awakened him to the harm being done to the monument by polluting factories nearby—by the early 1980s, the Taj Mahal's lustrous marble had been stained a dirty yellow. Mehta filed a public interest case against the offending industries. The following year, he turned his attention to the pollution of the Ganges. In a series of landmark cases, heard weekly over several years, Mehta succeeded in having three hundred factories closed and five thousand forced to install cleaner technology; the court ordered 250 municipalities to install sewage treatment plants. Mehta's were the most significant cases brought under India's Water Act of 1974; their proceedings revealed the extent of river pollution in India by the 1980s. In its 1988 judgment on a case Mehta brought against the owners of tanneries in the industrial town of Kanpur, the Supreme Court of India noted that "any further pollution of the river is likely to lead to a catastrophe." They noted the relentless discharge of sewage and chemical effluent into the river. In another case the same year, Mehta took on Kanpur Municipality. He brought as evidence a report from

Beginning in the 1980s, river pollution in India reached crisis proportions. CREDIT:
Dominique Faget/Getty Images

the Industrial Toxicology Research Centre, which showed that the
water of the Ganges was completely unfit for human consumption.[31]
Mehta won his cases; the polluting industries were ordered to amend
their ways. But in comparison with the scale of the problem of river
pollution in India, these were small victories in an enormous battle.

———

PROPELLED BY ECONOMIC TRANSFORMATION FAR MORE RAPID
than India's, China's rivers endured a comparable assault. In China,
too, under tighter political constraints, the 1980s saw the emergence
of an environmental movement—and there, too, water was a prime
concern. One of China's first private environmental organizations,
Friends of Nature, was formed by Liang Congjie in 1993. Liang Congjie's grandfather was Liang Qichao, the prominent late-nineteenth-
century reformer; his father was an architectural conservationist who

suffered brutal persecution during the Cultural Revolution. Liang Congjie took a keen interest in environmental issues from the 1980s. His approach initially was cautious; his activism began with seemingly innocuous targets, like his campaign to save the chiru, or Tibetan antelope. But Friends of Nature, like its counterparts in India, harnessed the power of information to illuminate a slow crisis. The organization began to publish the *China Environment Yearbook,* which was akin to the *State of the Indian Environment* reports, if less overtly critical of the government. Anxieties about water pollution and water shortages multiplied in China in the 1980s and 1990s, prompted by the breakneck pace of urbanization.[32]

In the 1990s, around the same time that P. Sainath undertook his investigative tour of rural India, journalist Ma Jun published a series of articles for Hong Kong's *South China Morning Post* on the state of China's waters, culminating in his influential 1999 book *China's Water Crisis.* He wrote of his realization that government officials and engineers were "trying to rob nature of the last drop of water to serve economic expansion." He noted that "while most people regarded the floods, dry spells, and sandstorms as some sort of evil force that demanded even larger engineering projects, I began to view them as nature's way of retaliating for man's reckless attempt to conquer and harness nature." He described how the flow of the Yellow River—the "mother river," cradle of Chinese civilization—began to decrease from 1972. In 1997, for a period of 330 days, the Yellow River failed altogether to reach the ocean. Ma Jun called the abuse of China's rivers a "heinous crime," as he made an emotional appeal to the power of the rivers and the reverence with which they had been treated for centuries. "To rescue the dying rivers with our devotion and work would be our most glorious effort," he urged.

The most shocking passages of his book described the pollution of the lower Yangzi River—we can see similarities in both tone and content with the writings of Indian environmentalists. Ma Jun described the river as "a vast open sea of garbage and sewage." It was clogged by rubbish from the dense traffic on the river: "Styrofoam lunch boxes and vegetable scraps, toilet waste, cooking oil, machine oil and industrial muck." Worst of all was the pollution from the cities. Ma Jun wrote that "day by day, week by week, month by month, in a monotonously inexorable fashion, they throw or pour the detritus of 400 million people into the waterway." In China, as in India, minor triumphs, small cleanups, have failed to stem a hurricane of waste.[33]

FACED WITH MULTIPLE WATER CRISES, THE INDIAN AND CHINESE states fell back on the strategy that they had favored since the 1940s—to turn, again and again, to large-scale hydraulic engineering. Ecological and social harm reinforced each other in the case of large dams, which continued to be constructed on an ever-expanding scale even after groundwater became a far more important source of water for irrigation. While they drowned forests and flooded fields, they also displaced millions of people. Concerns with the social suffering caused by large engineering schemes joined fears about water pollution to form a second major strand of environmental activism. Here, too, India was at the vanguard. The 1980s saw an escalation in the scale and reach of protest against large dams in India.

In 1978, India sought World Bank assistance for the mammoth Narmada project, which called for the construction of 30 large, 135 medium-sized, and 3,000 small dams along the Narmada River, which flowed west through the states of Madhya Pradesh, Maharashtra, and Gujarat to the Arabian Sea.[34] In 1985, the World Bank committed US$450 million to the project, around 10 percent of its total cost. Little had changed since the 1950s in the Indian government's iron certainty that the benefits of the project would outweigh its costs. On their initial estimate, seven thousand families would be displaced by the project. Plans were made for rehabilitation, but in keeping with common practice, only those with formal title to their lands were included. As the true scale of displacement and environmental destruction emerged, resistance grew. In the late 1980s, a cluster of nongovernmental organizations—a broad coalition that included human rights groups, environmentalists, students, and local people's associations—came together to form the Narmada Bachao Andolan, or NBA ("Save Narmada Movement"), led by the social activist Medha Patkar. Patkar, born in Mumbai in 1954 to parents who were active in the nationalist and labor movements, studied social work at the prestigious Tata Institute of Social Sciences, but abandoned the doctoral dissertation she had started as she became more involved with the struggles of marginalized communities along the Narmada valley. Under her leadership, the NBA harnessed the power of Gandhian nonviolent protest and drew on a rich vein of ideas that insisted on people's sovereignty over their lands and landscapes. Among their most resonant techniques was the "monsoon Satyagraha"—silent demonstrations held as the river's waters rose during the monsoon, slowly submerging the protesters until they stood waist-deep in water. It appealed symbolically to the power of climate and seasonality, which the dams sought to engineer away.[35]

The NBA succeeded in harnessing international support. In the United States, Lori Udall of the Environmental Defense Fund took up the fight. Patkar met with the World Bank in 1987, and pressure on the bank from international supporters of the Narmada movement led it, in 1991, to initiate an inquiry into the project. The bank's decision to withdraw funding for the project in 1993 was a victory for the Narmada movement, and marked a shift in the bank's previously uncritical support for large dams.[36] But the Indian government's response was defiant. Alongside nonviolent resistance, the Narmada movement took to the courts: they had some early success, and then, from the late 1990s, faced a series of defeats as the Supreme Court ruled in favor of the project's continuation. The World Bank's withdrawal served to harden the government's resolve to find private finance for the Narmada project. When the Sardar Sarovar Dam was finally declared open in autumn 2017 by Indian prime minister Narendra Modi—who had strongly supported the Narmada project when he was chief minister of Gujarat, condemning environmentalists

Environmental activist Medha Patkar, leader of the Narmada Bachao Andolan, joins a protest against the construction of a court complex at Pipliyahana Reservoir near Indore. CREDIT: *Hindustan Times*/Getty Images

as "anti-development" and purveyors of a "campaign of misinformation"—he took pains to point out that "with or without the World Bank, we completed this massive project on our own."[37]

Resistance to the Narmada Dam drew attention to the harm, both environmental and social, that arose from India's post-independence addiction to large dams. Research undertaken by scientists and activists in the 1980s and 1990s showed that these problems combined to devastating effect. In the first two decades after India's independence, an estimated half-million hectares of forest was submerged by dams; this loss of land accelerated in the 1970s and 1980s with ever-larger projects like the Narmada scheme and the equally controversial Polavaram Dam in Andhra, along the Godavari River. The dams themselves suffered from the failure of their designs to take into account the quantity of silt the rivers carried. Their architects had underestimated the extent of the problem, as silt-heavy rivers clogged up reservoirs. The lifespans of the great Bhakra and Hirakud Dams—two of the first to be built after independence—was significantly reduced by higher rates of siltation than planners anticipated. Large dams also caused a major problem with waterlogging—inundating agricultural land beyond its capacity to absorb moisture and so rendering it infertile. An estimated thirty-three thousand hectares of productive land were lost as a result of the Tungabhadra Dam. Here we see yet another contrast—as India's arid regions drew down their water tables by pumping groundwater, well-watered areas near the headworks of large dams suffered from excess. As they interfered with the ecology of water, dams also created the conditions for water- and insect-borne diseases to thrive. Many studies around large dams in India and elsewhere showed a significant rise in the incidence of malaria, as large reservoirs and canals provided conditions for the anopheles mosquito to flourish.[38]

All the while, the social disruption caused by large dams continued unabated. As we saw, the most comprehensive estimate for the number of people that have been displaced by dams in India since independence reaches 40 million people. The projects with the highest cost in lives disrupted were those of the 1980s: two of the largest dams in the Narmada project, the Sardar Sarovar and the Narmada Sagar, displaced two hundred thousand people each. A high proportion of those displaced were marginalized *adivasis* (tribal peoples), who had little power to negotiate adequate compensation from the state.[39] The fate of these internal refugees—refugees from water development projects—too often goes unrecorded. The work of journalists like P. Sainath and environmental campaigning organizations like the NBA has brought some of their stories to light. Novelist Arundhati Roy found a wide international audience with a visceral essay on the profound costs of India's addiction to large dams, though her polemical style also drew criticism.[40] Others have turned to fiction to depict their suffering. In 2001, Vairamuthu—a prolific lyricist who has written the words to more than seven thousand Tamil film songs—wrote a novel, *Kallikaatu Ithihaasam* ("Saga of the Drylands"), which won him the Sahitya Akademi award in 2003, India's highest literary honor. Vairamuthu chose a historical setting from his childhood to explore the suffering of those displaced in the name of progress. As a child, in the 1950s, Vairamuthu lived in one of fourteen villages flooded by the Vaigai Dam in Madurai, in southern Tamil Nadu. The wide attention his work received had a striking contemporary resonance in the 2000s, when debates about dams and displacement raged in India.[41]

AFTER ITS NARMADA DEBACLE, THE WORLD BANK SUPPORTED THE creation of a World Commission on Dams in 1997, charged with assessing the benefits and costs of dam building worldwide over the previous half century. The commission's membership included strong supporters as well as opponents of dams, among them Medha Patkar—but its report, when it appeared in 2000, was more critical of large dams than most critics expected it to be. The commissioners estimated that on average dams were 56 percent overbudget, and that they delivered less irrigation water and hydropower than they promised. The commission's assessment of their environmental impact was equally bleak. Challenging the view that hydropower was an ecologically preferable alternative to the use of fossil fuels, the commission's studies pointed to the large greenhouse gas emissions from rotting vegetation in the reservoirs of large dams. It also pointed to the ecological consequences that Indian scholars and environmentalists had long highlighted—dams altered river flow to the detriment of aquatic habitats; they interfered with the paths of migratory fish. By impounding silt, they robbed lands downstream of fertility.[42]

One of the consultants to the World Commission on Dams was Ramaswamy Iyer, a career civil servant who served as India's secretary of water resources in the mid-1980s. Iyer's intellectual rigor and honesty set him apart, reflected in his willingness to change his mind. As water secretary, Iyer had taken for granted the value of large dams. He played an important role in pushing through government approval for the Narmada project. But by the end of the 1980s, he began to be influenced by what he called "newly emerging concerns about environmental impacts and the displacement of large numbers of people." Environmental thinking began to influence government decisions in the late 1980s, he recalled, but the growing force of popular opposition to dams led to what he describes as a "retreat from enlightenment" in the 1990s. Indian administrators and policymakers came to view Medha Patkar and all that she represented with antagonism, particularly after the World Bank's withdrawal of funding for the Narmada project. The Indian government's response to the World Commission on Dams was brusque dismissal. The cavalcade of arguments about the harmful effects of large dams fell on deaf ears. Searching and thoughtful in his analysis, Iyer turned to history for illumination. The fundamental problem, he discerned, was the persistence of a deep legacy of water engineering, going back to Arthur Cotton; this had bequeathed to India a Western tradition of water engineering, to which Iyer had no objection, "but also the underlying Promethean attitude to nature," which he had started to see as more problematic. To that tradition was added a distinctively postcolonial addiction to what Iyer called the "magic spell of gigantism."[43]

Far from retreating from dam construction, the Indian state redoubled its efforts in the 2000s. It embarked on a scheme to link India's rivers, through one of the largest and most expensive construction projects in human history. It plans to spend at least US$80 billion on a project to link thirty-seven of its rivers through 14,000 kilometers of canals, transferring 170 billion cubic meters of water across India. Among the promised benefits of the scheme are an additional thirty to thirty-five gigawatts of electricity and better water supply for irrigation. The roots of the river linkage scheme lie in the nineteenth century, in the dreams of Arthur Cotton. More proximately, it was the brainchild of irrigation engineer K. L. Rao, who had worked alongside Kanwar Sain and A. N. Kholsa to launch India's dam-building revolution after independence. The idea gained traction in the 2000s, under the coalition government dominated by the Hindu nationalist

Bharatiya Janata Party. In 2012, the Indian Supreme Court decreed that it was a matter of "national interest" and that the project should be completed as quickly as possible. Environmentalists have raised concerns about the project's consequences and its disruption to already fragile hydrological systems.[44]

In the years leading up to his death in 2015, Ramaswamy Iyer remained an eloquent critic of the scheme. "The project is in essence an attempt to redesign the entire geography of the country," he wrote; "underlying it is the old hubristic idea of 'conquest of nature.'" He argued that the water diversion project was based on a simplistic and dangerous view of India's hydrology; even to divide India simply into "water surplus" and "water deficit" areas, in ignorance of local ecology, was absurd. The problem went deeper, Iyer thought: "Rivers are not human artefacts; they are natural phenomena, integral components of ecological systems, and inextricable parts of the cultural, social, economic and spiritual lives of the communities concerned."[45] Iyer gave voice to a view of water ecology that was at odds with the conceptions of the Indian state, but it was a view of which we have seen echoes throughout *Unruly Waters*. It was a view that sustained numerous local initiatives that pushed against the juggernaut of large dams, including careful local efforts to restore ancient irrigation systems, through a system of small and simple check dams, in arid regions of Rajasthan.[46]

But still, "gigantism" prevails. At the time of writing, the river-linking project has been given renewed emphasis, though it is years behind schedule, and it is far from clear that it will ever be realized. The counterpart project in China is much further advanced. China's own river diversion project seeks to redress the country's inequalities in the distribution of water by redistributing it on a massive scale. The South to North line, from Danjiangkou reservoir in central China to Beijing, opened in 2014; it is the most expensive infrastructure project the world has ever seen. Two-thirds of Beijing's tap water now comes from Danjiangkou, almost nine hundred miles away. Another arm of the diversion project, the "eastern route" that follows the old Grand

Canal, opened in 2013. Already, the diversion scheme has brought similar problems to those predicted in India—heightened water conflicts, wastage, social disruption, and substantial ecological harm to riverine ecosystems. The most ambitious part of the project—the western line, linking the headwaters of the Yellow and Yangzi rivers across the Tibetan Plateau—lies in the future, and it is the most likely to cause problems for China's neighbors.[47]

The ecological and social effects of dams have been well documented; over the past decade, new scientific research suggests that, cumulatively, the world's dams exercise a fundamental geological impact on Earth. The sheer scale of water engineering in the second half of the twentieth century is changing the shape of the world's most densely populated river deltas, which are now denied up to a third of the sediment from flowing rivers that have, over thousands of years, built up the deltas and replenished them. On one estimate, reservoirs have increased by 600 or 700 percent the amount of water held in the world's major rivers, but much less of it now reaches the sea. The once mighty Indus, like the Yellow River, is now a trickle by the time it reaches the Arabian Sea—dammed and diverted into a web of canals, many of them first built by the British in the late nineteenth century. The effect of hydraulic engineering has been to put coastal settlements—and mega-cities, above all—at greater risk of flooding, even before we take into account the effects of climate change and sea level rise.[48]

———

AS THE RISKS OF CLIMATE CHANGE BECOME INCREASINGLY EVIDENT, water becomes ever-more central to political and strategic conflicts at the heart of Asia. In the face of ecological uncertainty and strategic competition, the Himalayas are home to the greatest concentration of dam construction projects in the world. In historical perspective this marks the final frontier in a conquest of water that began in the nineteenth century. From the 1880s, as European and Indian explorers reached the source of Asia's great rivers in the Himalayas, it was known that the interaction of the Himalayan rivers and the monsoons held the key to Asia's water supply. As territorial borders sharpened up in that era—the border between British India and Tibet, for example, was marked by the McMahon line in 1914—there came the faint knowledge that struggles over water may lie in wait. Even when imperialism was overthrown in Asia after the end of the Second World War, the problem of transboundary rivers arose directly only in relation to the partition of India. As late as 1960, as we have seen, Indian intelligence agents dismissed reports that the Chinese were planning to dam the Brahmaputra, arguing that they had neither the labor nor the infrastructure to do so.

That Indian assessment was not mistaken. But things changed rapidly in the 1980s. It was in that decade that the Chinese state's dam-building ambitions fixated upon the Tibetan Plateau, source of Asia's rivers. By the 1980s, the large-scale settlement of Han Chinese in Tibet had changed the composition of the region's population; the construction of roads and railways made it less remote from the lowlands and river valleys. Above all, China's frenetic economic growth produced a demand for energy—and an uncomfortable dependence on imported oil—that the hydroelectric potential of the mountain rivers promised to meet.[49] As long as the Himalayan source of Asia's great rivers remained remote and forbidding, it mattered little who formally controlled them; but if the unruly waters were to be tamed, it mattered profoundly who brought them under control.

As I write this, more than four hundred large dams are planned in the Himalayan regions of India, Nepal, Bhutan, and Pakistan. Construction is already underway on many of these projects. A further one hundred dams are planned on the Chinese side, where so many of the rivers originate.

If these projects come to fruition, there will be a dam every thirty-two kilometers along the Himalayan rivers, making it the most heavily dammed region in the world. A secretive complex of public and private interests converge and compete to harness the waters of the high mountains. The hunger for energy is widely shared across the region, though demand is driven primarily by the voracious needs of China and India. Geopolitical rivalries play out in the negotiations over who will build the dams, and on whose terms. As multiple dams line up along the same river valleys, the risk to downstream users is grave. In the Indian states of Arunachal Pradesh and Assam, there are fears about Chinese plans for the Brahmaputra upstream in Tibet, where it is called the Yarlung Tsangpo. Downstream, Bangladesh is most vulnerable of all. Already in the 1980s, Bangladesh protested the effects of India's Farakka barrage, built in 1975 to divert water from the Ganges to the Bhagirathi-Hooghly, in part to revive the port of Calcutta that had, since the mid-twentieth century, suffered from severe silting. By reducing the river's flow to Bangladesh, the dam had an impact on soil fertility, irrigation, and health. With an increasing number of dam projects upstream, the risk to Bangladesh has multiplied.[50]

In one respect, the latest wave of dam construction departs from the precedents of the 1950s and 1960s—it is financed in a different way. Until the 1990s, large dams in India and China were financed primarily by the governments, with India receiving additional funding from international financial institutions like the World Bank, and China, until the split with the Soviet Union in 1961, benefiting from Soviet aid. The new rush to build dams depends more heavily on private capital. In India, public sector organizations like the National Hydroelectric Power Company and the North Eastern Electric Power Company play a major role in dam construction; but so too do private companies like Tata Power (architect of one of India's earliest hydroelectric dams, in the 1910s), and Reliance Energy. State governments have raised capital from domestic markets as international organizations have backed away from funding large dams. But the biggest shift is the role of China. China's dam-building industry, in the late 1990s and early 2000s emerged as a major force in the world. Given the scale of China's own dam building, the depth of engineering expertise in China rivals anywhere in the Western world; and that expertise has been matched by money. By 2008, ten Chinese companies were involved in thirteen dam projects in Nepal and nine in Pakistan, many of them financed by Chinese state-owned and private banks. When India's leading hydraulic engineers had visited China in 1954, they had found their Chinese counterparts dependent on Russian expertise, having to make do and improvise. By the end of the century, the Chinese dam industry led the world.[51]

Such is the rush for growth that warnings about the impact and the potential risks of these new Himalayan dams have been brushed aside. Environmental assessments on many of the projects have been cursory at best. Given that the dams are entwined with geopolitical and security considerations, given that governments around the region fear popular protest against the dams—which has been widespread not only in India, but increasingly in neighboring countries, too—considerable secrecy shrouds the plans. Even data about river flow across borders is guarded as a state secret. The Himalayan region is less densely populated than the river valleys, but the same problems that accompanied the large dams of the twentieth century are likely to follow here—drowned lands and displaced people. Large reservoirs are less common at these heights than in the lowlands, but diversions to the course of rivers affect life on the river. Mountain species are under threat from the loss of their habitats—already, the brown bear, the snow leopard, the musk deer, the golden mahseer, and the snow trout are imperiled. Much of the power generated by the large dams will be sold to large cities far away, while many local livelihoods are imperiled. The Lower Subansari Hydroelectric Project in the northeastern Indian state of Assam, one of India's most controversial—it has been stalled repeatedly by local protests—threatens the passage of country boats that carry a lot of local trade. The submergence of forest lands will deny local people their main source of firewood. Historian Rohan D'Souza describes the Brahmaputra as a "moving inland ocean" bound together by the rhythms of subsistence fishing and floodplain agriculture—a system that is under threat from the dam. The now familiar problem of siltation menaces many of the dams. But this is also one of Earth's most active seismic zones, with earthquakes of 8.0 or more on the Richter scale not uncommon. Fan Xiao, a geologist from Sichuan and a brave opponent of recent mountain dam projects in China, fears that dams will become "a source of permanent grief and regret for future generations yet unborn."[52]

———

THE GRAVEST RISK OF ALL—TO THE DAMS, TO THE HIMALAYAS, TO billions of people downstream—comes from climate change. Climate change affects the Himalayan glaciers two ways: by changing patterns of snowfall and by hastening the process of melting. Research findings are complex—not all glaciers are in retreat, and, more seriously, there are very few monitoring stations and few long-term studies. While research has been ongoing in the Chinese Himalayas since the 1990s, the Indian side has been virtually untouched by scientists. The inclusion (and later retraction) of a careless claim by the Intergovernmental Panel on Climate Change (IPCC) about the speed at which the Himalayan glaciers are melting was wielded by climate change skeptics to try to discredit the organization's work. But the consensus is overwhelming that the warming of the planet has led to a recession of the Himalayan glaciers since the mid-nineteenth century and at an accelerating pace in recent decades, if not uniformly everywhere across the mountain range. Most models predict that river flow will be augmented in the short term by the melting of the glaciers—bringing more frequent and severe floods, and even the risk of catastrophic dam collapses. Few observers believe that the designs for the large Himalayan dams have taken into account the uncertainties of climate change. The dangers are greater still given the heightened possibility of extreme rainfall—which, as we shall see, is likely. Around the middle of the twenty-first century, by 2050 or 2060, scientists predict that the dry season flow of the major Himalayan rivers will see significant declines. Not only will this diminution make many of the planned dams ineffective, it will put many lives and livelihoods at risk. More than 1.3 billion people rely directly on the Himalayan rivers for water; 3 billion people rely on the food, water, and energy the Himalayan rivers provide. Changes in the flow and behavior of the rivers as a direct result of the warming of the glaciers threatens a significant proportion of humanity.[53]

III

The monsoon has been a continuous thread through *Unruly Waters*—and it is with the monsoon that we conclude.

The breakthroughs in tropical meteorology of the late twentieth century shed new light on the scale and complexity of internal variability in the monsoon on multiple timescales—from the quasiperiodic impact of the ENSO system to the intraseasonal variations attributed to the Madden-Julian Oscillation. In recent years, the focus of scientific research has been on how the effects of anthropogenic climate change interact with the monsoon's natural variability in dangerous and unpredictable ways.

The most fundamental forces driving the monsoon, as we have seen, are the thermal contrast between the land and the ocean, and the availability of moisture. Climate change affects both of these drivers of wind and rain. The warming of the ocean's surface is likely to augment the amount of moisture the monsoon winds pick up on their journey toward the Indian subcontinent. But if the ocean surface warms more rapidly than the land, which appears to be happening in equatorial waters, this would narrow the temperature gradient that drives the winds, and so weaken circulation. Put simply, many climate models predict that the first of these processes will predominate: "wet gets wetter" as a result of greenhouse gas emissions. They predict, that is to say, that the moist monsoon lands will see an increase in rainfall. But the monsoon is an intricate phenomenon, as meteorologists have long known. It is increasingly clear that monsoon rainfall is affected not only by planetary warming but also by transformations on a regional scale, including the emission of aerosols—from vehicles, crop burning, and domestic fires—and changes in land use. The urgent challenge for climate science is to disentangle and to understand these global and regional influences on the behavior of the monsoon. And so far, the monsoon has proved much harder to capture in models than, say, global temperatures.[54]

The availability of detailed records of climate and rainfall in India—which themselves are a product of the history of Indian meteorology going back to the efforts of Henry Blanford and his colleagues in the late nineteenth century—have allowed scientists to reconstruct in detail the monsoon's behavior over the last sixty years. The picture these data present is complex, and in some ways surprising. Average summer rainfall over India has declined by around 7 percent since 1950. But what lies behind this trend?[55] The cause of the decline in rainfall lies in the pattern of India's development since independence. Its explanation, that is to say, lies in the province of economic history.

In the late 1990s, research vessels observed exceptionally high concentrations of aerosols in the northern Indian Ocean. Satellite images showed a stain that spread across the Gangetic plain and over the Indian Ocean—researchers called it the "brown cloud," an accurate if not a poetic description of the haze. Between January and March 1999, a large team of investigators set out to understand this brown cloud, taking readings from their base at the Kaashidhoo observatory on one of the most remote islands of the Maldives. The project was led by Veerabhadran Ramanathan, an Indian oceanographer based at the Scripps Institute in La Jolla, California. One of the scientists involved was Dutch atmospheric chemist Paul Crutzen, who around the same time also coined the phrase "the Anthropocene," referring to a new geological epoch in which human activity is the most important influence on Earth's physical processes.[56]

The project found that the haze was a noxious composite of sulfate, nitrate, black carbon, dust, and fly ash as well as naturally occurring aerosols including sea salt and mineral dust. Three-quarters of the composition of the brown cloud could be attributed directly to human activity especially concentrated along the densely populated Gangetic plain and northwestern India. In this region, where up to 80 percent of the population remains rural, and where many rural families continue to be deprived of electricity, much of the black carbon is produced by domestic burning of biomass—wood, crop residue, dung, and coal—used primarily for cooking. Open crop burning accounts for the rest. The stoves used in households are inefficient and combustion is incomplete, producing large amounts of soot. Apart from their likely effects on regional climate, these emissions also poison human bodies. On one estimate, more than four hundred thousand premature deaths each year in India can be attributed to indoor pollution. Black carbon combines, in the brown cloud, with sulfates and other aerosols—and the Gangetic plain bears an additional burden in this respect, as a result of pockets of intensive industrial and extractive activity. Since the late nineteenth century, the Indo-Gangetic plain has been the core region of India's extractive industries, built around the rich coal and mineral deposits in the Chota Nagpur region. Further along Yamuna River, the Delhi region is one of India's fastest-growing metropolitan areas, and its largest in absolute terms. Emissions have increased exponentially since the 1970s as India's population has grown, as its economy has expanded, as inequalities within and among regions have widened. The Gangetic plain suffers from a double pathology: the sulfur, carbon, and nitrogen dioxide emissions that accompany energy-intensive growth are combined with the black carbon that comes from the use of cheaper, dirtier fuels by millions without access to electricity. If India leads in black carbon, China, too, has a brown cloud problem, with sulfates from factory emissions dominating the mix there.[57]

All of this is shifting the monsoon's patterns. Aerosols absorb solar radiation, allowing less of it to reach Earth's surface. This cools the land, diminishes the temperature contrast between the land and sea, and weakens the atmospheric circulation that sustains the summer monsoon. Changes in circulation over the Indian subcontinent in turn affects the tightly integrated air-sea interaction that binds the Asian continent with the Indian Ocean, a system that already contains plenty of internal variability. Because of the way the Asian monsoon is linked to other parts of the planet's climate, it is possible that aerosols over South Asia and China have global consequences. When all of these effects are coupled with the impact of global warming on the ocean and the atmosphere, the instabilities multiply. Far from counteracting the effect of greenhouse gases in any simple sense, the impact of aerosols complicates them.[58]

A further driver of regional climate change is rapid changes in land use. Over the last 150 years, forest cover over most parts of Asia has declined dramatically. The intensification of agricultural production in India, and the use of more water for irrigation, have affected the moisture of the soil, its capacity to absorb or reflect heat. Crops reflect more solar radiation than forests, which tend to absorb it; the greater reflexivity of land planted with crops makes it cooler, once again weakening the temperature differentials that drive circulation and rainfall. Tropical meteorologist Deepti Singh points out that climate models have often failed to predict the monsoon's behavior in part because they are too abstract to take into account the "complex topography, temperature and moisture gradients in the region that can influence the monsoon circulation." The models omit, that is, precisely the details of landscape and microclimate that the meteorologists of a century earlier were so deeply interested in, which they depicted in their detailed local and regional maps of India's climate.[59]

We are left with the most bitter of ironies. Many of the measures taken to secure India against the vagaries of the monsoon in the second half of the twentieth century—intensive irrigation, the planting of new crops—have, through a cascade of unintended consequences, destabilized the monsoon itself. When the geographers of the early twentieth century wrote of "monsoon Asia," they saw the monsoon as sovereign—it shaped the lives of hundreds of millions of people, who waited on its every move. Monsoon Asia means something quite different now, when the monsoon's behavior, increasingly erratic, responds to human intervention.

AT ONE LEVEL, THE STORY OF HOW THE MONSOON HAS CHANGED
since the 1950s is a story of India's resilience. India has experienced
more droughts since the 1940s than in the half century before that,
a half century that saw so many devastating famines. Even on a
shorter timescale, there are signs of progress. In 2014 and 2015,
India experienced two successive years of drought that were as
severe as the monsoon failures of 1965 and 1966, which—as we
have seen—India could only ride out with massive external aid. In
2014–2015, there was no noticeable drop in agricultural production, which observers attribute to better planning—but also to much
better forecasts, enabled by the advances in meteorological understanding, and technology, that took root in the 1970s and 1980s.
Intraseasonal oscillations—the MJO and the Boreal Summer Intraseasonal Oscillation—have become more amenable to prediction,
improving forecasts on a timescale of two to four weeks. Alongside
a general drying trend, the monsoon has grown more prone to extremes over the past several decades. If India has received less rain
overall, more of it has come in torrents. Between 1981 and 2000,
wet spells have been more intense, while droughts have been more
frequent but less intense.[60] From the nineteenth century, understanding and predicting the fearsome cyclones that visit the Bay of Bengal
with regularity prompted the development of meteorology in India—
just as the menace of typhoons spurred research in the Philippines
and along the China coast. Predictions of the impact climate change
will have on the development of cyclones are as uncertain as those
that seek to model the overall behavior of the monsoon. The same
countervailing forces are at work: warming seas are, in theory, likely
to produce more cyclones—but not if the seas are warming faster
than the land. A more definite finding is that the Bay of Bengal's cyclones have grown in intensity in recent decades, as have hurricanes
in the Atlantic and tropical storms in other parts of the world. In
the Bay of Bengal, scientists predict that climate change will, in the
coming century, lead to fewer but more powerful cyclones—though

it is possible that the Arabian Sea, not known for cyclones, could see an increase.[61]

Nowhere in the world have tropical storms affected more lives than in Bangladesh. In the 1860s and 1870s, severe cyclones in that region of eastern Bengal, then part of British India, spurred the development of meteorological science. In the second half of the twentieth century, cyclones have been more frequent and just as devastating. Approximately 40 percent of global storm surges in the last fifty years have hit Bangladesh, including the two with the highest death tolls, in 1970 and in 1991. Five of the ten worst storms to affect any part of Asia in the twentieth century have struck Bangladesh.[62] But the past twenty years have witnessed a dramatic reduction in cyclone mortality in Bangladesh. Cyclone Sidr, which struck Bangladesh in 2007, was as severe—in terms of wind speed and rainfall—as cyclone Bhola of 1970, but the death toll was one hundred times smaller. An estimated five hundred thousand people died in the cyclone of 1970; in 2007, that number was below five thousand. In part this is a tribute to improvements in forecasting. The Bangladesh Meteorological Department's ability to track cyclones as they develop in the Bay of Bengal improved significantly, with assistance from a Japanese satellite as well as data from the US National Oceanic and Atmospheric Administration. After the fearsome cyclone of 1991, the Bangladesh government embarked on the construction of thousands of cyclone shelters, which have saved millions of lives. Cyclone warnings have become more effective, helped greatly by the spread of mobile phones to even the poorest villages. Changes in the landscape have also played a role. Coastal embankments have kept floodwaters out, although their impact on local ecology has been more controversial. An extensive program of mangrove reforestation has helped to restore one of the most effective natural flood defenses to parts of Bangladesh's low-lying coast.[63] But in Bangladesh, as in India and elsewhere in Southeast Asia, real gains in protecting people from tropical storms contend with a series of new risks—risks that the weather will become more extreme and less predictable, and manufactured risks in the form of unregulated coastal construction, rising population density, and galloping social inequality.

Research is underway into what governs the increasingly erratic, increasingly extreme behavior of the monsoon. It likely stems from the interaction, on multiple levels and over different timescales, of planetary warming, regional climate change, and natural variability. Recent advances in the oceanographic study of the Bay of Bengal make clear how much the sea's chemistry itself affects climate. The bay is less salty than most bodies of water because of the vast discharge of freshwater from the Himalayan rivers, and because it receives more rainfall than any other sea. This has implications for ocean circulation, temperature differentials, and the interaction of ocean and atmosphere, but the forces at work are still the subject of intensive research.[64] Some of that research looks to the deepest past for clues about the future. Satellites allowed for a new appreciation of climatic forces in the late twentieth century; now the seabed is the next frontier. In 2015, the 470-foot vessel *JOIDES Resolution,* equipped with a 200-foot drilling tower, collected sediment cores from the sea floor under the Bay of Bengal. Scientists seek a record of the monsoon's behavior going back 15 million years, embedded in the sunken fossils of microorganisms known as plankton foraminifera that once inhabited the surface water and now lie buried. The project aims to use that deep historical data to predict the monsoon's future behavior under conditions of global warming, by examining how the monsoon has responded to historic changes in temperature, salinity, sea level, and atmospheric carbon. It is ironic, perhaps fitting, that the *Resolution* was once an oil-drilling vessel, now converted to the more benign purpose of oceanographic investigation.[65]

ON JULY 26, 2005, 37.2 INCHES OF RAIN BATTERED THE CITY OF Mumbai, most of it between 2:30 p.m. and 7:30 p.m. A third of the city

Map of Mumbai, showing it flanked by the Arabian Sea to the west and Thane Creek to the east—much of contemporary Mumbai sits on reclaimed land. CREDIT: Illustration by Matilde Grimaldi

was flooded. Cellular phone networks crashed. The airport was shut when its runways flooded. Close to 150,000 people were stranded in stations on Mumbai's massive commuter rail network, which came to a standstill. Almost one thousand people died, tens of thousands were made homeless. The government was completely unprepared when faced with this extreme amount of rain, even though such downpours were not completely without precedent in Mumbai's history. To many observers, this was a freak of nature—or an act of God.

But a citizens' commission assembled by the city's nongovernmental organizations to report on the floods reached a different

A taxi under water during the Mumbai floods of July 26, 2005. CREDIT: *Hindustan Times*/Getty Images

conclusion. The commission argued that Mumbai had put itself in harm's way. After decades of relentless growth and expansion, Mumbai had few natural drainage channels left. They had been concreted over. There was nowhere for the water to go. Storm drains were clogged with waste, tidal flats had been built upon. What environmental regulations there were on paper could not rein in a boom in unauthorized construction—in a city where prime real estate was worth more than in New York or Hong Kong. The destruction of the mangroves of Mahim Creek—which stretched to seven hundred acres as late as 1930—for highway construction and urban development robbed the city of a natural buffer between land and sea.[66]

Beyond the suffering that the storm caused, it also provided a stark warning. If, as scientists predict, the "once-in-a-hundred-year" storm is likely, in the future, to materialize every ten or twenty years, or perhaps more regularly than that, Mumbai is acutely vulnerable, along with so many cities at the water's edge. The threat to the coasts, once again, comes from the sea—fueled no longer just by natural patterns but also by human activity that is once regional and planetary in its sources and its effects.

The prospect of the next big storm hitting Mumbai is the alarming picture that Amitav Ghosh sketches, powerfully, in his nonfiction work on climate change, *The Great Derangement*. Ghosh forces us to imagine Mumbai in a superstorm:

> At this point waves would be pouring into South Mumbai from both its sea-facing shorelines; it is not inconceivable that the two fronts of the storm surge would meet and merge. In that case the hills and promontories of South Mumbai would once again become islands, rising out of a wildly agitated expanse of water.

But in the face of catastrophe that is "inconceivably large," Ghosh argues that most states, like most human beings, are guided by "the inertia of habitual motion."[67]

In recent years, the view that we should live with and adapt to the natural hydraulic risks of littoral zones—to say nothing of how these risks are worsening with climate change—has infiltrated the worlds of architecture and design. Mumbai architects and urban theorists Anuradha Mathur and Dilip da Cunha insist that the colonial and postcolonial practice of drawing a firm boundary between land and water in Bombay stems from a fundamental misreading of the fluid coastal landscape. Mumbai during the monsoon demands to be seen "in cross-sectional depth," they argue, not in the two dimensions of maps and plans. When the monsoon comes, "there is too little time and too much water to make an orderly exit through courses delineated on maps." In their vision, the city in monsoon becomes a fluid, mutable organism at the boundary between land and water, shaped by the interplay of "the monsoon clouds above through the labyrinthine world of creeks, to the web of aquifers beneath." Only if we understand this, design for it, adapt to it, they argue, can we live with, rather than trying to engineer away, risk.[68]

Working within a tradition that goes back a century—to Blanford and Isis Pogson and Ruchi Ram Sahni—Indian meteorologists have a distinctive understanding of the climate risks facing India today. However much patterns of rainfall may be changing, they suggest, the monsoon has *always* presented a risk to South Asia: the fundamental source of escalating risk today lies in foolhardy policies.

This is what emerged in my conversations with S. Raghavan, formerly a senior officer in the Indian Meteorological Department, now retired in Chennai, where we met at his home. His father was a large farmer in an arid tract of rural Tamil Nadu. Raghavan grew up with an intimate knowledge of water and crops; long before he became a meteorologist, he recognized the rhythms of the monsoon. After taking a degree in physics at Madras University, Raghavan received three job offers: one from All India Radio, one from the auditor general's office, and one from the meteorological department. With little knowledge of meteorology, he took that option, excited in part by the chance to work with the latest technology, including the radiosondes he had seen on display at a stall in his college's engineering fair. At the height of the Cold War, Raghavan was sent to the United States on a government scholarship to study radar technology. When Delhi's Safdarjung Airport received its first radar in 1957, Raghavan was put in charge of its operation. In 1972, he returned to Madras to take charge of radar meteorology there, equipped with a cyclone warning radar purchased from Japan, which arrived only after a long tussle with the customs department at a time when India had stringent restrictions on imports and foreign exchange. That year, a serious cyclone struck the coastal town of Cuddalore. It was the first cyclone in India to be tracked by radar as it approached; with accurate information and early warnings, casualties were minimal. It was then, Raghavan said, that "I realized that I was doing some service to society."

But just as forecasting capacities improved, in the 1980s, the risk posed by extreme weather in India multiplied. The cause was manufactured, not climatic, he said. The reason so many millions who live in coastal India are in danger, he told me, was because governments, planners, developers, and citizens had completely neglected the ordinary climatic risks that coastal South Asia faces. "Time and time again, we put ourselves in harm's way," he said. Raghavan does not believe that the risks of cyclonic storms, for example, can be engineered away; at best they can be prepared for. He believes in early warning, and for that there is no substitute for the patient observation of weather fronts developing in the Bay of Bengal. His mission in retirement is the production of a Tamil lexicon of climatological terms, in the same way that the colonial meteorologists of the late nineteenth century were not averse to collecting local proverbs. He regularly gives talks to schools and residential societies about climate and weather. He described the destruction of cities' natural drainage and storm defenses. As late as the 1940s, he remembered seeing Chennai's Cooum River busy with traffic, including boats carrying salt from Andhra; it had become a "cesspit," he told me. Drains were blocked by a "plastics explosion"; the destruction of mangroves had taken away natural protection against storm surges. "Our own actions are responsible" for the crisis, he said.

The soft-spoken Mr. Raghavan was careful and precise in his judgments; as we talked, he often turned to his shelves to find a book, or to consult a folder of press clippings that he had maintained over many years. The day after we met he sent me a PowerPoint presentation he had made for a recent lecture. But there was no mistaking the emotion in what he said—he was both sad and angry at the way the risks of a monsoon climate had been disregarded. His was a view of the weather, and the climate, that was rarely about control—it was about adapting to known and felt risks. The long quest of India's meteorologists to understand the monsoon continues to shape their responses to a changing climate.

THESE ARE NOT, FOR THE MOST PART, THE LESSONS THAT ARE BEING learned. The map of cities at risk resembles a series of beads on a necklace threaded along the coastline of Asia. One study predicts that by 2070, nine out of the ten cities with the most people at risk from extreme weather will be in Asia—Miami is the only non-Asian inclusion. The list includes Kolkata and Mumbai in India, Dhaka in Bangladesh, Guangzhou and Shanghai in China, Ho Chi Minh City and Hai Phong in Vietnam, Bangkok in Thailand, and Yangon in Myanmar.[69] Each one of these cities will confront any change in the interaction of land and water, winds and rain over Asia's oceans. Just two years after the Mumbai floods, it was the turn of Jakarta, the Indonesian capital and the fastest-sinking city in the world, pulled down by the weight of construction, by the extraction of groundwater, and by the rising sea. Jakarta is sinking by between three and six inches every year. The storm in 2007 washed over the sea walls built to protect the city. Half the city was underwater, displacing 340,000 people from their homes. In Jakarta and in each of Asia's coastal megacities, climate change compounds a cavalcade of risks that are severe in and of themselves—hasty development driven by property speculation and new forms of middle-class consumption, crumbling health and sanitary infrastructures, and a lack of preparedness and precaution, are all symptoms of profound social and economic inequalities both among and within nations. Of all the countries in the world, few are more directly under threat than low-lying Bangladesh.[70]

IV

The struggle for water transcends Asia's borders. The Himalayan rivers, dammed and diverted and vulnerable to changes in glacier cover, flow through many nation-states on their descent to the sea. Planetary warming is a result of the historical emissions of fossil fuels—initially and cumulatively by the wealthy and industrialized countries of the world, but also, and increasingly since the 1980s, by China and India. Global warming interacts with and compounds

the effects of regional climate change. Aerosol emissions from the
Gangetic plain or from fast-industrializing areas of China have effects far beyond India's or China's borders, creating a series of brown
clouds that blanket the Indian Ocean and affect rainfall far away.
Climate change creates problems of distance—between the source
of pollution and its consequences—but it also creates new forms of
proximity in the form of shared risks and interdependence. The image of the Himalayas as "Asia's water tower" conveys both the scale
of the hydraulic system that binds much of Asia, and the scale of the
threat that they face from the destabilization of that source of so
much water. By the 1960s, the sea itself was a form of territory. The
Bay of Bengal was the crucible of the earliest monsoon science in the
nineteenth century; it remains the crucible of monsoon science today.
But it is a very different sort of space. It is crowded. It is contested.
It is walled off by borders in the sea as much as on land.[71] Even international cooperation in oceanographic research on the monsoon
has to confront the reality of borders at sea. A major project between
2013 and 2015 set out to investigate the Bay of Bengal and its role
in monsoon circulation; it brought together American, Indian, and
Sri Lankan scientists. Their research vessels roamed the Bay for two
years, taking an enormous number of measurements of ocean salinity, temperature, currents, and chemistry. Yet the map of their voyages, a dense set of tracks that the ships followed over those years,
is divided up by a thin line marking the extent of territorial waters
and exclusive economic zones; some, like the border of Myanmar, the
ships could not cross for political reasons.[72]

If borders at sea are forbidding, those on land are even more so.
Throughout Asia one of the ways in which communities have coped
with extreme weather has been to move—often temporarily, and not
necessarily over long distances. For regions that are threatened by
climate change and water-related risks, borders create barriers to
mobility. "Climate refugees" are much discussed in current legal and
political debates. But the Red Cross rightly stresses that the "popu-
list term 'climate refugees' is profoundly misleading": environmental

drivers of migration act "in conjunction with economic, social and political factors, and [are] linked to existing vulnerabilities," and it is "conceptually difficult to establish a precise category of environmental or climate migrant."[73] It would be a mistake to separate a discussion of "climate migration" from a broader consideration of regional patterns of mobility.

There is an odd historical resonance to some of the pronouncements about climate and migration. In the nineteenth century, too, many observers saw the movement of people across the Indian Ocean and the South China Sea as driven by climate—not by climate change but by climate's natural volatility.[74] The use of liquid metaphors to describe migration remains pervasive: a language of "floods" and "tides" and "waves" and "flows." Many of the region's migrants today come from places and from communities that have been mobile in the past. This is hardly a surprise. Some of the places most threatened by environmental catastrophe are also places—the coasts and the great river deltas—that have the longest histories of migration. But other affected regions lack the accumulated family connections, knowledge, experience, and access to credit to allow them to move. Forced immobility can be as dangerous, as traumatic, as forced migration. Controls on mobility have intensified since the middle of the twentieth century, and they are likely to harden: hysteria in India about "illegal migration" from Bangladesh, for instance, has led to the securitization and fortification of the border, though many people risk their lives to cross it out of desperation. The slow effects of climate change are as likely to leave people stranded, unable to move, as they are to spark a rush of "climate refugees."[75]

A recent study by the World Bank makes clear that although cross-border migration receives more attention, vastly more people migrate within their own countries than migrate internationally. The overwhelming majority of people who are displaced by climate change over the next three decades will move internally—an estimated 40 million people in South Asia alone, and 143 million people globally.[76] In the depersonalized language common to climate policy documents, the World Bank concludes that "several hotspots of climate in- and out-migration are in transboundary areas" of South Asia, and that these "must be explored for their opportunities and managed for their challenges."[77] But what does this really mean? It means the options facing people whose lives are threatened by drought or deluge will be constrained by borders as well as by poverty, gender, caste, or a lack of opportunity. It means that the closest refuge, if it should lie across a border, may not be a refuge at all. It means that many routes that make social, cultural, or ecological sense to people—routes embedded in family histories, routes across regions that have not always been divided by borders—will be blocked. It means that those who are compelled to cross closed frontiers in search of security will face unprecedented risks.

GIVEN THE WEIGHT OF BORDERS, ARE THERE PROSPECTS OF CLOSER regional cooperation to confront the problems of water and the threat of climate change? If so, these prospects are modest in scope and ambition. Existing regional institutions—the Association of Southeast Asian Nations (ASEAN), the newer and smaller Bay of Bengal Initiative for Multisectoral Technical Cooperation (BIMSTEC)—are focused overwhelmingly on the development of infrastructure and the promotion of trade. Though environmental protection is not absent from their concerns, it is not a high priority. When policy documents refer to climate, it is often as a metaphor, as in the often expressed hope of creating a "climate friendly to investment."[78] And when new infrastructure projects threaten ecological and social harm—as do so many of the port projects that proliferate along the Bay of Bengal's coasts—they have almost always proceeded regardless, except where they have met with significant public protest. Nevertheless the second half of the twentieth century did create agreements and institutions to manage water across borders, and these need to be strengthened wherever possible. Though flawed, the Indus Treaty signed in 1960 between India and Pakistan, with the World Bank's mediation, has largely worked. The two hostile neighbors have for the most part worked cooperatively to manage that shared river, though there have been periodic surges of tension between them. The Mekong River Commission, created in the 1950s by the UN, has outlasted the Cold War. Though the commission has often failed to prevent reckless development along the river, the growing involvement of China in its discussions suggests that states are taking more seriously the shared threats they face.

But the most promising initiatives to address shared risks may lie in the realms of science and civil society. From the start, climate science in Asia has been a cosmopolitan enterprise. In the late nineteenth century, observatories and scientists across imperial borders exchanged data and theory and reports. This is not to say that climate science stands apart from politics. It never has. In the nineteenth century, the development of meteorology was deeply entwined with imperial interests. But climate science provided a way of visualizing Asia beyond borders, as a vast and connected climatic space, bound together in every dimension—the oceans, the air, and the land. Meteorologists saw that the same storms menaced the Philippines as India. Growing knowledge of climatic connections inspired attempts to share warnings if not coordinate responses. In an era of nation-states, that level of cross-border cooperation among scientists has continued—and it is more vital than ever. Even as an organization like BIMSTEC is hampered by political tensions among its member states, it has made small but tangible gains in coordinating the sharing of early warnings to bolster disaster preparedness.[79] As meteorologists' ability to forecast storms has advanced with improvements in satellite technology as well as better models, that information is now more readily accessible to a wide public—mobile phones are ubiquitous across South Asia; even the smallest fishing vessels are now equipped with GPS technology.

Some of the most promising recent efforts to increase cooperation across borders to tackle Asia's water problems have focused on the sharing of information. As we have seen, data concerning the hydraulics of the Himalayan rivers are a closely guarded secret. The Third Pole, a nongovernmental organization based in London and New Delhi, dedicated to understanding and communicating the cross-border water issues faced by Asian states—with a focus on the Himalayan rivers—has compiled as much information as is available on river flow and on climatic trends. Using open source data, it has created a new mapping platform that allows for the sharing of data on river flow and hydropower, glaciers, and groundwater. This is now readily available to journalists, activists, and scholars. These maps of the Himalayan region transcend borders, emphasizing shared ecological challenges. The ability to visualize the risks holds the promise of stimulating a more coordinated response; it might even inspire new solidarities that come from a sense of shared vulnerability. These efforts to pool information have begun to mobilize public participation by so-called citizen scientists. Season Watch, an Indian organization, encourages its members, including schoolchildren, to submit detailed daily observations of climate, thereby linking very local experiences of changing seasonal cycles with changes on a regional and global scale.[80] This effort extends to the preservation of local archives. The World Meteorological Organization has urged the importance of "data rescue"—the recovery of records and logs of rainfall and temperature, often handwritten, preserved in local repositories and threatened by physical deterioration. These are potentially invaluable to climatologists looking for long-term patterns of change. Those very archives, as we have seen throughout this book, are full of evidence of the ways in which, in earlier times, not only storms and currents but scientific information crossed borders.

From the 1980s onward, there has been close cooperation among environmental activists across Asia. They have pooled information, campaigned together, and recognized that environmental degradation—including but not limited to climate change—is a menace they all face. Historian Prasenjit Duara sees reasons for hope in the organizations of what he calls "network Asia"—the web of NGOs, some of them religiously motivated and others resolutely secular, coming together to confront problems of water and climate. The environmental movement has at times been genuinely transregional, yet in both India and China, it has become clear in recent years that environmental organizations are vulnerable to crackdowns by the state. There is also an imaginative barrier to cross. As we have seen, the power of environmental activism very often comes from the ability to evoke a sense of emotional attachment to particular landscapes. Narratives about the past have been fundamental to the rise of environmentalism in India and elsewhere in Asia, but the pasts they have appealed to are profoundly local ones; their narratives juxtapose an earlier age of ecological innocence with the depredations of colonialism and modernity. An appeal to nationalism has been, and remains, one powerful way that environmentalists can mobilize public support. But this can make it more difficult to work across borders.[81]

At a time of environmental crisis, local histories and national responses are insufficient on their own. It is now easy to see on a map, or in an alarming graphic, the scale of water-related risks that Asia faces. It is clear that those risks pay no heed to borders. The promise of a new sort of environmental history, a more connected and expansive history of Asia's unruly waters, is to fill that space with cultural and political meaning—to show that the landscape of Asia's mountain rivers and its monsoons have also constituted a space of migration, a zone of trade, a path of pilgrimage.

Throughout history, water has both connected and divided Asia. The rivers and oceans have been thoroughfares of trade as well as zones of imperial domination. In the nineteenth century, when European empires dominated the world, Asia's hydrology underpinned many of the commodities that fueled global industrial capitalism. The storms that have always menaced coastal regions always crossed frontiers, but states have responded to them in different ways. As connections across Asia frayed in the mid-twentieth-century decades of nationalism and war, water, too, came under ever-tighter territorial control. One reason why almost all of Asia's new nation-states tried so boldly to harness water was to gain self-sufficiency in a postcolonial era in which their autonomy was nevertheless called into question by the machinations of the superpowers in the Cold War. They were spurred to do so by memories of water's lack—bitter memories of famine and suffering within living memory. They were spurred, especially in India, by a fear of the monsoon climate and the power it had over human life. "For us in India scarcity is only a missed monsoon away," Prime Minister Indira Gandhi said—and this sense of a battle against enormous natural forces inspired in her, as in so many others, a tug between despair and optimism that science and technology held the key to liberation.[82] Over time that insistence on self-sufficiency combined with a sense of perpetual crisis led to a narrowing of vision and a willful blindness to the consequences of repeated attempts to conquer nature. Today, the inability of states to think beyond their borders imperils lives and denudes the political imagination.

If there is one consistent lesson in *Unruly Waters,* it is that water management never has been, and can never be, a purely technical or a scientific question; neither can it be addressed on a purely national scale. Ideas about the distribution and management of water are deeply inflected with cultural values, with notions of justice, with ideas and fears about nature and climate—including very old fears about the monsoon, which grows more capricious. The battle continues to understand the monsoons and mountain rivers that shape Asia.

HISTORY AND MEMORY
AT THE WATER'S EDGE

"THERE IS THE SCIENTIFIC AND IDEOLOGICAL LANGUAGE FOR WHAT IS happening to the weather," writes novelist Zadie Smith, "but there are hardly any intimate words," no words that capture the sense of loss that climate change brings with it. "The weather has changed, is changing," Smith writes, "and with it so many seemingly small things . . . are being lost."[1] Faced with the forbidding scale of climate change, many responses are profoundly local. Indian farmers, deeply attuned to the tenor of the skies, are changing when they plant their seeds.[2] But changes in the weather also bring a sense of disorientation—a loss of one's bearings. Everywhere I traveled over the eight years I have been working on this book, I heard stories about the weather—stories of how it is not what it used to be. In many cases, these stories were prompted by a particular landscape that was familiar once, and is now unrecognizable. "Look there,"

I was told by a longtime resident of Thanjavur on a trip through Tamil Nadu in 2012, "when I was young, the river ran full, now it is completely dry."

There are many other kinds of loss that climate change threatens us with. A changing monsoon affects every form of life that depends on it. From the Gurukula botanical sanctuary in Wayanad, northern Kerala, Suprabha Seshan and her colleagues cultivate endangered plants native to the ecosystem of the Western Ghats, the western Indian mountain range that receives some of the most intensive rainfall during the summer monsoon. "We refer to these plants as refugees," she writes; many have been rescued from areas where forests have already been cut down. "The weather features regularly in our speech," she writes. The gardeners' work depends on an intuitive knowledge of the weather. But these patterns are changing. Seshan observes that "ever since I have been here, about 24 years now, I have heard people talking about how the monsoon has gone awry, that it is no longer what it used to be. We also know this from scientific data, but crucially for us, we know this from the behaviour of the plants and animals in our sanctuary." Here, meteorological research and local perceptions match. Everyone is sure that the southwest monsoon has weakened—and has become more unpredictable. Local species are "confused," Seshan writes, by the weather's signals. Temperatures are too high for some mountain species to thrive, and rising temperatures bring new diseases. "I worry," Seshan concludes, "that the monsoon, with its moods and savage powers, might altogether cease."[3]

———

ALONG THE COAST OF SOUTHEASTERN INDIA, TOO, ARE MANY SIGNS of irreversible change. In a small village near Pondicherry, earlier this decade, I met Mr. Rathnam, a fisherman in his fifties whose family have been fishers in the area for generations. On both sides of the narrow strip of beach on which we sat were granite sea walls. "If not for these walls," he said, "the sea would have taken this settlement long

The remains of a house: an archive on the landscape of coastal erosion. CREDIT: Sunil Amrith

ago." The beach has been eaten away over the past twenty years, most noticeably by the construction of a large new port in Pondicherry, a few minutes down the coast. The Pondicherry port marked the beginning of an explosion in port construction in India, with dozens of ports currently planned for India's eastern and western seaboards. They eye the newly flourishing commercial opportunities of the Indian Ocean's littoral, which is vibrant again after falling into decline for the second half of the twentieth century. The ports cause enormous upheaval to the coastline. "Where these boats are now," Mr. Rathnam said, pointing to the beach, "those were all houses. Look, you can see the remains of the floors of houses." I saw little fragments jutting out from the soil, a small archive of coastal environmental history.

He is convinced that the sea is changing in ways beyond what is visible to the eye, beyond the visibly changing shape and extent of the beach. The weather is "unpredictable," he said; "the seasons seem to mean nothing now." He was convinced the monsoons are shifting, and in his narrative the Indian Ocean tsunami of 2004 was the

A narrowing stretch of beach near Pondicherry, eroded by the construction of a new port. CREDIT: Sunil Amrith

moment when "everything changed." The tsunami was a geological phenomenon, caused by an undersea earthquake, but to Mr. Rathnam it seemed a portent of fundamental change. He paused his story. "I don't understand the sea anymore," he said, suddenly: the sea that he has known, intimately and instinctively, for a lifetime. I asked him what does the future holds. "Nothing," he said; "there will be no fish left to catch."

Climate change is not the most obvious or proximate cause of his distress. Here, as elsewhere in Asia, the effects of climate change compound a crisis already far advanced—a product of reckless development and galloping inequality. Mr. Rathnam's livelihood has been threatened by the concentration of power in the hands of a small number of highly capitalized owners of large trawlers. There are fewer fish to catch because of what a recent report calls "an uncontrolled addition of fishing boats between 1965 and 1998." The size of the catch has collapsed, and its composition has changed: fewer large predators, fewer fish that command high prices on the market. The dramatic fall in their incomes has pushed many small fishers ever deeper into debt. Development along the large highway down the coast from Chennai has led to a spike in property speculation, fueling a construction boom that flouts coastal zone regulations. A tidal wave of plastic, and effluent from factories and power plants, floats out to sea. Compounding each of those challenges, climate change is also now making itself felt. Rising sea surface temperatures in the Bay of Bengal have exceeded the boundaries that can sustain many forms of aquatic life.[4]

One of the questions I asked Mr. Rathnam that day on the beach, was, "What happened to the family that lived in this house, and others like them?" One part of the answer, I expected—there had been a large movement of younger people to the growing cities, and to Chennai in particular.

But there was another part of the answer that I did not expect. In light of the work I had spent the previous decade doing on migration across the Indian Ocean, I had the feeling of a very familiar map of migration being drawn before my eyes. All but the very poorest households in the village, Mr. Rathnam told me, has at least one family member overseas. A similar story emerged in the neighboring hamlet. Older routes of migration have been reinvigorated—plenty of sons and nephews in the village were in Singapore and Malaysia, working in construction. Others had taken more recent paths. Many work on fishing fleets in the Persian Gulf. Colonial connections, too, continue to shape people's trajectories in Pondicherry, which was a French-ruled enclave within British India—one older fisher turned to his memories of the "French time," and then enumerated his family members now living in Paris. Old geographies still matter. In this part of South India, people experience and imagine climate change at home in relation to a constellation of distant places; family histories of mobility are reactivated as a means of support or insurance. But borders are harder than ever to cross. Every day, South Indian fishermen, struggling to make a living, stray into Sri Lankan territorial waters in search of fish; many have been arrested and detained by the Sri Lankan coast guard.

People experience climate change in space as well as in time. They mark change in terms of their memories of the seasons as they used to be, or of epochal storms that now seem portents of the future. But they also mark it through traces on the landscape, through memories of old houses and old neighbors. Traces of those earlier times lie embedded as debris at the water's edge.

ACKNOWLEDGMENTS

THE RESEARCH I DID FOR THIS BOOK BETWEEN 2012 AND 2015 received funding under the European Union's Seventh Framework Programme (FP/2007–2013/ERC Grant Agreement 284053) from the European Research Council, held at Birkbeck College, London. Since 2015, my work has been supported by Harvard University's Faculty of Arts and Sciences; I am grateful to deans Michael Smith, Nina Zipser, and Laura Fisher for providing me with the resources that make my work possible. More recently I have benefited from the support of the Infosys Science Foundation and the MacArthur Foundation, and their extraordinary generosity has helped me bring this project to a conclusion. I would also like to thank Carol Richards and the late David Richards for their generous support of the Center for History and Economics.

The research for this book would not have been possible without the help of many fine archivists and librarians. Within the wonderful Harvard library system, I would especially like to thank Fred

Burchstead, Ramona Islam and Richard Lesage, Laura Linard at the Baker Library, and everyone at the Map Collection. The staff at the Asian and African Studies reading room at the British Library have helped me over many years and many projects. In India I would like to thank the staff of the National Archives of India and the Nehru Memorial Museum in Delhi; Kiran Pandey at the Centre for Science and Environment's library; the staff of the Tamil Nadu State Archives in Chennai; and the Maharashtra State Archives Department in Mumbai. Elsewhere, I would like to thank the staff of the National Maritime Museum in Greenwich, the World Bank Archives in Washington, DC, the National Archives of Myanmar in Yangon, and the National Archives of Sri Lanka, Colombo.

Over the years I have been working on this book, there has been a revolution in digitization. Sources that I spent weeks or months locating are now, as I write this, freely available online. A particular treasure for historians of South Asia has been the digitization of the library of the Gokhale Institute of Politics and Economics. That remarkable institution's collection of reports, occasional papers, and theses span a century and are now available to historians everywhere; this is truly a public service. The National Archives of India, too, has embarked on an ambitious program of digitization; access came too late for me to make use of these digitized materials in this book, but it will make new kinds of work possible in the future.

I have benefited from excellent research assistance over the course of writing this book. I would like to thank Sneha, who was indefatigable in finding sources at the National Archives in Delhi, and to two Harvard undergraduates, Aaisha Shah and Ellie Lasater-Guttman, who tracked down many textual and visual materials for me.

I am grateful to the many people who have shared their stories and memories with me over the course of writing this book. From among the extraordinary community of Indian meteorologists who have dedicated their careers to understanding the monsoon, I am especially grateful to Ranjan Kelkar, S. Raghavan, and S. R. Ramanan, who were generous with their time and spent many hours talking to me. They also shared unpublished material with me, and facilitated access to the wealth of material in the libraries of the Regional Meteorology Centre, Chennai, and the India Meteorological Department in Pune. There are many more people to thank who preferred to speak informally or to remain anonymous: my understanding of water and climate would have been poorer without the opportunity to learn from many fishers and cultivators and government officials in Tamil Nadu and Pondicherry. I would particularly like to thank R. Rajamanickam for an illuminating introduction to coastal communities around Pondicherry, and for setting up some initial interviews.

Writing about the history of meteorology has taken me far from my areas of expertise, and I am deeply grateful for the advice of colleagues in the fields of tropical meteorology and climate science. Professor Peter Webster of Emory University and Marena Lin of Harvard's Department of Earth and Planetary Sciences were generous in answering questions and sharing unpublished work with me; I am particularly grateful to Professor Adam Sobel of Columbia University who read drafts of two chapters, providing invaluable feedback, clarifying my understanding, and saving me from errors.

My wonderful literary agent, Don Fehr, has been supportive and encouraging throughout the process. Don made it possible for me to work with an editorial "dream team" of Brian Distelberg at Basic Books and Simon Winder at Penguin. Brian's thorough and deeply insightful comments on a ramshackle draft pushed me to sharpen the argument and tighten the structure. Simon's edits were brilliant, honing in on where I had been evasive and suggesting many fruitful comparisons and connections. The final version has benefited enormously from the meticulous and thoughtful line editing of Roger Labrie, and from Bill Warhop's sensitive copyediting. Melissa Veronesi oversaw the production process with grace and efficiency.

LOOKING BACK AT MY CAREER OVER THE PAST FIFTEEN YEARS, nothing makes me happier than that Emma Rothschild, the scholar from whom I have learned more than from anyone else, is now my closest colleague. Nobody has done more to support and inspire my work. I depend on her advice even more now than when I first arrived on her doorstep as a graduate student.

I began this book while I was at Birkbeck College, University of London, where I taught for nine years. I would like to thank my former colleagues at Birkbeck, especially Chandak Sengoopta and Hilary Sapire. Since arriving at Harvard in 2015, I have accumulated many debts. I have been lucky to have the support of three exceptional deans: Diana Sorensen followed by Robin Kelsey in the Division of Arts and Humanities, and Claudine Gay in the Division of Social Science. Thomas Skerry eased my move to Harvard in many ways. At the Department of South Asian Studies, Cheryl Henderson's kindness and efficiency make everything possible; I would also like to thank Parimal Patil for his support during his time as the department's chair. At the Department of History, I have been fortunate to work with Rob Chung, Kimberly Richards O'Hagan, and their team. I am grateful to David Armitage and Dan Smail for their kindness during their respective terms as chair. At the Center for History and Economics, it is truly a pleasure to work with Emily Gauthier and Jennifer Nickerson—their dedication sustains the amazing intellectual community of the center. I would also like to express my thanks to the staff of the Harvard Asia Center, the Weatherhead Center for International Affairs, the Harvard Academy for International and Area Studies, and the Harvard University Center for the Environment.

For their warmth and camaraderie and for many kinds of help, I would like to thank friends and colleagues at Harvard: Sugata Bose, Allan Brandt, Richard Delacy, Arunabh Ghosh, David Jones, Gabriela Soto Laveaga, Kenneth Mack, Durba Mitra, Jonathan Ripley, Charles Rosenberg, Amartya Sen, Ajantha Subramanian, Michael Szonyi, and Karen Thornber.

The greatest joy of moving to Harvard has been the opportunity to work with a remarkable group of students. My undergraduate students' commitment to positive change, their courage, and their talent give me optimism for the future in these dark times. I have also been privileged to work with many outstanding graduate students: Mou Banerjee, Aniket De, Yuting Dong, Shireen Hamza, Neelam Khoja, Kiran Kumbhar, Lei Lin, Amulya Mandava, Tsitsi Mangosho, Mircea Raianu, Priyasha Saksena, Hannah Shepherd, and, beyond Harvard, Jack Loveridge and Lucas Mueller. I am especially grateful to the four students I've worked with most closely over the last three years: Divya Chandramouli, Hardeep Dhillon, Sarah Kennedy Bates, and Iris Yellum—I have learned far more from them than they have from me.

Dispersed though they are across the world, I am always grateful for the friendship of Isabel Hofmeyr, Maya Jasanoff, Diana Kim, Sumit Mandal, Mahesh Rangarajan, Taylor Sherman, Naoko Shimazu, Benjamin Siegel, Kavita Sivaramakrishnan, Glenda Sluga, Eric Tagliacozzo, and A. R. Venkatachalapathy—each of them is a model of scholarly integrity, generosity, and kindness. This time again, I owe a particular debt to Tim Harper.

For helpful conversations, ideas, or invitations to present my work, I would like to thank: Seema Alavi, Michiel Baas, Abhijit Banerjee, Ritu Birla, Anne Blackburn, Dipesh Chakrabarty, Joya Chatterji, Rohit De, Prasenjit Duara, David Engerman, Amitav Ghosh, Ramachandra Guha, Anne Hansen, Namrata Kala, Akash Kapur, Adil Hasan Khan, Sunil Khilnani, T. M. Krishna, Michael Laffan, Melissa Lane, David Ludden, Amala Mahadevan, Rochona Majumdar, Farina Mir, Kazuya Nakamizo, Michael Ondaatje, Prasannan Parthasarathi, Jahnavi Phalkey, Gyan Prakash, Srinath Raghavan, Bhavani Raman, Jonathan Rigg, Harriet Ritvo, Tansen Sen, Tomoko Shiroyama, Mrinalini Sinha, Vineeta Sinha, Helen Siu, K. Sivaramakrishnan, Smriti Srinivas, Julia Stephens, Kohei Wakimura, Roland Wenzlhumer, and Nira Wickramasinghe. I join many people around the world in mourning the loss of Christopher Bayly, who remains a guiding light.

THIS IS THE FIRST BOOK I HAVE WRITTEN SINCE BECOMING A PARENT, and I could not do my work without the work of many others. I am full of admiration for the creative and nurturing teachers at Radcliffe Child Care Center in Cambridge, Massachusetts, and I appreciate deeply the care of Marlene Boyette, Pearl Kerber, and Uyen-Nguyen Tran. I am grateful to the many people who have made Cambridge, Massachusetts, feel like home: Ian Miller and Crate Herbert, who have welcomed us from the start; Priyanka Shankar; the kind neighbors on our street; the community of Radcliffe parents, and especially Laura Muir and Danny Pallin; all at Cambridge Friends Meeting.

My family has sustained me through this time of many transitions. Barbara Phillips has been a caring presence and she has often put her own plans on hold and traveled a long way to help look after the children. Megha Amrith has been many things—an intellectual inspiration, a listening ear, a trusted source of advice, a wonderful travel companion, and a loving and enthusiastic aunt to my children. Over these years we are lucky to have added Andreas Werner and his parents to our extended family. Jairam and Shantha Amrith have given me everything—and they continue to support me in everything that I do.

Nothing I have ever achieved would have been possible without the love and generosity of Ruth Coffey. The origins of this book lie in conversations I had with Ruth a decade ago, when she was studying for a master's degree in environmental management in London; those conversations have developed over many travels together in South and Southeast Asia. Over the past few years my absence on research trips has placed an additional burden of childcare on her, and she has taken it on with grace while embarking on a judicial career in England, teaching law at Harvard, and pursuing many other projects. She is my anchor and my guiding light.

Theodore was born when this book was in its early stages, and he has traveled many miles with me in search of water stories. I am blessed by the exuberant joy he brings to every day. He started to read just as I was writing the last sentences of this book—he asked me the other day if it was finished, and suggested that my next book should be written for children. Lydia arrived as this project neared its conclusion, and she has made my life richer and more full of wonder. I dedicate this book to them both, with love and gratitude.

ARCHIVES AND SPECIAL COLLECTIONS

BRITISH LIBRARY, LONDON

India Office Records
East India Company Board of Control Records
East India Company Factory Records
Economic Department Records
Marine Records
Public & Judicial Department Records
Political & Secret Records
Official Publications Series

NATIONAL MARITIME MUSEUM, GREENWICH
British India Steam Navigation Company papers
Irrawaddy Flotilla Company papers [uncataloged]

NATIONAL ARCHIVES OF INDIA, NEW DELHI

Department of Revenue and Agriculture

Department of Revenue, Agriculture and Commerce

Department of Education, Health and Lands

Department of Industries and Labour: Meteorology

Home Department, Judicial Branch & Public Branch

Meteorological Department

Ministry of External Affairs

Ministry of Irrigation

Ministry of States

Political Department

TAMIL NADU STATE ARCHIVES, CHENNAI

Fisheries Department

Public Works Department

MAHARASHTRA STATE ARCHIVES DEPARTMENT, MUMBAI

Public Works Department: Irrigation, 1868–1909

INDIA METEOROLOGICAL DEPARTMENT, PUNE

Miscellaneous reports, charts, and memoirs

WORLD BANK GROUP ARCHIVES, WASHINGTON, DC

Damodar Multi-Purpose Project, India: Administration, Correspondence, and Negotiations, 1953–1957

Drought Prone Areas Project, India: Correspondence, 1973–1985

Indus Basin Dispute, General Negotiations and Correspondence, 1949–1960

Uttar Pradesh Tube Wells Projects, India: Correspondence, 1961–1992

NOTES

CHAPTER ONE: THE SHAPE OF MODERN ASIA

1. E. M. Forster, *A Passage to India* (London: Edward Arnold, 1924), 116; Pranay Lal, *Indica: A Deep Natural History of the Indian Subcontinent* (New Delhi: Allen Lane, 2017), 258.

2. Norton Ginsburg, ed., *The Pattern of Asia* (New York: Prentice-Hall, 1958), 5–6.

3. V. Ramanathan et al., "Atmospheric Brown Clouds: Impact on South Asian Climate and Hydrological Cycle," *Proceedings of the National Academy of Sciences* 102 (2005): 5326–5333.

4. Asia Society, *Asia's Next Challenge: Securing the Region's Water Future, a Report by the Leadership Group on Water Security in Asia* (New York: Asia Society, 2009), 9; C. J. Vörösmarty et al., "Global Threats to Human Water Security and River Biodiversity," *Nature* 467 (September 30, 2010): 555–561; Chris Buckley and Vanessa Piao, "Rural Water, Not City Smog, May be China's Pollution Nightmare," *New York Times*, April 11, 2016; Malavika Vyawahare, "Not Just Scarcity, Groundwater Contamination Is India's Hidden Crisis," *Hindustan Times*, March 22, 2017.

5. Intergovernmental Panel on Climate Change, *Climate Change 2014: Impacts, Adaptation, and Vulnerability* (Geneva: IPCC, 2014); World Bank, *Turn Down the Heat: Climate Extremes, Regional Impacts, and the Case for Resilience* (Washington, DC: World Bank, 2013); Deepti Singh et al., "Observed Changes in Extreme Wet and Dry Spells During the South Asian Summer Monsoon," *Nature Climate Change* 4 (2014): 456–461.

6. Benjamin Strauss, "Coastal Nations, Megacities, Face 20 Feet of Sea Rise," Climate Central, July 9, 2015, accessed January 12, 2018, www.climate central.org/news/nations-megacities-face-20-feet-of-sea-level-rise-19217.

7 Mike Davis, *Ecology of Fear: Los Angeles and the Imagination of Disaster* (New York: Metropolitan Books, 1998); David Blackbourn, *The Conquest of Nature: Water, Landscape, and the Making of Modern Germany* (New York: W.W. Norton, 2006).

8. Dipesh Chakrabarty, "The Climate of History, Four Theses," *Critical Inquiry* 35 (2009): 197–222.

9. Amitav Ghosh, *The Great Derangement: Climate Change and the Unthinkable* (Chicago: University of Chicago Press, 2016).

10. Karl Wittfogel, *Oriental Despotism: A Comparative Study of Total Power* (New Haven, CT: Yale University Press, 1957).

11. Much of the scholarship on Chinese environmental history over the long term is surveyed in Mark Elvin, *The Retreat of the Elephants: An Environmental History of China* (New Haven, CT: Yale University Press, 2004); see also Peter Perdue, *Exhausting the Earth: State and Peasant in Hunan, 1550–1850* (Cambridge, MA: Harvard University Press, 1986), and Kenneth Pomeranz, *The Making of a Hinterland: State, Society, and Economy in Inland North China, 1853–1937* (Berkeley: University of California Press, 1993). On India, key works include Dharma Kumar, *Land and Caste in South India: Agricultural Labour in the Madras Presidency During the Nineteenth Century* (Cambridge: Cambridge University Press, 1965); C. J. Baker, *An Indian Rural Economy: The Tamilnad Countryside, 1880–1955* (Oxford: Clarendon Press, 1984), and Sugata Bose, *Agrarian Bengal: Economy, Social Structure and Politics, 1919–1947* (Cambridge: Cambridge University Press, 1986).

12. Marc Bloch, *The Historian's Craft,* trans. Peter Putnam (New York: Knopf, 1953), 26.

13. Fernand Braudel, "Histoire et sciences sociales: la longue durée," *Annales, economies, sociétés, civilisations* 13 (1958), 725–753; K. N. Chaudhuri, *Trade and Civilisation in the Indian Ocean: An Economic History from the Rise of Islam to 1750* (Cambridge: Cambridge University Press, 1985), quotation from Braudel on p. 23.

14. "Sampling device," from Charles E. Rosenberg, *Explaining Epidemics and Other Studies in the History of Medicine* (Cambridge: Cambridge University Press, 1992), 279.

15. Kenneth Pomeranz, "The Great Himalayan Watershed: Water Shortages, Mega-Projects and Environmental Politics in China, India, and Southeast Asia," *Asia Pacific Journal: Japan Focus* 7 (2009): 1–29.

16. Raj Patel and Jason W. Moore, *A History of the World in Seven Cheap Things: A Guide to Capitalism, Nature, and the Future of the Planet* (Berkeley: University of California Press, 2017), 44–63.

17. Eric Hobsbawm, *The Age of Capital, 1848–1875* (London: Weidenfeld and Nicolson, 1975), 48.

18. Ramachandra Guha, *India After Gandhi: The History of the World's Largest Democracy* (London: HarperCollins, 2007).

19. Peter D. Clift and R. Alan Plumb, *The Asian Monsoon: Causes, History and Effects* (Cambridge: Cambridge University Press, 2008), vii.

20. Sunita Narain, Science and Democracy Lecture, Harvard University, December 4, 2017.

21. Gilbert T. Walker, "The Meteorology of India," *Journal of the Royal Society of Arts* 73 (1925): 838–855, quotation on p. 839; Charles Normand, "Monsoon Seasonal Forecasting," *Quarterly Journal of the Royal Meteorological Society* 79 (1953): 463–473, 469; A. Turner and H. Annamalai, "Climate Change and the South Asian Monsoon," *Nature Climate Change* 2 (2012): 587–595.

22. Bob Yirka, "Earliest Example of Large Hydraulic Enterprise Excavated in China," Phys.org, December 5, 2017, accessed December 15, 2017, phys.org/news/2017-12-earliest-large-hydraulic-enterprise-excavated.amp.

CHAPTER TWO: WATER AND EMPIRE

1. "Madras Government request the Court of Directors' sanction for the expenditure of 5000 rupees on deepening the Pamban Channel between India and Ceylon," October 1833–March 1835, Board's Collections: British Library [hereafter BL] India Office Records [hereafter IOR], F/4/1523/60207.

2. H. Morris, "A Descriptive and Historical Account of the Godavery District in the Presidency of Madras," (1878): BL, IOR, V/27/66/18.

3. E. Halley, "An Historical Account of the Trade Winds and the Monsoons, Observable in the Seas Between and Near the Tropicks, with an attempt to assign the physical cause of the sail winds," *Philosophical Transactions of the Royal Society of London* 16 (1686): 153–168.

4. The clearest explanations of the monsoon can be found in Peter J. Webster, "Monsoons," *Scientific American* 245 (1981): 108–119; and Peter D. Clift and R. Alan Plumb, *The Asian Monsoon: Causes, History and Effects* (Cambridge: Cambridge University Press, 2008).

5. Jos Gommans, *Mughal Warfare: Indian Frontiers and Highroads to Empire, 1500–1700* (London: Routledge, 2003), chapter 1; Jos Gommans,

"The Silent Frontier of South Asia, c. AD 1100–1800," *Journal of World History* 9, no. 1 (1998): 1–23.

6. Victor Lieberman, *Strange Parallels: Southeast Asia in Global Context, c. 800–1830*, vol. 2, *Mainland Mirrors: Europe, Japan, China, South Asia, and the Islands* (Cambridge: Cambridge University Press, 2009), 632–636.

7. Diana Eck, "Ganga: The Goddess in Hindu Sacred Geography," in *The Divine Consort: Radha and the Goddesses of India*, ed. John Hawley and Donna Wulff (Boston: Beacon Press, 1982), 166–183; Diana Eck, *India: A Sacred Geography* (New York: Harmony, 2012); Anne Feldhaus, *Connected Places: Religion, Pilgrimage and the Geographical Imagination in India* (New York: Palgrave Macmillan, 2003).

8. Karl Wittfogel, *Oriental Despotism: A Comparative Study of Total Power* (New Haven, CT: Yale University Press, 1957); Kathleen D. Morrison, "Dharmic Projects, Imperial Reservoirs, and New Temples of India: An Historical Perspective on Dams in India," *Conservation and Society* 8 (2010): 182–195; Kathleen D. Morrison, *Daroji Valley: Landscape, Place, and the Making of a Dryland Reservoir System* (New Delhi: Manohar Press, 2009).

9. Peter Jackson, *The Delhi Sultanate* (Cambridge: Cambridge University Press, 1999); Sunil Kumar, *The Emergence of the Delhi Sultanate, 1192–1286* (New Delhi: Oxford University Press, 2007).

10. James L. Wescoat Jr., "Early Water Systems in Mughal India," *Environmental Design: Journal of the Islamic Environmental Design Research Centre* 2, ed. Attilo Petruccioli (Rome: Carucci Editions, 1985), 51–57.

11. *Babur Nama*, trans. Annette Susannah Beveridge (New Delhi: Penguin, 2006), 93, 264–265. For later discussion of Mughal-era irrigation, see Irfan Habib, *The Agrarian System of Mughal India (1556–1707)* (London: Asia Publishing House, 1963), 24–36.

12. Muzaffar Alam and Sanjay Subrahmanyam, eds., *The Mughal State, 1526–1750* (New Delhi: Oxford University Press, 1998); Lieberman, *Strange Parallels*, 636–637.

13. John F. Richards, *The Mughal Empire* (Cambridge: Cambridge University Press, 1995); Lieberman, *Strange Parallels*.

14. Gommans, *Mughal Warfare*.

15. Gommans, *Mughal Warfare*; *The Akbarnama of Abu'l Fazl*, vol. 3, trans. H. Beveridge (Calcutta: Asiatic Society, 1910), 135–136.

16. Irfan Habib, *An Atlas of the Mughal Empire* (New Delhi: Oxford University Press, 1982), reference to Kanauj on plate 8B.

17. Prasannan Parthasarathi and Giorgio Riello, "The Indian Ocean in the Long Eighteenth Century," *Eighteenth-Century Studies* 48 (Fall 2014): 1–19.

18. Anthony Reid, "Southeast Asian Consumption of Indian and British Cotton Cloth, 1600–1850," in *How India Clothed the World: The World*

of South Asian Textiles, 1500–1850, ed. Giorgio Riello and Tirthankar Roy (Leiden: Brill, 2009), 31–52.

19. Armando Coresao, trans., *The Suma Oriental of Tomé Pires* (London: Hakluyt Society, 1944), 3:92–93.

20. Sinappah Arasaratnam, *Merchants, Companies and Commerce on the Coromandel Coast 1650–1740* (New Delhi: Oxford University Press, 1986), 98–99.

21. Sanjay Subrahmanyam and C. A. Bayly, "Portfolio Capitalists and the Political Economy of Early Modern India," *Indian Economic and Social History Review* 25 (1988): 401–424; Richards, *The Mughal Empire*; Alam and Subrahmanyam, eds., *The Mughal State*; "giant pump" from Lieberman, *Strange Parallels,* 694–696.

22. David Ludden, "History Outside Civilisation and the Mobility of South Asia," *South Asia* 17 (1994): 1–23.

23. H. V. Bowen, John McAleer, and Robert J. Blyth, *Monsoon Traders: The Maritime World of the East India Company* (London: Scala, 2011).

24. C. A. Bayly, *Indian Society and the Making of the British Empire* (Cambridge: Cambridge University Press, 1988).

25. C. A. Bayly, *Rulers, Townsmen and Bazaars: North Indian Society in the Age of British Expansion* (Cambridge: Cambridge University Press, 1983), 1; Murari Kumar Jha, "The Rhythms of the Economy and Navigation along the Ganga River," in *From the Bay of Bengal to the South China Sea,* ed. Satish Chandra and Himanshu Prabha Ray (New Delhi: Manohar, 2013), 221–247.

26. James Rennell, *Memoir of a Map of Hindoostan; or, The Mogul Empire* (London: M. Brown, 1788), 280.

27. T. F. Robinson, "William Roxburgh, 1751–1815: The Founding Father of Indian Botany" (PhD dissertation, University of Edinburgh, 2003).

28. "A Meteorological Diary, & c. Kept at Fort St. George in the East Indies. By Mr William Roxburgh, Assistant-Surgeon to the Hospital at Said Fort. Communicated by Sir John Pringle, Bart. P.R.S.," *Philosophical Transactions of the Royal Society of London* 68 (1778): 180–193.

29. Robinson, "William Roxburgh."

30. Alexander Dalrymple, ed., *Oriental Repertory* (London: G. Biggs, 1793–1797), 2:58–59.

31. Record of proceedings at Fort Saint George, February 8, 1793, Madras Public Consultations, January 28–March 8, 1793, BL IOR, P/241/37.

32. Letter from William Roxburgh to Joseph Banks, August 30, 1791: BL, IOR, European Manuscripts, EUR/K148, ff. 243–47; Andrew Ross cited in Robinson, "William Roxburgh," 224n4.

33. William Roxburgh, "Remarks on the Land Winds and their Causes," *Transactions of the Medical Society of London* (1810), 189–211.

34. Richard H. Grove, *Green Imperialism: Colonial Expansion, Tropical Island Edens and the Origins of Environmentalism, 1600–1860* (Cambridge: Cambridge University Press, 1995), 399–400.

35. Richard Drayton, *Nature's Government: Science, Imperial Britain, and the "Improvement" of the World* (New Haven, CT: Yale University Press, 2000).

36. Letter from William Roxburgh to Andrew Ross, February 14, 1793, in Dalrymple, *Oriental Repertory,* 73.

37. Dalrymple, *Oriental Repertory,* 56; Robinson, "William Roxburgh," quotations on pp. 237–238, 241.

38. Jurgen Osterhammel, *The Transformation of the World: A Global History of the Nineteenth Century*, trans. Patrick Camiller (Princeton, NJ: Princeton University Press, 2014), 656; E. A. Wrigley, *Energy and the English Industrial Revolution* (Cambridge: Cambridge University Press, 2010), 91–112; Terje Tvedt, *Water and Society: Changing Perceptions of Societal and Historical Development* (London: I.B. Tauris 2016), 19–44.

39. Joseph Dalton Hooker, *Himalayan Journals: Notes of a Naturalist in Bengal, the Sikkim and Nepal Himalayas, the Khasia Mountains & c.* (London: J. Murray, 1854), 1:87.

40. James Ranald Martin, *Notes on the Medical Topography of Calcutta* (Calcutta: G.H. Huttmann, 1837), 90–93; on climate and racial thinking, see David Arnold, *The Tropics and the Traveling Gaze: India, Landscape and Science, 1800–1856* (Seattle: University of Washington Press, 2005).

41. Arthur Thomas Cotton, "Report on the Irrigation, & c., of Rajahmundry District" [1844], House of Commons Sessional Papers, XLI (1850), quotation on pp. 4–5 of Cotton's report.

42. Cotton, "Report on the Irrigation, & c., of Rajahmundry District," 13.

43. Arthur Cotton, "On a Communication between India and China by the line of the Burhampooter and Yang-tsze," *Journal of the Royal Geographical Society* 37 (1867): 231–239, quotation on p. 232.

44. For a detailed account of their rivalry, see Alan Robertson, *Epic Engineering: Great Canals and Barrages of Victorian India,* ed. Jeremy Berkoff (Melrose, UK: Beechwood Melrose Publishing, 2013).

45. Anthony Acciavatti, *Ganges Water Machine: Designing New India's Ancient River* (New York: Applied Research and Design Publishing, 2015), 120.

46. James L. Wescoat Jr., "The Water and Landscape Heritage of Mughal Delhi," accessed June 22, 2016, www.delhiheritagecity.org/pdfhtml/mughal/JW-the-water-and-landscape-heritage-of-mughal-delhi-Oct8.pdf.

47. Henry Yule, "A Canal Act of the Emperor Akbar, with some notes and remarks on the History of the Western Jumna Canals," *Journal of the Asiatic Society of Bengal* 15 (1846).

48. Proby Cautley, "On the Use of Wells, etc. in Foundations as Practiced by the Natives of the Northern Doab," *Journal of the Asiatic Society of Bengal* 8 (1839): 327–340.

49. Proby Cautley, *Report on the Central Doab Canal*, BL, IOR, V/27 /733/3/1.

50. G. W. MacGeorge, *Ways and Works in India: Being an Account of Public Works in that Country from the Earliest Times up to the Present Day* (London: Archibald Constable & Company, 1894), 153.

51. B. H. Tremenheere, "On Public Works in the Bengal Presidency," *Minutes of Proceedings of the Institute of Civil Engineers* 17 (1858): 483–513.

52. Proby T. Cautley, *Report on the Ganges Canal Works: from their Commencement until the Opening of the Canal in 1854,* 3 vols. (London: Smith, Elder & Co, 1860), 3:2.

53. Jan Lucassen, "The Brickmakers' Strike on the Ganges Canal in 1848–1849," *International Review of Social History* 51 (2006) supplement: 47–83.

54. Ganges Canal Committee, *A Short Account of the Ganges Canal* (Calcutta: Ganges Canal Committee, 1854), 3; the Hindi version was published as *Ganga Ki Nahar Ka Sankshepa Varnana* (Agra: Ganges Canal Committee, 1854).

55. "Short Account of the Ganges Canal," *North American Review,* October 1855, 81.

56. David Washbrook, "Law, State and Agrarian Society in Colonial India," *Modern Asian Studies* 15, no. 3 (1981): 648–721; Mayo quoted in David Ludden, *India and South Asia: A Short History* (London: Oneworld, 2014), 150.

57. David Mosse (with assistance from M. Sivan), *The Rule of Water: Statecraft, Ecology, and Collective Action in South India* (New Delhi: Oxford University Press, 2003), 29; Terje Tvedt, "'Water Systems': Environmental History and the Deconstruction of Nature," *Environment and History* 16 (2010): 143–166, quotation on p. 160.

58. Amitav Ghosh, "Of Fanas and Forecastles: The Indian Ocean and Some Lost Languages of the Age of Sail," *Economic and Political Weekly,* June 21, 2008, 56–62.

59. Henry T. Bernstein, *Steamboats on the Ganges: An Exploration in the History of India's Modernization through Science and Technology* (Bombay: Orient Longmans, 1960), 7–8, 13–16.

60. "Impediments to the Traffic on the Ganges and Jumna, arising from the number of customs chokeys," (February 5, 1833), BL, IOR, F/4/1506.

61. Bernstein, *Steamboats*, 28–31.

62. Bernstein, *Steamboats*, 84–99.

63. David Arnold, *Science, Technology and Medicine in Colonial India* (Cambridge: Cambridge University Press, 2000).

64. Bernstein, *Steamboats*, 99.

65. William Cronon, *Nature's Metropolis: Chicago and the Great West* (New York: W.W. Norton, 1991), 74; Richard White, *Railroaded: The Transcontinentals and the Making of Modern America* (New York: W.W. Norton, 2011).

66. Ian Kerr, *Engines of Change: The Railroads That Made India* (Westport, CT: Praeger, 2007).

67. Minute by Lord Dalhousie to the Court of Directors, April 20, 1853, in *Railway Construction in India: Select Documents,* ed. S. Settar (New Delhi: Indian Council of Historical Research, 1999), 2:23–57.

68. MacGeorge, *Ways and Works*, 221.

69. Karl Marx, "The Future Results of the British Rule in India," *New York Daily Tribune*, August 8, 1853; Edwin Merrall, *A Letter to Col. Arthur Cotton, upon the Introduction of Railways in India upon the English Plan* (London: E. Wilson, 1860), quotations on pp. 8 and 47.

70. MacGeorge, *Ways and Works*, 422–426, quotation on p. 426; Ian J. Kerr, *Engines of Change: The Railroads That Made India* (Westport, CT: Praeger, 2007).

71. C. H. Lushington, quoted in Tarasankar Banerjee, *Internal Market of India, 1834–1900* (Calcutta: Academic Publishers, 1966), 90–91.

72. Quotation from Banerjee, *Internal Market*, 323; Dave Donaldson, "Railroads of the Raj: Estimating the Impact of Transportation Infrastructure," (working paper, MIT/NBER, 2010); Robin Burgess and Dave Donaldson, "Railroads the Demise of Famine in Colonial India," (working paper, LSE/MIT/NBER, 2012).

73. On immobility see Joya Chatterji, "On Being Stuck in Bengal: Immobility in the 'Age of Mobility,'" *Modern Asian Studies* 51 (2017): 511–541; quotation from Arnold, *Science, Technology and Medicine*, 110.

74. MacGeorge, *Ways and Works*, 220–221, 358; Madhav Rao, cited in Kerr, *Engines of Change*, 4.

75. Arnold, *Science, Technology, and Medicine*.

76. MacGeorge, *Ways and Works*, 328–331; Kerr, *Engines of Change*, 47–51.

77. Rudyard Kipling, "The Bridge Builders," in *The Day's Work* (New York: Doubleday & McClure, 1899), 3–50.

78. W. W. Hunter, *Statistical Account of Bengal* (London: Trubner & Co., 1877), 14:31. On railways and the ecology of malaria: Iftekhar Iqbal, *The Bengal Delta: Ecology, State, and Social Change, 1840–1943* (Basingstoke: Palgrave/Macmillan, 2010), 117–139.

79. *New York Observer and Chronicle*, December 8, 1864.

80. J. E. Gastrell and Henry F. Blanford, *Report on the Calcutta Cyclone of the 5th of October 1864* (Calcutta: O.T. Cutter, 1866), 11, 31–32.

81. Gastrell and Blanford, *Calcutta Cyclone*, 139.

82. Gastrell and Blanford, *Calcutta Cyclone*, 109, 127.

83. Henry Piddington, *The Sailor's Horn-Book for the Law of Storms* (London: John Wiley, 1848); description of "storm wave" in Henry Piddington, *The Horn-Book of Storms for the Indian and China Seas* (Calcutta: Bishop's College Press, 1844), 20; Piddington's inspiration was William Reid, *An Attempt to Develop the Law of Storms By Means of Facts* (London: John Weale, 1838).

84. Gastrell and Blanford, *Calcutta Cyclone*, 11–13.

85. Gastrell and Blanford, *Calcutta Cyclone*, 4.

86. Gastrell and Blanford, *Calcutta Cyclone*, 70.

87. Gastrell and Blanford, *Calcutta Cyclone*, 14–15.

88. Gastrell and Blanford, *Calcutta Cyclone*, 108.

89. Hooker, *Himalayan Journals*, 1:97.

CHAPTER THREE: THIS PARCHED LAND

1. Mike Davis, *Late Victorian Holocausts: El Niño Famines and the Making of the Third World* (London: Verso, 2001); on the China famine, see Kathryn Edgerton-Tarpley, *Tears from Iron: Cultural Responses to Famine in Nineteenth-Century China* (Berkeley: University of California Press, 2008).

2. *The Constitution of the Poona Sarvajanik Sabha and its Rules* (Poona, 1870); quote from S. R. Mehrotra, "The Poona Sarvajanik Sabha: The Early Phase (1870–1880)," *Indian Economic and Social History Review* 9 (1969): 293–321; C. A. Bayly, *Recovering Liberties: Indian Thought in the Age of Liberalism and Empire* (Cambridge: Cambridge University Press, 2012).

3. Poona Sarvajanik Sabha, "Famine Narrative, No. 1," October 21, 1876.

4. *Medical and Sanitary Report of the Native Army of Madras, for the Year 1875* (Madras: Government Press, 1876), 49; *Report of the Indian Famine Commission*, 2 vols. (London: HM Stationery Office, 1880); William Digby, *The Famine Campaign in Southern India* (London: Longmans, Green & Co., 1878), 1:6; W. W. Hunter, *The Indian Empire: Its People, History and Products* (London: Trubner & Co., 1886), 542. For a perspective from current climate science: Edward R. Cook et al., "Asian Monsoon Failure and Megadrought over the Last Millennium," *Science* 328 (2010): 486–489.

5. Arup Maharatna, "Regional Variation in Demographic Consequences of Famines in Late 19th Century and Early 20th Century India," *Economic and Political Weekly* 29 (June 4, 1994): 1399–1410; Arup Maharatna, *The Demography of Famines: An Indian Historical Perspective* (New Delhi: Oxford University Press, 1996); Tim Dyson, "On the Demography of South Asian Famines, I," *Population Studies* 45 (1991): 5–25.

6. Richard Strachey, "Physical Causes of Indian Famines," May 18, 1877, *Notices of the Proceedings of the Meetings of the Members of the Royal Institution of Great Britain* 8 (1879): 407–426.

7. "The Famine, Letter from the Affected Districts," *The Examiner,* March 24, 1877, 363.

8. Digby, *Famine Campaign,* 1:67–68, 1:155–156.

9. Digby, *Famine Campaign,* 1:174–175.

10. J. Norman Lockyer and W. Hunter, "Sun-Spots and Famines," *The Nineteenth Century,* (November 1877), 601.

11. Strachey, "Physical Causes," 411.

12. Mark Elvin, "Who Was Responsible for the Weather? Moral Meteorology in Late Imperial China," *Osiris* 13 (1998): 213–237; Richard White, *The Republic for Which It Stands: The United States During Reconstruction and the Gilded Age, 1865–1896* (New York: Oxford University Press, 2017), 425–427.

13. "The Causes of Famine in India," *New York Times (NYT),* August 25, 1878, 6.

14. "The Famine: Letter from the Affected Districts," *The Examiner,* March 24, 1877, 363.

15. "Causes of Famine," *NYT,* August 25, 1878.

16. Villiyappa Pillai, *Panchalakshana Thirumukavilasam* [1899] (Madurai: Sri Ramachandra Press, 1932).

17. W. G. Pedder, "Famine and Debt in India," *The Nineteenth Century* (September 1877).

18. Digby, *Famine Campaign,* 1:172–174.

19. Letter from Sarvajanik Sabha Rooms to S. C. Bayley, Additional Secretary to the Government of India, April 1, 1878, in Poona Sarvajanik Sabha, *Journal* 1 (1878).

20. "Letter from the Affected Districts" (1877), 363.

21. Richard H. Grove, *Green Imperialism: Colonial Expansion, Tropical Island Edens and the Origins of Environmentalism, 1600–1860* (Cambridge: Cambridge University Press, 1995); Diana K. Davis, *The Arid Lands: History, Power, Knowledge* (Cambridge, MA: MIT Press [2016]).

22. Cited in Davis, *Arid Lands,* 83.

23. "Causes of Famine," *NYT,* August 25, 1878.

24. Philindus, "Famines and Floods in India," *Macmillan's Magazine,* November 1, 1877, 236–256; George Perkins Marsh, *Man and Nature, or, Physical Geography As Modified by Human Action* (New York: Scribner, 1865).

25. "Famine Narrative no. 1," October 21, 1876, in Poona Sarvajanik Sabha, *Journal* 1.

26. Ramachandra Guha, "An Early Environmental Debate: The Making of the 1878 Forest Act," *Indian Economic and Social History Review* 27 (1990): 65–84.

27. Valentine Ball, "On Jungle Products Used as Articles of Food in Chota Nagpur," in *Jungle Life in India: Or, the Journeys and Journals of an Indian Geologist* (London: Thos. De La Rue & Co., 1880), 695–699.

28. George Chesney, "Indian Famines," *Nineteenth Century* 2 (November 1877): 603–620.

29. Digby, *Famine Campaign,* 1:148–150.

30. Florence Nightingale, "A Missionary Health Officer in India," *Good Words*, January 20, 1879, 492–496.

31. "Causes of Famine," *NYT,* August 25, 1878.

32. Dadabhai Naoroji, *Poverty of India* (London: Vincent Brooks, Day and Son, 1878), 42–43, 66.

33. See especially Davis, *Late Victorian Holocausts.*

34. Chandrika Kaul, "Digby, William (1849–1904)," *Oxford Dictionary of National Biography.*

35. *Report of the Indian Famine Commission, Part 1, Famine Relief* (London: Stationery Office, 1880): 9–10; Jean Drèze, "Famine Prevention in India" (working paper 45, WIDER: United Nations University, Helsinki, 1988), 45.

36. "Wasting Public Money," *The Economist*, July 4, 1874.

37. *The Black Pamphlet of Calcutta. The Famine of 1874. By a Bengal Civilian* (Calcutta: William Ridgeway, 1876).

38. Digby, *Famine Campaign,* 1:48.

39. "Letter from the Affected Districts," (1877), 363.

40. Edgerton-Tarpley, *Tears from Iron,* 152–153.

41. Lance Brennan, "The Development of the Indian Famine Code," in *Famine as a Geographical Phenomenon,* ed. Bruce Currey and Graeme Hugo (Dordrecht: D. Reidel, 1984), 91–112.

42. Drèze, "Famine Prevention in India."

43. On Caird, see Peter J. Gray, "Famine and Land in Ireland and India, 1845–1880: James Caird and the Political Economy of Hunger," *Historical Journal* 49 (2006): 193–215.

44. W. Stanley Jevons, "Sun-Spots and Commercial Crises," *Nature* 19 (1879): 588–590; Lockyer and Hunter, "Sun-Spots and Famines."

45. *Report of the Indian Famine Commission, Part 1,* 7.

46. *Report of the Indian Famine Commission, Part 2, Measures of Protection and Prevention* (London: Stationery Office, 1880), 9.

47. *Report of the Indian Famine Commission, Part 2,* 150–151.

48. Ira Klein, "When the Rains Failed: Famine, Relief, and Mortality in British India," *Indian Economic and Social History Review* 21 (1984): 185–214, quotation on p. 185.

49. *Report of the Indian Famine Commission, 1898* (London: Stationery Office, 1898); *Report of the Indian Famine Commission, 1901* (London: Stationery Office, 1901).

50. George Lambert, *India, the Horror-Stricken Empire: Containing a Full Account of the Famine, Plague, and Earthquake of 1896–7; including a Complete Narrative of the Relief Work through the Home and Foreign Commission* (Elkhard, IN: Mennonite Publishing Co., 1898).

51. Cited in C. S. Ramage, *The Great Indian Drought of 1899* (Boulder, CO: Aspen Institute for Humanistic Studies, 1977), 4.

52. Vaughan Nash, *The Great Famine and Its Causes* (London: Longmans, Green and Co., 1900), 11–14, 18–19, 27, 47.

53. Jon Wilson, *The Chaos of Empire: The British Raj and the Conquest of India* (New York: Public Affairs, 2016), 341–347; Georgina Brewis, "'Fill Full the Mouth of Famine': Voluntary Action in Famine Relief in India, 1896–1901," *Modern Asian Studies* 44 (2010): 887–918.

54. Davis, *Late Victorian Holocausts*, 22, 9.

55. Sanjoy Chakravorty, *The Price of Land: Acquisition, Conflict, Consequence* (New Delhi: Oxford University Press, 2013), 88.

56. Robin Burgess and Dave Donaldson, "Railroads and the Demise of Famine in Colonial India" (working paper, 2012), available at http://dave-donaldson.com/wp-content/uploads/2015/12/Burgess_Donaldson_Volatility_Paper.pdf.

57. Jurgen Osterhammel, *The Transformation of the World: A Global History of the Nineteenth Century*, trans. Patrick Camiller (Princeton: Princeton University Press, 2014), 208–209.

58. Mike Davis, *Ecology of Fear: Los Angeles and the Imagination of Disaster* (New York: Metropolitan Books, 1998).

CHAPTER FOUR: THE AQUEOUS ATMOSPHERE

1. J. Elliott, *Vizagapatam and Backergunge Cyclones* (Calcutta: Bengal Secretariat Press, 1877), 165–167. The most common spelling of his name is Eliot, which is the variant I use in the text, but in this publication it appears as Elliott.

2. Elliott, *Vizagapatam*, 158, 182.

3. Elliott, *Vizagapatam*, 159.

4. Elliott, *Vizagapatam*, 183.

5. Paul N. Edwards, "Meteorology as Infrastructural Globalism," *Osiris* 21 (2006): 229–250; on Britain's role, see Katharine Anderson, *Predicting the Weather: Victorians and the Science of Weather Prediction* (Chicago: University of Chicago Press, 2005).

6. Luke Howard, *Essay On the Modification of Clouds* [1803], 3rd ed. (London: John Churchill & Sons, 1865); Jean-Baptiste Lamarck, "Nouvelle définition des termes que j'emploie pour exprimer certaines formes des nuages qu'il importe de distinguer dans l'annotation de l'état du ciel," *Annuaire Météorologique pour l'an XIII de la République Française* 3 (1805): 112–133; H. Hildebrandsson, A. Riggenbach, and L. Teisserenc de Bort, eds., *Atlas International des Nuages* (Paris: IMO, 1896); for further discussion, see Lorraine Daston, "Cloud Physiognomy: Describing the In-

describable," *Representations* 135 (2016): 45–71, and Richard Hamblyn, *Clouds: Nature and Culture* (London: Reaktion, 2017).

7. University of Madras, *Tamil Lexicon* (Madras: University of Madras, 1924–1936), 219, 1680; William Crooke, *A Glossary of North Indian Peasant Life*, ed. Shahid Amin (New Delhi: Oxford University Press, 1989), Appendix D, "A Calendar of Agricultural Sayings"; C. A. Benson, "Tamil Sayings and Proverbs on Agriculture," *Bulletin*, Department of Agriculture, Madras No. 29, New Series (1933), paragraphs 144, 163, 168, 213, 311. I have modified some of the translations from the Tamil.

8. Henry F. Blanford, "Winds of Northern India, in Relation to the Temperature and Vapour-Constituent of the Atmosphere," *Philosophical Transactions of the Royal Society* 164 (1874): 563.

9. India, Meteorological Department, *Report on the Meteorology of India in 1876* (Calcutta: Office of the Superintendent of Government, 1877).

10. India, Meteorological Department, *Report on the Meteorology of India in 1877* (Calcutta: Office of the Superintendent of the Government, 1878).

11. "Administrative Report of the Meteorological Reporter to the Government of India, 1884–85," BL, IOR, V/24/3022, quotations on pp. 5–14.

12. Kapil Raj, *Relocating Modern Science: Circulation and the Construction of Knowledge in South Asia and Europe, 1650–1900* (Basingstoke: Palgrave/Macmillan, 2007); Mandy Bailey, "Women and the RAS: 100 Years of Fellowship," *Astronomy & Geophysics* 57 (February 2016): 19–21.

13. *Memoirs of Ruchi Ram Sahni: Pioneer of Science Popularisation in Punjab*, ed. Narender K. Sehgal and Subodh Mahanti (New Delhi: Vigyan Prasar, 1997), 15–17.

14. *Memoirs of Ruchi Ram Sahni*, 16.

15. Henry F. Blanford, *Meteorology of India: Being the Second Part of the Indian Meteorologist's Vade-Mecum* (Calcutta: Government Printer, 1877), 48.

16. Henry F. Blanford, *A Practical Guide to the Climates and Weather of India, Ceylon and Burmah and the Storms of the Indian Seas* (London: MacMillan and Co., 1889), 42.

17. Blanford, *Meteorology of India*, 144–145.

18. Blanford, *Meteorology of India*, 48.

19. Blanford, *Practical Guide*, 64.

20. Henry F. Blanford, *The Rainfall of India*, India Meteorological Memoirs vol. 3 (Calcutta: Government Printer, 1886–1888), 76.

21. Blanford, *Rainfall of India*, 79–81.

22. Henry F. Blanford, "On the Connexion of the Himalaya Snowfall with Dry Winds and Seasons of Drought in India," *Proceedings of the Royal Society of London* 37 (1884): 3–22.

23. List of library holdings in "Administration Report of the Meteorological Department in Western India for the year 1880–81," BL, IOR, V/24/3023.

24. Blanford, "On the Connexion of the Himalaya Snowfall."

25. Richard Grove, "The East India Company, the Raj and El Niño: The Critical Role Played by Colonial Scientists in Establishing the Mechanisms of Global Climate Teleconnections, 1770–1930," in *Nature and the Orient: The Environmental History of South and Southeast Asia,* ed. Richard Grove, Vineeta Damodaran, and Satpal Sangwan (New Delhi: Oxford University Press, 1998), 301–323.

26. John Eliot, *Climatological Atlas of India* (Edinburgh: J. Bartholomew & Co., 1906), xi–xii.

27. *Dictionary of National Biography,* 1912 supplement (London: Smith, Elder & Co., 1912).

28. *Memoirs of Ruchi Ram Sahni,* 23.

29. John Eliot, *Handbook of Cyclonic Storms in the Bay of Bengal* (Calcutta: Bengal Secretariat Press, 1890).

30. Rev. Jose Algué, *The Cyclones of the Far East,* 2nd ed. (Manila: Philippines Weather Bureau, 1904), 219.

31. Robert Hart, "Documents Relating to 1. The Establishment of Meteorological Stations in China; and 2. Proposals for Co-operation in the Publication of Meteorological Observations and Exchange of Weather News by Telegraph along the Pacific Coast of Asia" [1874], published in *Chinese Maritime Customs Project Occasional Papers,* no. 3, ed. Robert Bickers and Catherine Ladds (Bristol: University of Bristol, 2008); for further discussion, see Robert Bickers, "'Throwing Light on Natural Laws': Meteorology on the China Coast, 1869–1912," in *Treaty Ports in Modern China: Law, Land, and Power,* ed. Robert Bickers and Isabella Jackson (London: Routledge, 2016), 179–200.

32. Agustín Udías, "Meteorology of the Observatories of the Society of Jesus," *Archivum Historicum Societatis Iesu* 65 (1996): 157–170; James Francis Warren, "Scientific Superman: Father José Algué, Jesuit Meteorology, and the Philippines under American Rule, 1897–1924," in *Colonial Crucible: Empire in the Making of the Modern American State,* ed. Alfred W. McCoy and Francisco A. Scarano (Madison: University of Wisconsin Press, 2009), 508–522.

33. *Cosmos,* no. 1091 (1906), 717–719: cited in Warren, "Scientific Superman," 515.

34. Algué, *Cyclones,* 3.

35. Algué, *Cyclones,* 219–229; Eliot, *Handbook of Cyclonic Storms.*

36. Charles Normand, "Seasonal Monsoon Forecasting," *Quarterly Journal of the Royal Meteorological Society* 79 (1953): 463–473; Eliot, *Climatological Atlas,* xiii.

37. Eliot, *Climatological Atlas.*

38. *The Imperial Gazetteer of India,* rev. ed. (Oxford: Clarendon Press, 1901), 5–6, 19–22.

39. Sven Hedin, *Trans-Himalaya: Discoveries and Adventures in Tibet* (New York: Macmillan, 1909), 1:279, 1:284.

40. Halford Mackinder, "The Geographical Pivot of History," *Geographical Journal* 4 (1904): 421–444.

CHAPTER FIVE: THE STRUGGLE FOR WATER

1. Italo Calvino, *Invisible Cities,* trans. William Weaver [1974] (London: Vintage, 1997), 17.

2. Dadabhai Naoroji, *Poverty and Un-British Rule in India* (London: Swan, Sonnenschein & Co., 1901), 648–653.

3. M. G. Ranade, *Essays in Indian Economics: A Collection of Essays and Speeches* (Bombay: Thacker & Co., 1899), quotation on p. 66.

4. R. C. Dutt, *Open Letters to Lord Curzon on Famines and Land Assessments in India* (London: K. Paul, Trench & Trübner, 1900), quotations on pp. 1, 17; R. C. Dutt, *The Economic History of India in the Victorian Age* (London: K. Paul, Trench & Trübner, 1904), 172.

5. Mary Albright Hollings, *The Life of Colin Scott-Moncrieff* (London: J. Murray, 1917), 298.

6. Bernard S. Cohn, *Colonialism and Its Forms of Knowledge: The British in India* (Princeton, NJ: Princeton University Press, 1996); Nicholas B. Dirks, *Castes of Mind: Colonialism and the Making of Modern India* (Princeton, NJ: Princeton University Press, 2001).

7. *Report of the Indian Irrigation Commission, 1901–1903* (London: HM Stationery Office, 1903), 1:2–4; Hollings, *Colin Scott-Moncrieff,* 299.

8. *Report of the Indian Irrigation Commission,* 1:5–14.

9. *Report of the Indian Irrigation Commission,* 1:16, 1:124–125.

10. Letter from W. C. Bennett, Director of the Department of Agriculture and Commerce, North-West Provinces and Oudh, to the Secretary, Board of Revenue, NW Provinces, May 27, 1883, Maharashtra State Archives Department, Mumbai [hereafter MSA], Public Works Department: Irrigation Branch [hereafter PWD: Irrigation], v. 406 (1868–1890), M167–169.

11. Bennett to Board of Revenue, NW Provinces, May 27, 1883, MSA, PWD: Irrigation, v. 406, M167–169.

12. Letter from W. W. Goodfellow, Superintending Engineer, Belgaum to the Secretary to the Government, Public Works Department, Bombay, October 17, 1883, MSA, PWD: Irrigation, v. 406, M199.

13. V. Sriram, "Made in Madras," *The Hindu,* November 16, 2014.

14. Letter from A. Chatterton to the Secretary to the Commissioner of Revenue Settlement, Department of Land Records and Agriculture, May 23, 1905, MSA, PWD: Irrigation, v. 272 (1904–1909), M164–165.

15. Letter from A. Chatterton to the Director of Agriculture, Poona, July 15, 1906, MSA, PWD: Irrigation, M199–215.

16. "in reality part of the great desert": in James Douie, "The Punjab Canal Colonies," lecture delivered at the Royal Society of Arts on May 7, 1914, *Journal of the Royal Society of Arts* 62 (1914): 611–623, quotation on p. 612; "irrigation was not designed": in E. H. Calvert, *The Wealth and Welfare of the Punjab* (Lahore, 1922), 123.

17. Douie, "Punjab Canal Colonies," 614.

18. Mrinalini Sinha, *Colonial Masculinity: The 'Manly Englishman' and the 'Effeminate Bengali' in the Late Nineteenth Century* (Manchester: Manchester University Press, 1995).

19. Douie, "Punjab Canal Colonies," 615–616.

20. M.W. Fenton, Financial Commissioner, in 1915: quoted in Indu Agnihotri, "Ecology, Land Use, and Colonisation: The Canal Colonies of Punjab," *Indian Economic and Social History Review* 33 (1996): 37–58.

21. Quoted in David Gilmartin, *Blood and Water: The Indus River Basin in Modern History* (Berkeley: University of California Press, 2015), 168.

22. Gilmartin, *Blood and Water,* 175–176.

23. Thomas Gottschang and Diana Lary, *Swallows and Settlers: The Great Migration from North China to Manchuria* (Ann Arbor: University of Michigan Press, 2000).

24. John F. Richards, *Unending Frontier: An Environmental History of the Early Modern World* (Berkeley: University of California Press, 2003).

25. Hung Chung Chang, "Crop Production in China, with Special Reference to Production in Manchuria" (Master of Science thesis, University of Michigan Agricultural College, 1922).

26. Indu Agnihotri, "Ecology, Land Use, and Colonisation: The Canal Colonies of Punjab," *Indian Economic and Social History Review* 33 (1996): 37–58.

27. Petition from Sakharam Balaji, undated (ca. 1903), MSA, PWD: Irrigation, v. 124, "Petitions" (1899–1903).

28. "The humble memorial of the inhabitants of the within mentioned villages in the Belgaum Taluka, of the Belgaum District," [undated, 1903], MSA, PWD: Irrigation, v. 124, "Petitions" (1899–1903).

29. J. Sion, *Asie Des Moussons,* book 9 of the *Géographie Universelle,* ed. P. Vidal De La Blanche (Paris: Librairie Armand Colin, 1928), 2:363.

30. Matthew Gandy, *The Fabric of Space: Water, Modernity and the Urban Imagination* (Cambridge, MA: MIT Press, 2014), 114–119; Ira Klein, "Urban Development and Death: Bombay City, 1870–1914," *Modern Asian Studies* 20 (1986): 725–754; Hector Tulloch, *The Water Supply of Bombay* (Roorkee: Thomason College Press, 1873).

31. Indian Industrial Commission, *Report* (Calcutta: Government Printing, 1918).

32. Indian Industrial Commission, *Report,* 57–62.

33. M. Visvesvaraya, *Memoirs of My Working Life* [1951] (New Delhi: Government of India Publications Division, 1960), 9.

34. M. Visvesvaraya, *Reconstructing India* (London: P.S. King & Son, 1920), 127.

35. Visvesvaraya, *Memoirs,* 115–124.

36. S. Muthiah, "Madras Miscellany," *The Hindu,* November 24, 2014.

37. Extract from an Official Note of 1899 on the Desirability of Developing the Agricultural Department, *Madras Fisheries Bureau,* Bulletin No. 1; F. A. Nicholson, "The Marine Fisheries of the Madras Presidency," paper read at Lahore Industrial Conference, 1909: contained in BL, IOR, V/25/550/3.

38. F. A. Nicholson, *Note on Fisheries in Japan* (Madras: Government Press, 1907).

39. James Hornell, *A Statistical Analysis of the Fishing Industry of Tuticorin,* Madras Fisheries Bulletin vol. 11, report no. 3 (Madras: Government Press, 1917). For another perspective, see the insightful discussion of Nicholson and Hornell in Ajantha Subramanian, *Shorelines: Space and Rights in South Asia* (Stanford: Stanford University Press, 2009), 107–124.

40. Edward Buck, "Report on the Control and Utilization of Rivers and Draignage for the Fertilization of the Land and Mitigation of Malaria" (1907), MSA, PWD: Irrigation, v. 267 (1904–1909).

41. Christopher J. Baker, *An Indian Rural Economy: The Tamilnad Countryside, 1880–1955* (Oxford: Clarendon Press, 1984); Sugata Bose, *Peasant Labour and Colonial Capital: Rural Bengal Since 1770* (Cambridge: Cambridge University Press, 1993); the "tide of indebtedness" was a phrase used by a colonial official in Dhaka, quoted by Bose; Haruka Yanagisawa, *A Century of Change: Caste and Irrigated Lands in Tamilnadu, 1860s–1970s* (New Delhi: Manohar, 1996). On low yields in rain-fed agriculture, see Latika Chaudhary, Bishnupriya Gupta, Tirthankar Roy, and Anand V. Swamy, eds., *A New Economic History of Colonial India* (New York: Routledge, 2016), 100–116.

42. Royal Commission on Agriculture in India, *Abridged Report* (Bombay: Government Central Press, 1928), 5.

43. *Report of the United Provinces Provincial Banking Enquiry Committee, 1929–30* (Allahabad: Government Press, 1930), 2:119, 234.

44. *Report of the United Provinces Provincial Banking Enquiry Committee, 1929–30* (Allahabad: Government Press, 1930), 3:137.

45. Royal Commission on Agriculture in India, *Evidence Taken in the Bombay Presidency,* vol. 2, part 1 (London: Stationery Office, 1927), 342.

46. J. S. Chakravarti, "Agricultural Insurance," *Agricultural Journal of India* 12 (1917): 436–441, quotations on pp. 436–437; J. S. Chakravarti, *Agricultural Insurance: A Practical Scheme Suited to Indian Conditions* (Bangalore: Government Press of Mysore, 1920). The comment on Chakravarti's

prescience is from P. K. Mishra, "Is Rainfall Insurance a New Idea? Pioneering Work Revisited," *Economic and Political Weekly* 30 (1995): A84–A88.

47. Indian Industrial Commission, *Report*, 4.

48. P. A. Sheppard, revised by Isabel Falconer, "Walker, Gilbert Thomas," *Oxford Dictionary of National Biography*, https://doi.org/10.1093/ref:odnb /36692.

49. G. I. Taylor, "Gilbert Thomas Walker, 1868–1958," *Biographical Memoirs of Fellows of the Royal Society* 8 (1962): 166–174.

50. D. R. Sikka, "The Role of the India Meteorological Department, 1875–1947," in *Science and Modern India: An Institutional History, c.1784–1947,* ed. Uma Das Gupta (New Delhi: Pearson, 2010), chapter 14.

51. Gilbert T. Walker, "The Meteorology of India," *Journal of the Royal Society of Arts* 73 (July 1925): 838–855, quotation on p. 839.

52. Gilbert T. Walker, "Correlation in Seasonal Variations of Weather, VIII. A Preliminary Study of World-Weather," *Memoirs of the Indian Meteorological Department* 24, part 4 (1923): 75–131, quotation on p. 75.

53. Michael Bardecki, "Walker Circulation," in *Encyclopedia of Global Warming and Climate Change*, ed. S. George Philander, 2nd ed. (New York: Sage, 2005), 1:1073.

54. Gilbert T. Walker, "On the Meteorological Evidence for Supposed Changes of Climate in India," *Indian Meteorological Memoirs* 21, part 1 (1910): 1–21.

55. Gilbert T. Walker, "Correlation in Seasonal Variations of Weather, II," *Indian Meteorological Memoirs* 21, part 2 (1910): 21–45, quoted in J. M. Walker, "Pen Portraits of Past Presidents—Sir Gilbert Walker, CSI, ScD, MA, FRS," *Weather* 52 (1997): 217–220, quotation on p. 219.

56. Richard W. Katz, "Sir Gilbert Walker and a Connection Between El Niño and Statistics," *Statistical Science* 17 (2002): 97–112.

57. Walker, "Correlation, VIII" (1923), 109.

58. Walker, "Meteorology of India," 843.

59. Gilbert T. Walker, "Correlation in the Seasonal Variations of Weather," *Quarterly Journal of the Royal Meteorological Society* 44 (1918): 223–234; Gilbert T. Walker, "The Atlantic Ocean," *Quarterly Journal of the Royal Meteorological Society* 53 (1927): 71–113, quotations on p. 113.

60. Priya Satia, "Developing Iraq: Britain, India and the Redemption of Technology in the First World War," *Past and Present* 197 (2007): 211–255.

61. Walker, "Meteorology of India," 849.

62. Taylor, "Gilbert Thomas Walker," 171.

63. Gilbert T. Walker, Review of *Climate Through the Ages: A Study of Climatic Factors and Climatic Variations* by C.W.P. Brooks, *Quarterly Journal of the Royal Meteorological Society* 53 (1927): 321–323.

64. Gilbert T. Walker, "On Monsoon Forecasting in India," *Bulletin of the American Meteorological Society* 19 (1938): 297–299.

65. Charles Normand, "Monsoon Seasonal Forecasting," *Quarterly Journal of the Royal Meteorological Society* 79 (1953): 463–73, quotations on p. 469.

66. Walker, "Meteorology of India," 838–855, quotation on p. 848.

67. Sikka, "India Meteorological Department"; the quotations from Walker and Field draw on their papers deposited in the office of the director-general of meteorology in India, which were not available for consultation by researchers.

68. Calvino, *Invisible Cities*, 17.

CHAPTER SIX: WATER AND FREEDOM

1. M. K. Gandhi, *Hind Swaraj and Other Writings*, ed. Anthony J. Parel (Cambridge: Cambridge University Press, 1997), 131.

2. Jawaharlal Nehru to B. J. K. Hallowes (Deputy Commissioner, Allahabad, and President of the Famine Relief Fund of Gonda), June 26, 1929, in *The Essential Writings of Jawaharlal Nehru*, ed. S. Gopal and Uma Iyengar (Delhi: Oxford University Press, 2003), 12.

3. Jawaharlal Nehru, "The Basis of Society," Presidential Address to Bombay Youth Congress, Poona, December 12, 1928, in *Essential Writings of Jawaharlal Nehru*, 1:8–10.

4. Sun Yat-sen, *The International Development of China* (New York: Knickerbocker Press, 1922).

5. Sun Yat-sen, "Third Principle of the People: People's Livelihood," cited in Deirdre Chetham, *Before the Deluge: The Vanishing World of the Yangtze's Three Gorges* (New York: Palgrave Macmillan, 2002), 117.

6. David A. Pietz, *The Yellow River: The Problem of Water in Modern China* (Cambridge, MA: Harvard University Press, 2015), chapter 3; quotations on pp. 93–94.

7. This account of the Mahad protest draws on Christophe Jaffrelot, *Dr Ambedkar and Untouchability: Analysing and Fighting Caste* (London: Hurst and Company, 2005), 47–48.

8. Sudipta Kaviraj, "Ideas of Freedom in Modern India," in *The Idea of Freedom in Asia and Africa*, ed. Robert H. Taylor (Stanford, CA: Stanford University Press, 2002), 120–121.

9. M. K. Gandhi, "Salt Tax," *Young India*, February 27, 1930, in *Collected Works of Mahatma Gandhi* (New Delhi: Government of India Publications Division, 1970), 48:499–500.

10. Robert Carter and Erin McCarthy, "Watsuji Tetsurô," in *The Stanford Encyclopedia of Philosophy*, ed. Edward N. Zalta (Winter 2014 Edition), http://plato.stanford.edu/archives/win2014/entries/watsuji-tetsuro/.

11. Watsuji Tetsuro, *A Climate: A Philosophical Study*, trans. Geoffrey Bownas (Tokyo: Ministry of Education, 1961), 18–20.

12. Watsuji, *Climate*, 25–26.

13. Watsuji, *Climate*, 22–23, 38.

14. Watsuji, *Climate*, 39.

15. Mukerjee makes an appearance in, among others: C. A. Bayly, *Recovering Liberties: Indian Thought in the Age of Liberalism and Empire* (Cambridge: Cambridge University Press, 2012), and Alison Bashford, *Global Population: History, Geopolitics, and Life on Earth* (New York: Columbia University Press, 2014).

16. Radhakamal Mukerjee, "Social Ecology of a River Valley," *Sociology and Social Research* 12 (1927): 341–347, quotations on p. 342; Radhakamal Mukerjee, *Regional Sociology* (New York and London: Century and Co., 1926); Radhakamal Mukerjee, *The Changing Face of Bengal: A Study in Riverine Economy* (Calcutta: University of Calcutta Press, 1938).

17. Léon Metchnikoff, *La Civilisation et Les Grands Fleuves Historiques* (Paris: Hachette, 1889); Mukerjee, "Social Ecology," quotation on 342; William Willcocks, *Ancient System of Irrigation in Bengal and its Application to Modern Problems* (Calcutta: University of Calcutta Press, 1930).

18. Mukerjee, "Social Ecology," 345–347.

19. C. J. Baker, "Economic Reorganization and the Slump in South and Southeast Asia," *Comparative Studies in Society and History* 23 (1981): 325–349.

20. I explore these migrations in detail in Sunil S. Amrith, *Crossing the Bay of Bengal: The Furies of Nature and the Fortunes of Migrants* (Cambridge, MA: Harvard University Press, 2013), especially chapters 4 and 5.

21. Confidential letter from the Agent of the Government of India in British Malaya to the Government of India, April 3, 1933: NAI, Department of Education, Health and Lands: Overseas, file no. 206-2/32—L&O.

22. J. S. Furnivall, *Netherlands India* (Cambridge: Cambridge University Press, 1939), 428.

23. "World's Largest Dam Opened," *The Statesman*, August 22, 1934.

24. Handwritten memo by "SA," April 30, 1938, appended to the file of correspondence following the Chief Engineer's "Note on the Beneficial Effects of the Stanley Reservoir to Cauvery Delta Irrigation," Tamil Nadu State Archives, Chennai, Government Order 547-I, 27/2/1936.

25. Handwritten note in Government of Madras Public Works Department, Government Order number 375, February 24, 1938. Tamil Nadu State Archives, Chennai [TNSA].

26. Pietz, *Yellow River*, chapter 3.

27. *National Planning Committee No. 2: Being an Abstract of the Proceedings and other Particulars Relating to the National Planning Committee* (Bombay: K.T. Shah, 1940), 43.

28. "Burma-China Frontier: Chinese Claim to the Irrawaddy Triangle" (1933), BL, IOR, L/P&S/12/2231, enclosing William Credner's article in *Eastern Miscellany* (Shanghai), January 10, 1931; P. M. R. Leonard and

V. G. Robert, *Report on the Fourth Expedition to the "Triangle" for the Liberation of Slaves* (Rangoon: Government Printing, 1930), quotation in text from notes in archival file.

29. *Madras Fisheries Bulletin, 1918–1937* (Madras: Government Printing, 1938), 2; see Ajantha Subramanian, *Shorelines: Space and Rights in South Asia* (Stanford: Stanford University Press, 2009), 120–124.

30. Micah Muscolino, "Yellow River Flood, 1938–47," DisasterHistory .org, accessed March 3, 2018, www.disasterhistory.org/yellow-river-flood -1938-47; Micah S. Muscolino, *The Ecology of War in China: Henan Province, the Yellow River, and Beyond, 1938–1947* (Cambridge: Cambridge University Press, 2015).

31. Christopher Bayly and Tim Harper, *Forgotten Armies: The Fall of British Asia, 1941–45* (London: Allen Lane, 2004).

32. Srinath Raghavan, *India's War: The Making of Modern South Asia, 1939–45* (London: Allen Lane, 2016).

33. India Meteorological Department, *Hundred Years of Weather Service (1875–1975)*, bound typescript in the library of the Regional Meteorological Centre, Chennai, consulted in February 2015.

34. Sunil S. Amrith, "Food and Welfare in India, c. 1900–1950," *Comparative Studies in Society and History* 50 (2008): 1010–1035; the quotation from Nehru is in a letter to B. J. K. Hallowes, June 26, 1929, see note 2 above.

35. *The Ramakrishna Mission: Bengal and Orissa Cyclone Relief, 1942–44* (Howrah: Ramakrishna Mission, 1944), 1–2.

36. Bayly and Harper, *Forgotten Armies*, 282–291.

37. Note to Famine Commission (1944): Papers of L. G. Pinnell, British Library, Asian and African Studies Collection, European Manuscripts: MSS Eur D 911/7.

38. On ecological decline, see Iftekhar Iqbal, *The Bengal Delta: Ecology, State, and Social Change, 1840–1943* (Basingstoke: Palgrave/MacMillan, 2010), chapter 8. On famine, see Sugata Bose, "Starvation Amidst Plenty: The Making of Famine in Bengal, Honan and Tonkin, 1942–45," *Modern Asian Studies* 24, no. 4 (1990): 699–727; Amartya Sen, *Poverty and Famines: An Essay on Entitlement and Deprivation* (Oxford: Oxford University Press, 1981); Paul Greenough, *Prosperity and Misery in Modern Bengal: The Famine of 1943–4* (New York: Oxford University Press, 1982); and Bayly and Harper, *Forgotten Armies*, 282–291.

39. Jawaharlal Nehru, *The Discovery of India* [1946] (New Delhi: Oxford University Press, 2003), 496–498; S. G. Sardesai, *Food in the United Provinces* (Bombay: People's Publishing House, 1944), 19, 36–37.

40. V. D. Wickizer and M. K. Bennett, *The Rice Economy of Monsoon Asia* (Stanford, CA: Stanford University Press, 1941), 1, 189.

41. Nehru, *Discovery of India*, 535; Gyan Chand, *Problem of Population* (London: Oxford University Press, 1944), 10.

42. File note on Bhakra Dam Project, February 23, 1945, NAI, Political Department, I A Branch: file no. 21(22)—IA/45.

43. File note by T.A.W. Foy, October 31, 1946, NAI, 21(22)—IA/45.

44. Meghnad Saha, editorial, *Science and Culture* 1 (1935): 3–4.

45. Meghnad Saha, "Flood," *Science and Culture* 9 (September 1943): 95–97.

46. Meghnad Saha and Kamalesh Ray, "Planning for the Damodar Valley" (originally published in *Science and Culture*, 10 [1944]), in *Collected Works of Meghnad Saha*, ed. Santimay Chatterjee (Bombay: Orient Longman, 1987), 2:115–144, quotations on pp. 116, 132, 135.

47. Saha and Ray, "Planning for the Damodar Valley," 132, 135.

CHAPTER SEVEN: RIVERS DIVIDED, RIVERS DAMMED

1. The best narrative account of the period is in Christopher Bayly and Tim Harper, *Forgotten Wars: The End of Britain's Asian Empire* (London: Allen Lane, 2007); on the early Cold War in Asia, see Odd Arne Westad, *The Cold War: A World History* (New York: Basic Books, 2017), chapter 5.

2. For a concise overview, see Yasmin Khan, *The Great Partition: The Making of India and Pakistan* (New Haven, CT: Yale University Press, 2007).

3. Vazira Fazila-Yacoobali Zamindar, *The Long Partition and the Making of Modern South Asia: Refugees, Boundaries, Histories* (New York: Columbia University Press, 2007); Joya Chatterji, *The Spoils of Partition: Bengal and India, 1947–67* (Cambridge: Cambridge University Press, 2007); for a moving set of testimonies, see Urvashi Butalia, *The Other Side of Silence: Voices from the Partition of India* (London: Hurst, 2000).

4. David Gilmartin, *Blood and Water: The Indus River Basin in Modern History* (Berkeley: University of California Press, 2015), 206.

5. Government of India, Press Information Bureau, "Facts about Canal Dispute"—enclosure in a letter from S. V. Sampath to all Indian Missions abroad, September 27, 1949: National Archives of India [hereafter NAI], Ministry of External Affairs [hereafter MEA], File 6/1/7-XP (P)/49.

6. Joya Chatterji, "The Fashioning of a Frontier: The Radcliffe Line and Bengal's Border Landscape, 1947–52," *Modern Asian Studies* 33, no. 1 (1999): 185–242.

7. Rammanohar Lohia, *The Guilty Men of India's Partition* (Allahabad: Kitabistan, 1960).

8. Ayesha Jalal, *The Pity of Partition: Manto's Life, Times, and Work Across the India-Pakistan Divide* (Princeton: Princeton University Press, 2013).

9. Saadat Hasan Manto, "Yazid," in *Naked Voices: Stories and Sketches*, trans. Rakhshanda Jalil (New Delhi: Roli Books, 2008), 106.

10. India Meteorological Department, *Hundred Years of Weather Service (1875–1975)*, bound typescript in the library of the Regional Meteorological Centre, Chennai, consulted in February 2015, p. 55.

11. C. N. Vakil, *Economic Consequences of the Partition*, 2nd ed. (Bombay: National Information and Publications, 1949), 3–4.

12. Gilmartin, *Blood and Water*, 206.

13. Daniel Haines, *Rivers Divided: Indus Basin Waters in the Making of India and Pakistan* (Oxford: Oxford University Press, 2016), chapter 3.

14. Cited in Haines, *Rivers Divided*, 51.

15. India, Ministry of External Affairs, Directive on Canal Water Dispute Between India and Pakistan, NAI, MEA, File 6/1/7-XP (P)/49.

16. Directive on Canal Water Dispute, NAI, MEA, File 6/1/7-XP (P)/49.

17. Manu Goswami, *Producing India: From Colonial Economy to National Space* (Chicago: University of Chicago Press, 2004).

18. Gilmartin, *Blood and Water*, 212; Bashir A. Malik, *Indus Waters Treaty in Retrospect* (Lahore: Brite Books, 2005), 104.

19. David E. Lilienthal, "Kashmir: Another 'Korea' in the Making?," *Collier's* 128, no. 5 (1951): 58.

20. "Today in Earthquake History: Assam, 1950," Seismo Blog, Berkeley Seismology Lab, accessed March 1, 2018, http://seismo.berkeley.edu /blog/2017/08/15/today-in-earthquake-history-assam-1950.html; M. C. Podder, "Preliminary Report on the Assam Earthquake of 15th August 1950," *Bulletin of the Geological Survey of India*, Series B 2 (1950): 1–40; Francis Kingdon Ward, "Aftermath of the Assam Earthquake of 1950," *The Geographical Journal* 121 (1955): 290–303.

21. *Census of India, 1951*, vol. 1, Part 1A (New Delhi: Government Press, 1953).

22. Georges Canguilhem, *The Normal and the Pathological*, trans. Carolyn R. Fawcett (New York: Zone Books, 1989), 161.

23. *Census of India 1951*, vol. 1, Part 1A: 126–131.

24. *Census of India 1951*, vol. 1, Part 1A: 150.

25. Sanjoy Chakravorty, *The Price of Land: Acquisition, Conflict, Consequence* (New Delhi: Oxford University Press, 2013), see especially chapter 7.

26. Report of American Famine Mission to India, led by T. W. Schultz, cited in Henry Knight, *Food Administration in India, 1939–47* (Stanford, CA: Stanford University Press, 1954), 253; Government of India, Foodgrains Policy Committee, *Interim Report* (New Delhi, 1948); Government of India, Foodgrains Policy Committee, *Final Report* (New Delhi, 1948); *Report of the Foodgrains Enquiry Committee, 1957* (New Delhi: Ministry of Food & Agriculture, 1957), 26–27.

27. Jawaharlal Nehru's letter to India's chief ministers, April 15, 1948, in *Letters for a Nation: From Jawaharlal Nehru to His Chief Ministers,*

1947–1963, ed. Madhav Khosla (New Delhi: Allen Lane, 2014), 147–148. On the Hirakud Dam, see Rohan D'Souza, "Damming the Mahanadi River: The Emergence of Multi-Purpose River Valley Development in India (1943–46)," *Indian Economic and Social History Review* 40 (2003): 81–105.

28. Henry C. Hart, *New India's Rivers* (Bombay: Orient Longmans, 1956), 250.

29. India, Central Water-Power, Irrigation & Navigation Commission, *Quinquennial Report, April 1945–March 1950*, p. 2: BL, IOR, V/24/4496.

30. There is an illuminating account of Khosla's career in Daniel Klingensmith, *One Valley and a Thousand: Dams, Nationalism, and Development* (New Delhi: Oxford University Press, 2007).

31. A. N. Khosla, "Our Plans," *Indian Journal of Power and River Valley Development [IJPRVD]*, June 1951: 1–4.

32. India, Central Water & Power Commission, *Major Water & Power Projects of India* (Bhagirath Pamphlet 1, June 1957).

33. Jawharlal Nehru, speech at the opening of the Nangal Canal, July 8, 1954, in *Jawaharlal Nehru: An Anthology*, ed. Sarvepalli Gopal (New Delhi: Oxford University Press, 1980), 213–215.

34. "Nehru Shows Chou India's Dam Project," *New York Times*, January 1, 1957, 4.

35. On the history of the Films Division, see Peter Sutoris, *Visions of Development: Films Division of India and the Imagination of Progress, 1948–75* (London: Hurst, 2016), which includes a discussion of Ezra Mir's career; see also, Judith Pernin et al., "The Documentary Film in India, 1948–1975," undated, Hong Kong Baptist University, last accessed May 13, 2018, http://digital.lib.hkbu.edu.hk/documentary-film/india.php#footnote.

36. *Bhakra Nangal*, dir. N.S. Thapa, Government of India Films Division (1958).

37. On Lilienthal, see Klingensmith, *One Valley and a Thousand*.

38. Hart, *New India's Rivers*, 97.

39. Hugh Tinker, "A Forgotten Long March: The Indian Exodus from Burma, 1942," *Journal of Southeast Asian Studies* 6 (1975): 1–15; Sunil S. Amrith, *Crossing the Bay of Bengal: The Furies of Nature and the Fortunes of Migrants* (Cambridge, MA: Harvard University Press, 2013), chapter 6.

40. Letter from the Secretary to the Government of Punjab, Public Works Department, to the Secretary to the Governor General, June 18, 1945, NAI, file no. 21(22)—IA/45; Proceedings of a meeting held on January 31, 1946, between representatives of Madras and Hyderabad: Government of India, Political Branch: Hyderabad Residency; NAI, file no. 92(2), 1946.

41. Hart, *New India's Rivers*, 115.

42. "Lathi Charge on Strikers," *Times of India*, January 31, 1954, 9.

43. Hart, *New India's Rivers*, 178–184.

44. Ashis Nandy, "Dams and Dissent: India's First Modern Environmental Activist and His Critique of the DVC Project," *Futures* 33 (2001): 709–731.

45. World Bank Group Archives, Washington DC [hereafter WBA], File no. 1787276, Indus Basin Dispute, General Negotiations, 1949–52, Correspondence; File no, 1787280, Notes of Mission, September 1–16, 1954; File no. 1787263, Chronology of Indus Waters Dispute. Files 1787269 and 1787270 (Indus Basin Dispute, Working Party, Correspondence vol. 3 & 4), for example, both had several items that had been removed as unsuitable for declassification.

46. M. V. V. Ramana, *Inter-State River Water Disputes in India* (Hyderabad: Orient Longman, 1992), chapter 4.

47. Sumathi Ramaswamy, *The Goddess and the Nation: Mapping Mother India* (Durham, NC: Duke University Press, 2009), 244.

48. *Mother India,* dir. Mehboob (1957).

49. Quoted in Ramaswamy, *Goddess and the Nation,* 243.

50. Sangita Gopal and Sujata Moorti, *Global Bollywood: Travels of Hindi Song and Dance* (Minneapolis: University of Minnesota Press, 2008), 60.

51. Brian Larkin, "Bollywood Comes to Nigeria," accessed November 14, 2017, www.samarmagazine.org/archive/articles/21.

52. James C. Scott, *Seeing Like a State: Why Certain Schemes to Improve the Human Condition Have Failed* (New Haven, CT: Yale University Press, 1998); Arturo Escobar, *Encountering Development: The Making and Unmaking of the Third World* (Princeton, NJ: Princeton University Press, 1995).

53. Chakravorty, *Price of Land,* 113–114; Rohan D'Souza, "Framing India's Hydraulic Crises: The Politics of the Modern Large Dam," *Monthly Review* 60 (2008): 112–124.

54. File note, February 23, 1945, Bhakra Dam Project, Government of India, Political Department, IA Branch, NAI, file no. 21(22)—IA/45.

55. File note, Anon., March 12, 1945, Bhakra Dam Project, Government of India, Political Department, IA Branch: NAI, file no. 21(22)—IA/45.

56. File note, Anon., March 12, 1945, NAI, file no. 21(22)—IA/45.

57. Proceedings of a meeting held on January 31, 1946, between representatives of Madras and Hyderabad: Government of India, Political Branch: Hyderabad Residency; NAI, file no. 92(2), 1946, quotations from this file.

58. P. Chaturvedi and A. Dalal, *Law of Special Economic Zone: National and International Perspective* (Kolkata: Eastern Law House, 2009), 342, cited in Chakravorty, *Price of Land,* 115.

59. Walter Fernandes and Enakshi Ganguly Thukral, eds., *Development, Displacement and Rehabilitation* (New Delhi: Indian Social Institute, 1989); Esther Duflo and Rohini Pande, "Dams," *Quarterly Journal of Economics* 122 (2007): 601–646; Satyajit Singh, *Taming the Waters: The Political Economy of*

Large Dams in India (New Delhi: Oxford University Press, 1997), 182–203; Chakravorty, *Price of Land*, quotations on pp. 123–130.

60. Singh, *Taming the Waters*, 133–158.

61. Jawaharlal Nehru, "Social Aspects of Small and Big Projects," Inaugural address at the 29th annual meeting of the Central Board of Irrigation and Power, New Delhi, November 17, 1958, in Baldev Singh, *Jawaharlal Nehru on Science and Society: A Collection of His Writings and Speeches* (New Delhi: Nehru Memorial Museum and Library, 1990), 172–175.

62. United Nations, Economic Commission for Asia and the Far East (ECAFE), *Economic Survey of Asia and the Far East, 1948* (Bangkok: ECAFE, 1949); C. Hart Schaaf, "The United Nations Economic Commission for Asia and the Far East," *International Organization* 7 (1953): 463–481, quotation on p. 468.

63. Hart Schaaf, "Economic Commission for Asia" (1953).

64. Hart Schaaf, "Economic Commission for Asia" (1953), 481.

65. UN, ECAFE, *Economic Survey of Asia and the Far East, 1954* (Bangkok: ECAFE, 1955); chapter 10 covers the People's Republic of China.

66. Kanwar Sain and K. L. Rao, *Report on the Recent River Valley Projects in China* (New Delhi: Government of India Central Water and Power Commission, 1955), their full itinerary appears in Appendix F; "Mao is our Buddha" reported in Kanwar Sain, *Reminiscences of an Engineer* (New Delhi: Young Asia Publications, 1978), 208–209.

67. Sain, *Reminiscences of an Engineer*, 208–209.

68. Sain and Rao, *River Valley Projects in China*, 206–207; list of Chinese officials in Appendix G.

69. Sain and Rao, *River Valley Projects in China*, 154.

70. Judith Shapiro, *Mao's War Against Nature: Politics and Environment in Revolutionary China* (Cambridge: Cambridge University Press, 2001), chapter 1.

71. Cited in Philip Ball, *Water Kingdom: A Secret History of China* (Oxford: The Bodley Head, 2016), 225.

72. Sain and Rao, *River Valley Projects in China*, 162.

73. For an elaboration of the continuity argument, see Ball, *Water Kingdom*, chapter 8.

74. Sain and Rao, *River Valley Projects in China*. Hao's speech is reproduced in Appendix A.

75. Sain, *Reminiscences of an Engineer*, 210.

76. Christopher Sneddon, *Concrete Revolution: Large Dams, Cold War Geopolitics, and the US Bureau of Reclamation* (Chicago: University of Chicago Press, 2015); David Biggs, *Quagmire: Nation-Building and Nature in the Mekong Delta* (Seattle: University of Washington Press, 2010).

77. Biggs, *Quagmire*, 172.

78. Sneddon, *Concrete Revolution*.

79. Sain, *Reminiscences of an Engineer,* 388–392.

80. Nehru's letter to Zhou Enlai, September 26, 1959, published in *India-China Conflict* (New Delhi: Indian Ministry of External Affairs, 1964).

81. Letter from B. C. Mishra, Ministry of External Affairs to Apa B. Pant, Political Officer of the Government of India, Gangtok, Sikkim, October 7, 1960, in India, Ministry of External Affairs, "Construction of dam on Brahmaputra and Indus group of rivers by the Chinese": NAI, MEA, file F no. 4(75)—T 60.

82. Letter from R. S. Kapoor, Indian trade agent, Gyantse, Tibet to Apa Pant, Political Officer of the Government of India, Gangtok, Sikkim, December 15, 1960: NAI, MEA, file F no. 4(75)—T 60.

83. Letter marked "top secret," from K. K. Framji, Chief Engineer & Joint Secretary, Ministry of Irrigation and Power to B. C. Mishra, DS (China), Ministry of External Affairs, January 5, 1961: NAI, MEA, file F no. 4(75)—T 60.

84. Rohinton Mistry, *Such a Long Journey* (London: Faber & Faber, 1991), 10.

CHAPTER EIGHT: THE OCEAN AND THE UNDERGROUND

1. Neel Mukherjee, *The Lives of Others* (London: Vintage, 2014), 195–197.

2. Indian National Committee on Oceanic Research [hereafter INCOR], *International Indian Ocean Expedition: Indian Scientific Programmes, 1962–1965* (New Delhi: Council of Scientific and Industrial Research, 1962), 15.

3. Bernard Bailyn, "The Challenge of Modern Historiography," *American Historical Review* 87 (1982): 1–24, quotations on pp. 10–11.

4. Sunil S. Amrith, *Crossing the Bay of Bengal: The Furies of Nature and the Fortunes of Migrants* (Cambridge, MA: Harvard University Press, 2013), chapter 7.

5. India, Ministry of External Affairs, Memorandum on the International Conference of Plenipotentiaries on the Law of the Sea [undated, probably late 1957]. NAI, MEA: UN II Section, file no. 9(6) UN II/57.

6. Daniel Behrman, *Assault on the Largest Unknown: The International Indian Ocean Expedition* (Paris: UNESCO Press, 1981), 10–11; G. E. R. Deacon, "The Indian Ocean Expedition," *Nature* 187 (August 13, 1960): 561–562.

7. Warren S. Wooster, "Indian Ocean Expedition," *Science,* n.s., 150 (October 15, 1965): 290–292.

8. INCOR, *International Indian Ocean Expedition,* 15.

9. *The Indian Ocean Bubble,* issue 5, March 1, 1960, Woods Hole Oceanographic Institution Open Access Server, last accessed March 10, 2018, https://darchive.mblwhoilibrary.org/handle/1912/218.

10. Behrman, *Assault,* 27.

11. Behrman, *Assault,* 52.

12. INCOR, *International Indian Ocean Expedition,* 1–5.

13. Klaus Wyrtki, *Oceanographic Atlas of the International Indian Ocean Expedition* (Washington, DC: National Science Foundation, 1971).

14. Behrman, *Assault,* 64.

15. INCOR, *International Indian Ocean Expedition,* 44.

16. Gilbert T. Walker, "The Atlantic Ocean," *Quarterly Journal of the Royal Meteorological Society* 53 (1927): 113.

17. Deacon, "Indian Ocean Expedition"; INCOR, *International Indian Ocean Expedition,* 12; Wyrtki, *Oceanographic Atlas,* 7.

18. C. S. Ramage, *Monsoon Meteorology* (London: Academic Press, 1971), 1; Thomas A. Schroeder, "A Personal View of the History of the Department of Meteorology, University of Hawaii at Manoa" (2006), accessed June 1, 2016, www.soest.hawaii.edu/met/history.pdf.

19. Sanchari Pal, "Anna Mani Is One of India's Greatest Woman Scientists," *The Better India,* January 21, 2017, accessed May 1, 2018, https://www.thebetterindia.com/83063/anna-mani-scientist-meteorology-ozone-wind-energy/.

20. C. S. Ramage, *Meteorology in the Indian Ocean* (Geneva: World Meteorological Association, 1965).

21. Ramage, *Meteorology in the Indian Ocean.*

22. Behrman, *Assault,* 67.

23. Behrman, *Assault,* 65; C. S. Ramage and C. R. Raman, *Meteorological Atlas of the International Indian Ocean Expedition* (Washington, DC: US Government Printer, 1972).

24. Behrman, *Assault,* 66.

25. Ramage, *Meteorology in the Indian Ocean.*

26. Roger Revelle and H. E. Suess, "Carbon Dioxide Exchange Between Atmosphere and Ocean and the Question of an Increase of Atmospheric CO^2 During the Past Decades, " *Tellus* 9 (1957): 18–27, quotation on pp. 19–20. On oceanography and the discovery of climate change, see Naomi Oreskes, "Changing the Mission: From the Cold War to Climate Change," in *Science and Technology in the Global Cold War,* ed. Naomi Oreskes and John Krige (Cambridge, MA: MIT Press, 2014), 141–187.

27. Behrman, *Assault,* 11–12.

28. P. K. Das, *The Monsoons* (New Delhi: National Book Trust, 1968), 6.

29. Francine Frankel, *India's Political Economy, 1947–1977: The Gradual Revolution* (Princeton, NJ: Princeton University Press, 1980), 247–248.

30. Statistics on Indian wheat imports from Nick Cullather, *The Hungry World: America's Cold War Battle Against Poverty in Asia* (Cambridge, MA: Harvard University Press, 2010), 144; *Economic Survey of Indian Agriculture for 1966–67* (New Delhi: Government of India, 1969); Frankel, *India's Political Economy,* 293.

31. David Ludden, *An Agrarian History of South Asia* (Cambridge: Cambridge University Press, 1999), especially chapter 4.

32. David C. Engerman, *The Price of Aid: The Economic Cold War in India* (Cambridge, MA: Harvard University Press, 2018); Cullather, *Hungry World*.

33. Cullather, *Hungry World*, 207.

34. C. Subramaniam, *Hand of Destiny*, vol. 2, *The Green Revolution* (Bombay: Bharatiya Vidya Bhavan, 1993), 137–138.

35. Frankel, *India's Political Economy*, 270.

36. Subramaniam, *Hand of Destiny*, vol. 2, 154, 165–167.

37. "Years Before a Revolution," *Times of India*, August 22, 1965.

38. Quoted in Mahesh Rangarajan, "Striving for a Balance: Nature, Power, Science and Indira Gandhi's India, 1917–1984," *Conservation and Society* 7 (2009): 299–312.

39. Cullather, *Hungry World*, 223.

40. Paul R. Brass, "The Political Uses of Crisis: The Bihar Famine of 1966–1967," *Journal of Asian Studies* 45 (1986): 245–267, 249.

41. Ronald E. Doel and Kristine C. Harper, "Prometheus Unleashed: Science as a Diplomatic Weapon in the Lyndon B. Johnson Administration," *Osiris* 21 (2006): 66–85.

42. James R. Fleming, *Fixing the Sky: The Checkered History of Weather and Climate Control* (New York: Columbia University Press, 2010); Kristine C. Harper, *Make It Rain: State Control of the Atmosphere in Twentieth Century America* (Chicago: University of Chicago Press, 2017).

43. Lyndon B. Johnson, *The Vantage Point: Perspectives of the Presidency, 1963–1969* (New York: Holt, Rinehart and Winston, 1971), 226.

44. Doel and Harper, "Prometheus Unleashed," 80, 83.

45. Rajni Kothari, "The Congress 'System' in India," *Asian Survey* 4 (1964): 1161–1173; Confidential Despatch from British High Commission, Delhi to London, 3 March 1967, in United Kingdom National Archives (UKNA), "India—Political Affairs—Internal" FO 37/35.

46. Ashutosh Varshney, *Democracy, Development, and the Countryside: Urban-Rural Struggles in India* (Cambridge: Cambridge University Press, 1998), 57.

47. Geoffrey Parker, *Global Crisis: War, Climate Change & Catastrophe in the Seventeenth Century* (New Haven, CT: Yale University Press, 2013); Sam White, *The Climate of Rebellion in the Early Modern Ottoman Empire* (Cambridge: Cambridge University Press, 2011).

48. Indira Gandhi, "Man and Environment," speech at the Plenary Session of United Nations Conference on Human Environment, Stockholm, June 14, 1972: full text available at http://lasulawsenvironmental.blogspot .com/2012/07/indira-gandhis-speech-at-stockholm.html, last accessed May 14, 2018.

49. Paul Ehrlich, *The Population Bomb* (New York: Ballantine Books, 1968), 15–16.

50. Indira Gandhi, "Man and Environment," speech at the Plenary Session of United Nations Conference on Human Environment, Stockholm, June 14, 1972; Jairam Ramesh, "Poverty Is the Greatest Polluter: Remembering Indira Gandhi's Stirring Speech in Stockholm," *The Wire*, June 7, 2017, accessed November 30, 2017, https://thewire.in/144555/indira-gandhi-nature-pollution/.

51. On coercive population control in India, see Matthew Connelly, *Fatal Misconception: The Struggle to Control World Population* (Cambridge, MA: Harvard University Press, 2008); Emma Tarlo, *Unsettling Memories: Narratives of the Emergency in Delhi* (London: Hurst, 2003).

52. Shyam Divan and Armin Rosencranz, eds., *Environmental Law and Policy in India: Cases, Materials and Statutes* (New Delhi: Oxford University Press, 2001), 167–241; the case cited is Aggarwal Textile Industries v. State of Rajasthan, S.B.C. Writ Petition No. 1375/80, March 2, 1981, presented in Divan and Rosencranz, *Environmental Law*, 187.

53. Anthony Acciavatti, "Re-imagining the Indian Underground: A Biography of the Tubewell," in *Places of Nature in Ecologies of Urbanism*, ed. Anne Rademacher and K. Sivaramakrishnan (Hong Kong: Hong Kong University Press, 2017), 206–237, quotation on p. 207.

54. Tushaar Shah, *Taming the Anarchy: Groundwater Governance in South Asia* (New York: Routledge, 2008); Tushaar Shah, "Climate Change and Groundwater: India's Opportunities for Mitigation and Adaptation," *Environmental Research Letters* 4 (2009): 1–13.

55. Roger Revelle and V. Lakshminarayana, "Ganges Water Machine," *Science*, n.s., 188 (1975): 611–616, quotation on p. 611; K. L. Rao, *India's Water Wealth: Its Assessment, Uses, and Projections* (New Delhi: Orient Longman, 1975).

56. Joshua Eisenman, "Building China's 1970s Green Revolution: Responding to Population Growth, Decreasing Arable Land, and Capital Depreciation," in *China, Hong Kong, and the Long 1970s: Global Perspectives*, ed. Priscilla Roberts and Odd Arne Westad (New York: Palgrave Macmillan, 2017), 55–86.

57. Sigrid Schmalzer, *Red Revolution, Green Revolution: Scientific Farming in Socialist China* (Chicago: University of Chicago Press, 2016), quotation on p. 13.

58. Francine Frankel, *India's Green Revolution: Economic Gains and Political Costs* (Princeton, NJ: Princeton University Press, 1971); N. K. Dubash, *Tubewell Capitalism: Groundwater Development and Agrarian Change in Gujarat* (New Delhi: Oxford University Press, 2002); Shah, "Climate Change and Groundwater."

59. L. J. Walinsky, ed., *Agrarian Reform As Unfinished Business: The Selected Papers of Wolf Ladejinsky* (Washington, DC: World Bank, 1977); G. Rosen, "Obituary: Wolf Ladejinsky (1899–1975)," *Journal of Asian Studies* 36 (1976): 327–328.

60. Wolf Ladejinsky, "Drought in Maharashtra (Not in a Hundred Years)," typescript contained in World Bank Archives (WBA), file number 1167800, Drought Prone Areas Project—India—Correspondence vol. 1.

61. Jean Drèze, "Famine Prevention in India" (working paper 45, WIDER: United Nations University, Helsinki, 1988), 69–75.

62. John A. Young, "Physics of the Monsoon: The Current View," in *Monsoons*, ed. Jay S. Fein and Pamela L. Stephens (New York: John Wiley & Sons, 1987), 211–243, quotation on p. 211; see the discussion of "moist processes" in Peter J. Webster, "Monsoons," *Scientific American* 245 (1981): 108–119; on modeling, see Kirsten Hastrup and Martin Skrydstrup, eds., *The Social Life of Climate Change Models: Anticipating Nature* (London: Routledge, 2012).

63. Jacob Bjerknes, "A Possible Response of the Atmospheric Hadley Circulation to Equatorial Anomalies of Ocean Temperature," *Tellus* 18 (1966): 820–829; Jacob Bjerknes, "Atmospheric Teleconnections from the Equatorial Pacific," *Journal of Physical Oceanography* 97 (1969): 163–172.

64. P. J. Webster, H. R. Chang, and V. E. Toma, *Tropical Meteorology and Climate* (Oxford: Wiley Blackwell, in press), chapter 14.

65. R. A. Madden and P. R. Julian, "Detection of a 40–50 Day Oscillation in the Zonal Wind in the Tropical Pacific," *Journal of the Atmospheric Sciences* 28 (1971): 702–770; R. A. Madden and P. R. Julian, "Description of Global-Scale Circulation Cells in the Tropics with a 40–50 Day Period," *Journal of the Atmospheric Sciences* 29 (1972): 1109–1123.

66. David M. Lawrence and Peter J. Webster, "The Boreal Summer Intraseasonal Oscillation: Relationship between Northward and Eastward Movement of Convection," *Journal of the Atmopheric Sciences* 59 (2002): 1593–1606.

67. Adam Sobel, *Storm Surge: Hurricane Sandy, Our Changing Climate, and Extreme Weather of the Past and Future* (New York: Harper Wave, 2014), 9–20.

68. On MONEX, see Behrman, *Assault*, 64; Webster, Chang, and Toma, *Tropical Meteorology*, chapter 14.

69. C. S. Ramage, *The Great Indian Drought of 1899*, Occasional Paper, Aspen Instiute for Humanistic Studies, Program on Science, Technology, and Humanism (1977), quotations on pp. 4, 6.

70. *Declaration of the Climate Conference* (Geneva: World Meteorological Organization, 1979), 1.

CHAPTER NINE: STORMY HORIZONS

1. World Bank Group, "China Overview," accessed February 12, 2018, www.worldbank.org/en/country/china/overview.

2. Sumit Ganguly and Rahul Mukherjee, *India Since 1980* (Cambridge: Cambridge University Press, 2012), chapter 3.

3. Anil Agarwal, Kalpana Sharma, and Ravi Chopra, *The State of India's Environment, 1982: A Citizens' Report* (New Delhi: Centre for Science and Environment, 1982), 20.

4. Naomi Oreskes, "The Scientific Consensus on Climate Change," *Science* 306 (December 2004): 1686.

5. Intergovernmental Panel on Climate Change, *Climate Change 2014: Impacts, Adaptation, and Vulnerability* (Geneva: IPCC, 2014); World Bank, *Turn Down the Heat: Climate Extremes, Regional Impacts, and the Case for Resilience* (Washington, DC: World Bank, 2013).

6. Andreas Malm, *The Progress of This Storm: Nature and Society in a Warming World* (London: Verso, 2018), 5.

7. Angus Deaton, *The Great Escape: Health, Wealth, and the Origins of Inequality* (Princeton, NJ: Princeton University Press, 2013).

8. Hannah Ritchie, "Yields vs. Land Use: How the Green Revolution Enabled Us to Feed a Growing Population," *Our World in Data*, August 22, 2017, accessed February 10, 2018, https://ourworldindata.org/yields-vs-land-use-how-has-the-world-produced-enough-food-for-a-growing-population.

9. Khushwant Singh, "The Indian Monsoon in Literature," in *Monsoons*, ed. Jay S. Fein and Pamela L. Stephens (New York: John Wiley & Sons, 1987), 35–50, quotations on p. 48.

10. Jyoti Bhatt, "Divination of Rainy Days: An Annual Festival in Gujarat" [1987], in Asia Art Archive, Hong Kong: Jyoti Bhatt Archive, last accessed April 24, 2018, https://aaa.org.hk/en/collection/search/archive/jyoti-bhatt-archive-english/object/divination-of-rainy-days-an-annual-festival-in-gujarat.

11. University of Hawaii at Manoa Economics Department, "Harry T. Oshima (1918–1998)," accessed February 16, 2018, www.economics.hawaii.edu/history/faculty/oshima.html.

12. Harry T. Oshima, *Economic Growth in Monsoon Asia: A Comparative Survey* (Tokyo: University of Tokyo Press, 1987); Harry T. Oshima, "Seasonality and Underemployment in Monsoon Asia," *Philippine Economic Journal* 19 (1971): 63–97.

13. Statistics from A. Vaidyanathan, *Water Resources of India* (New Delhi: Oxford University Press, 2013); T. Shah, "Climate Change and Groundwater: India's Opportunities for Mitigation and Adaptation," *Environmental Research Letters* 4 (2009): 1–13, quotation on p. 3.

14. Meera Subramanian, *A River Runs Again: India's Natural World in Crisis, from the Barren Cliffs of Rajasthan to the Farmlands of Karnataka* (New York: PublicAffairs, 2015), 9–66, on Punjab; on Gujarat, see David Hardiman, "The Politics of Water Scarcity in Gujarat," in Amita Baviskar ed. *Waterscapes: The Cultural Politics of a Natural Resource* (New Delhi: Permanent Black, 2006), 39–62.

15. Daniyal Mueenuddin, "Nawabdin Electrician," in *In Other Rooms, Other Wonders* (New York: W.W. Norton, 2009), 13–28, quotation on p. 13.

16. Jane Qiu, "China Faces Up to Groundwater Crisis," *Nature* 466 (2010): 308.

17. David A. Pietz, *The Yellow River: The Problem of Water in Modern China* (Cambridge, MA: Harvard University Press, 2015), 264–265; M. Webber et al., "The Yellow River in Transition," *Environmental Science and Policy* 11 (2008): 422–429.

18. M. Rodell, I. Velicogna, and J. S. Famiglietti, "Satellite-Based Estimates of Groundwater Depletion in India," *Nature* 460 (2009): 999–1002.

19. M. K. Gandhi, "Some Mussooree Reminiscences," *Harijan*, June 23, 1946, 198; Ramachandra Guha, *The Unquiet Woods: Ecological Change and Political Protest in the Himalaya* (New Delhi: Oxford University Press, 1989).

20. Kathleen D. Morrison, "Dharmic Projects, Imperial Reservoirs, and New Temples of India: An Historical Perspective on Dams in India," *Conservation and Society* 8 (2010): 184.

21. Ambedkar's statement was delivered in India's Constituent Assembly on November 4, 1948; for research on the complexity of water management in pre-modern India, see David Mosse, *The Rule of Water: Statecraft, Ecology, and Collective Action in South Asia* (New Delhi: Oxford University Press, 2003); Haruka Yanagisawa, *A Century of Change: Caste and Irrigated Lands in Tamil Nadu, 1860s–1970s* (New Delhi: Manohar, 1996); for an overview that is skeptical of the idea that colonial rule was an absolute ecological watershed, see Mahesh Rangarajan, "Environmental Histories of India: Of States, Landscapes, and Ecologies," in *The Environment and World History*, ed. Kenneth Pomeranz and Edmund Burke III (Berkeley: University of California Press, 2009), 229–254.

22. Agarwal, Chopra, and Sharma, *The State of India's Environment, 1982*; A. Agarwal and Sunita Narain, eds., *The State of India's Environment, 1984–85: A Second Citizens' Report* (New Delhi: Centre for Science and Environment, 1985), quotation from "Statement of Shared Concern"; The Centre for Science and Environment also produced a documentary film on water harvesting: *Harvest of Rain*, dir. Sanjay Kak (1995), Centre for Science and Environment, 1995; Tim Forsyth, "Anil Agarwal," in *Fifty Key Thinkers on Development*, ed. D. Simon (London: Routledge, 2005), 9–14.

23. Vandana Shiva, *The Violence of the Green Revolution: Third World Agriculture, Ecology and Politics* (London: Zed Books, 1991), 11.

24. The authors of the first Indian report cite the inspiration of Penang in their preface: Agarwal, Chopra, and Sharma, *State of India's Environment, 1982*; on the Third World Network, see its website, accessed February 1, 2018, www.twn.my/twnintro.htm; on the Consumers' Association of Penang, see Matthew Hilton, *Prosperity for All: Consumer Activism in an Era of Globalization* (Ithaca, NY: Cornell University Press, 2009); on the rise and fall of the New International Economic Order, see Nils Gilman, "The New International Economic Order: A Reintroduction," *Humanity* (Spring 2015): 1–16.

25. Anil Agarwal and Sunita Narain, *Global Warming in an Unequal World: A Case of Environmental Colonialism* (Delhi: Centre for Science and Environment, 1991).

26. P. Sainath, *Everybody Loves a Good Drought: Stories from India's Poorest Districts* (New Delhi: Penguin, 1996), quotations on pp. 319–320.

27. Aseem Shrivastava and Ashish Kothari, *Churning the Earth: The Making of Global India* (New Delhi: Viking, 2012), 176–183; P. Sainath, "Farm Suicides: A 12-Year Saga," *The Hindu*, January 25, 2010; P. Sainath, "The Largest Wave of Suicides in History," *The Hindu*, February 16, 2009; Akta Kaushal, "Confronting Farmer Suicides in India," *Alternatives* 40 (2016): 46–62.

28. Gyansham Shah, Harsh Mander, Sukhadeo Thorat, Satish Deshpande, and Amita Baviskar, *Untouchability in Rural India* (New Delhi: Sage, 2006), 75.

29. Agarwal, Chopra, and Sharma, *State of India's Environment, 1982*, 20–23.

30. Darryl D'Monte, *Temples of Tombs? Industry versus Environment, Three Controversies* (New Delhi: Centre for Science and Environment, 1985), 15.

31. M. C. Mehta v. Union of India (Kanpur Tanneries), *All India Reporter* (1988), SC 1037; M. C. Mehta v. Union of India (Municipalities), *All India Reporter* (1988), SC 1115: cases cited in *Environmental Law and Policy in India: Cases, Materials and Statutes*, ed. Shyam Divan and Armin Rosencranz (New Delhi: Oxford University Press, 2001), 210–225. Mehta's biography from the M. C. Mehta Foundation, http://mcmef.org/m-c-mehta/; details of his career are available in his citation for the Goldman Prize, which he won in 1996, www.goldmanprize.org/recipient/mc-mehta/.

32. Judith Shapiro, *China's Environmental Challenges* (London: Polity, 2012), 112–118.

33. Ma Jun, *China's Water Crisis*, trans. Nancy Yang Liu and Lawrence R. Sullivan (Norwalk, CT: EastBridge/International Rivers, 2004), quota-

tions on pp. vii–xi, 79–80; first published in Chinese as *Zhongguo shui weiji* (Beijing: China Environmental Sciences Publishing House, 1999).

34. "India—Mr McNamara's Meeting with the Indian Finance Minister," Memorandum of September 27, 1978: WBA, Contacts with Member Countries: India—Correspondence 09, folder 1771081.

35. See the entry on Medha Patkar on the website of the Goldman Environmental Prize, which she won in 1992. Accessed March 1, 2018, www .goldmanprize.org/recipient/medha-patkar/.

36. "Bankwide Lessons Learned from the Experience with the India Sardar Sarovar (Narmada) Project," World Bank report, May 19, 1993, accessed March 19, 2018, http://documents.worldbank.org/curated /en/221941467991015938/Lessons-learned-from-Narmada.

37. Smita Narula, "The Story of Narmada Bachao Andolan: Human Rights in the Global Economy and the Struggle against the World Bank," *New York University Public Law and Legal Theory Working Papers* 106 (2008); Balakrishnan Rajagopal, "The Role of Law in Counter-hegemonic Globalization and Global Legal Pluralism: Lessons from the Narmada Valley Struggle in India," *Leiden Journal of International Law* 18 (2005): 345–355; Alf Gunvald Nilsen, *Dispossession and Resistance in India: The River and the Rage* (London: Routledge, 2010). Modi's comments quoted in "54 Years On, Modi Opens Sardar Sarovar Dam," *FirstPost,* September 18, 2017, accessed March 20, 2018, www.firstpost.com/politics/sardar-sarovar-dam -inaugurated-narendra-modi-alleges-conspiracy-to-stop-project-congress -calls-it-election-gimmick-4054251.html.

38. A synthesis of the research of the 1980s and 1990s appears in Satyajit Singh, *Taming the Waters: The Political Economy of Large Dams in India* (New Delhi: Oxford University Press, 1997), especially 133–163 on ecological consequences.

39. On displacement, see Sanjoy Chakravorty, *The Price of Land: Acquisition, Conflict, Consequence* (New Delhi: Oxford University Press, 2013), especially Appendix A9.2; and Singh, *Taming the Rivers,* 182–203. For a global perspective on dam displacement, see International Committee of the Red Cross, *World Disasters Report 2012: Focus on Forced Migration and Displacement* (Geneva: Red Cross, 2012). On the disproportionate impact of dams on marginalized communities, see Esther Duflo and Rohini Pande, "Dams," *Quarterly Journal of Economics* 122 (2007): 601–646.

40. Arundhati Roy, "The Greater Common Good," *Outlook,* May 24, 1999; for a critique, see Ramachandra Guha, "The Arun Shourie of the Left," *The Hindu,* November 26, 2000.

41. Vairamuthu, *Kallikaatu Ithihaasam* (Chennai: Thirumagal, 2001).

42. *Dams and Development: A New Framework for Decision Making. The Report of the World Commission on Dams* (London: Earthscan, 2000).

43. Ramaswamy R. Iyer, "The Story of a Troubled Relationship," *Water Alternatives* 6 (2013): 168–176, quotations on pp. 169, 175; on Iyer's career, see Amita Baviskar's obituary: "He Watered the Arid Fields of Administration with Intellectual Rigour and Honesty," *The Wire,* September 11, 2015, last accessed May 2, 2018, https://thewire.in/environment/watering-the-arid -fields-of-administration-with-intellectual-rigour-and-honesty.

44. Ravi S. Jha, "India's River Linking Project Mired in Cost Squabbles and Politics," *The Guardian,* February 5, 2013, accessed May 4, 2018, www .theguardian.com/environment/2013/feb/05/india-river-link-plan-progress -slow; Supreme Court of India. Writ Petition (Civil) No. 512 of 2002 in Re. Networking of Rivers, judgment, accessed May 14, 2018, http://courtnic .nic.in/supremecourt/temp/512200232722012p.txt; Y. A. Alagh, G. Pangare, and B. Gujja, *Interlinking of Rivers in India: Overview and Ken-Betwa Link* (New Delhi: Academic Foundation, 2006).

45. Ramaswamy R. Iyer, "River Linking Project: A Disquieting Judgment," *Economic and Political Weekly,* April 7, 2012, 33–40, quotations on p. 37.

46. Meera Subramanian, *A River Runs Again: India's Natural World in Crisis, from the Barren Cliffs of Rajasthan to the Farmlands of Karnataka* (New York: PublicAffairs, 2015).

47. "China Has Built the World's Largest Water Diversion Project," *The Economist,* April 5, 2018; on the scheme in historical perspective, see Kenneth Pomeranz, "The Great Himalayan Watershed: Water Shortages, Mega-Projects and Environmental Politics in China, India, and Southeast Asia," *Asia Pacific Journal: Japan Focus* 7 (2009), accessed February 1, 2018, https://apjjf .org/-Kenneth-Pomeranz/3195/article.html.

48. C. J. Vörösmarty et al, "Battling to Save the World's River Deltas," *Bulletin of the Atomic Scientists,* 65, 2 (2009): 31–43; James Syvitski, "Sinking Deltas Due to Human Activities," *Nature Geoscience* 2 (2009): 681–686; Roger L. Hooke, "On the History of Humans as Geomorphic Agents," *Geology* 28 (2000): 843–846.

49. Pomeranz, "The Great Himalayan Watershed."

50. Shripad Dharmadhikary, *Mountains of Concrete: Dam Building in the Himalayas* (Berkeley: International Rivers, 2008); Douglas Hill, "Trans-boundary Water Resources and Uneven Development: Crisis Within and Beyond Contemporary India," *South Asia: Journal of South Asian Studies* 36 (2013): 243–257; John Vidal, "China and India 'Water Grab' Dams Put Ecology of the Himalayas in Danger," *The Observer,* August 10, 2013.

51. Dharmadhikary, *Mountains of Concrete.*

52. Dharmadhikary, *Mountains of Concrete;* R. Grumbine and M. Pandit, "Threats from India's Himalaya Dams," *Science* 339 (2013): 36–37; Rohan D'Souza, "Pulses Against Volumes: Transboundary Rivers and Pan-Asian

Connectivity," in *Heading East: Security, Trade, and Environment Between India and Southeast Asia,* ed. Karen Stoll Farrell and Sumit Ganguly (Oxford: Oxford University Press, 2016); Jane Qiu, "Flood of Protest Hits Indian Dams," *Nature* 492 (2012): 15–16; Fan Xiao quoted in Charlton Lewis, "China's Great Dam Boom: A Major Assault on Its Rivers," *Yale Environment 360,* November 4, 2013, accessed March 1, 2018, https://e360.yale.edu/features /chinas_great_dam_boom_an_assault_on_its_river_systems.

53. T. Bolch et al., "The State and Fate of the Himalayan Glaciers," *Science* 336 (2012): 310–314; World Bank, *Turn Down the Heat;* Dexter Filkins, "The End of Ice: Exploring a Himalayan Glacier," *New Yorker,* April 4, 2016.

54. S. P. Xie et al., "Towards Predictive Understanding of Regional Climate Change," *Nature Climate Change 5* (2015): 921–930.

55. A. Turner and H. Annamalai, "Climate Change and the South Asian Monsoon," *Nature Climate Change* 2 (2012): 587–595; M. Bollasina, Y. Ming, and V. Ramaswamy, "Anthropogenic Aerosols and the Weakening of the South Asian Summer Monsoon," *Science* 334 (2011): 502–505; Deepti Singh, "South Asian Monsoon: Tug of War on Rainfall Changes," *Nature Climate Change* 6 (2016): 20–22; R. Krishnan et al., "Deciphering the Desiccation Trend of the South Asian Monsoon Hydroclimate in a Warming World," *Climate Dynamics* 47 (2016): 1007–1027.

56. J. Lelieveld, P. J. Crutzen, V. Ramanathan et al., "The Indian Ocean Experiment: Widespread Air Pollution from South and Southeast Asia," *Science* 291 (2001): 1031–1036; P. J. Crutzen and E. F. Stoermer, "The Anthropocene," *Global Change Newsletter* 41 (2000): 17–18.

57. V. Ramanathan, "Atmospheric Brown Clouds: Impact on South Asian Climate and Hydrological Cycle," *Proceedings of the National Academy of Science* 102 (2005): 5326–5333; H. V. Henriksson et al., "Spatial Distributions and Seasonal Cycles of Aerosols in India and China seen in Global Climate-Aerosol Model," *Atmospheric Chemistry and Physics* 11 (2011): 7975–7990.

58. Bollasina, Ming, and Ramaswamy, "Anthropogenic Aerosols"; Theodore G. Shepherd, "Atmospheric Circulation as a Source of Uncertainty in Climate Change Projections," *Nature Geoscience* 7 (2014): 703–708.

59. Singh, "South Asian Monsoon," 21; Krishnan et al., "Deciphering the Desiccation Trend"; D. Niyogi, C. Kishtawal, S. Tripathi, and R. Govindaraju, "Observational Evidence that Agricultural Intensification and Land Use Change May Be Reducing the Indian Summer Monsoon Rainfall," *Water Resources Research* 46 (2010), https://doi.org/10.1029/2008WR007082.

60. D. Singh, M. Tsiang, B. Rajaratnam, and N. Diffenbaugh, "Observed Changes in Extreme Wet and Dry Spells During the South Asian Summer Monsoon," *Nature Climate Change* 4 (2014): 456–461; Krishnan et al., "Deciphering the Desiccation Trend"; B. N. Goswami, S. A. Rao, D. Sengupta,

and S. Chakravorty, "Monsoons to Mixing in the Bay of Bengal: Multiscale Air-Sea Interactions and Monsoon Predictability," *Oceanography* 29 (2016): 28–37.

61. Adam Sobel, *Storm Surge: Hurricane Sandy, Our Changing Climate, and Extreme Weather of the Past and Future* (New York: Harper Wave, 2014), 203–232; Amitav Ghosh, *The Great Derangement: Climate Change and the Unthinkable* (Chicago: University of Chicago Press, 2016), 41–43.

62. Data from the EM-DAT International Disaster Database, last accessed April 22, 2018, www.emdat.be.

63. Ubydul Haque et al., "Reduced Deaths from Cyclones in Bangladesh: What More Needs to Be Done?," *Bulletin of the World Health Organization* 90 (2012): 150–156, doi: 10.2471/BLT.11.088302.

64. A. Mahadevan et al., "Freshwater in the Bay of Bengal: Its Fate and Role in Air-Sea Heat Exchange," *Oceanography* 29 (2016): 72–81.

65. "Seafloor Holds 15 Million Years of Monsoon History," accessed April 10, 2018, https://news.brown.edu/articles/2015/02/monsoons.

66. Concerned Citizens' Commission, *Mumbai Marooned: An Inquiry into the Mumbai Floods, 2005* (Mumbai: Conservation Action Trust, 2006).

67. Ghosh, *Great Derangement*, 50–51.

68. Anuradha Mathur and Dilip da Cunha, "The Sea and Monsoon Within: A Mumbai Manifesto," in *Ecological Urbanism*, ed. Mohsen Mostafavi with Gareth Doherty (Cambridge, MA: Harvard University Graduate School of Design/Lars Müller, 2010), 194–207.

69. Susan Hanson et al., "A Global Ranking of Port Cities with High Exposure to Climate Extremes," *Climatic Change* 104 (2011): 89–111; Orrin H. Pilkey, Linda Pilkey-Jarvis, and Keith C. Pilkey, *Retreat from a Rising Sea: Hard Choices in an Age of Climate Change* (New York: Columbia University Press, 2016), 65–74.

70. International Federation of Red Cross and Red Crescent Societies, "Indonesia: Jakarta Floods," Information Bulletin 4.2007, September 26, 2007; Pilkey, Pilkey-Jarvis, and Pilkey, *Retreat from a Rising Sea*, 70–71.

71. On the contemporary geopolitics of the Bay, see Sunil S. Amrith *Crossing the Bay of Bengal: The Furies of Nature and the Fortunes of Migrants* (Cambridge, MA: Harvard University Press, 2013), chapter 8.

72. A. Mahadevan et al., "Bay of Bengal: From Monsoons to Mixing," *Oceanography* 29 (2016): 14–17, map on p. 16.

73. International Federation of Red Cross Societies, *World Disasters Report 2012: Focus on Forced Migration and Displacement* (Geneva: IFRC, 2013), 231.

74. Amrith, *Crossing the Bay of Bengal*, chapter 8.

75. Joya Chatterji, "Dispositions and Destinations: Refugee Agency and 'Mobility Capital' in the Bengal Diaspora, 1947–2007," *Comparative Studies*

in Society and History 55 (2013): 273–304; IFRC, *World Disasters Report 2012*, 38.

76. *Groundswell: Preparing for Internal Climate Migration* (Washington, DC: World Bank, 2018).

77. World Bank, "Policy Note #2: Internal Climate Migration in South Asia" (2018), accessed May 14, 2018, https://openknowledge.worldbank .org/bitstream/handle/10986/29461/GroundswellPN2.pdf?sequence =7&isAllowed=y.

78. ASEAN, *Master Plan on ASEAN Connectivity 2025* (Jakarta: Asean Secretariat, 2016); Constantino Xavier, *Bridging the Bay of Bengal: Towards a Stronger BIMSTEC* (New Delhi: Carnegie India, February 2018).

79. Aparna Roy, "Bay of Bengal Diplomacy," *The Hindu,* October 10, 2017.

80. Season Watch, accessed May 1, 2018, www.seasonwatch.in/.

81. Prasenjit Duara, *The Crisis of Global Modernity: Asian Traditions and a Sustainable Future* (Cambridge: Cambridge University Press, 2014).

82. Gandhi quoted in Singh, "Indian Monsoon in Literature," 50.

EPILOGUE: HISTORY AND MEMORY AT THE WATER'S EDGE

1. Zadie Smith, "Elegy for a Country's Seasons," *New York Review of Books,* April 3, 2014.

2. Namrata Kala, "Learning, Adaptation, and Climate Uncertainty: Evidence from Indian Agriculture," working paper, August 2017, accessed March 3, 2018, https://namratakala.files.wordpress.com/2017/08/kala_learning_aug 2017_final.pdf, last.

3. Suprabha Seshan, "Once, the Monsoon," June 22, 2017, accessed March 10, 2018, https://countercurrents.org/2017/06/22/once-the-monsoon/.

4. M. Rajshekhar, "Why Tamil Nadu's Fisherfolk Can No Longer Find Fish," *Scroll,* July 8, 2016, accessed April 15, 2018, https://scroll.in/article /808960/why-tamil-nadus-fisherfolk-can-no-longer-find-fish, the quotation in the text is from Rajshekhar's report; E. Vivekanandan, "Impact of Climate Change in the Indian Marine Fisheries and Potential Adaptation Options," in *Coastal Fishery Resources of India: Conservation and Sustainable Utilisation* (Cochin, India: Society of Fisheries Technologists, 2010), 169–184; Amitav Ghosh and Aaron Savio Lobo, "Bay of Bengal: Depleted Fish Stocks and Huge Dead Zone Signal Tipping Point," *The Guardian,* January 31, 2017.

INDEX